THE DYNAMICAL IONOSPHERE

THE DYNAMICAL IONOSPHERE

A Systems Approach to Ionospheric Irregularity

Edited by

MASSIMO MATERASSI

Researcher, Institute for Complex Systems of the National Research Council (ISC-CNR), Florence, Italy

BIAGIO FORTE

Research Fellow, Department of Electronic and Electrical Engineering, University of Bath, Bath, United Kingdom

ANTHEA J. COSTER

Assistant Director and Principal Research Scientist, MIT Haystack Observatory, Westford, MA, United States

SUSAN SKONE

Associate Professor, The University of Calgary|HBI Department of Geomatics Engineering, Calgary, AB, Canada

ELSEVIER

Elsevier
Radarweg 29, PO Box 211, 1000 AE Amsterdam, Netherlands
The Boulevard, Langford Lane, Kidlington, Oxford OX5 1GB, United Kingdom
50 Hampshire Street, 5th Floor, Cambridge, MA 02139, United States

Notices
Knowledge and best practice in this field are constantly changing. As new research and experience broaden
our understanding, changes in research methods, professional practices, or medical treatment may become
necessary.

Practitioners and researchers must always rely on their own experience and knowledge in evaluating and
using any information, methods, compounds, or experiments described herein. In using such information
or methods they should be mindful of their own safety and the safety of others, including parties for whom
they have a professional responsibility.

To the fullest extent of the law, neither the Publisher nor the authors, contributors, or editors, assume any liability
for any injury and/or damage to persons or property as a matter of products liability, negligence or otherwise, or
from any use or operation of any methods, products, instructions, or ideas contained in the material herein.

Library of Congress Cataloging-in-Publication Data
A catalog record for this book is available from the Library of Congress

British Library Cataloguing-in-Publication Data
A catalogue record for this book is available from the British Library

ISBN: 978-0-12-814782-5

For information on all Elsevier publications
visit our website at https://www.elsevier.com/books-and-journals

Publisher: Candice Janco
Acquisition Editor: Marisa LaFleur
Editorial Project Manager: Sara Pianavilla
Production Project Manager: Selvaraj Raviraj
Cover Designer: Matthew Limbert

Typeset by SPi Global, India

Contents

Contributors

Managlathayil Ali Abdu National Institute for Space Research (Instituto Nacional de Pesquisas Espaciais-INPE), Sao Jose dos Campos, Brazil

Lucilla Alfonsi Istituto Nazionale di Geofisica e Vulcanologia, Rome, Italy

Asti Bhatt Center for Geospace Studies, SRI International, Menlo Park, CA, United States

Brett Carter SPACE Research Center, RMIT University, Melbourne, VIC, Australia

Christopher J. Coleman The University of Adelaide, Adelaide, SA, Australia

Giuseppe Consolini INAF-Institute for Space Astrophysics and Planetology, Rome, Italy

Anthea J. Coster MIT Haystack Observatory, Westford, MA, United States

Philip J. Erickson Atmospheric and Geospace Sciences Group, MIT Haystack Observatory, Westford, MA, United States

Biagio Forte Department of Electronic and Electrical Engineering, University of Bath, Bath, United Kingdom

D.L. Hysell Earth and Atmospheric Sciences, Cornell University, Ithaca, NY, United States

Dennis L. Knepp NorthWest Research Associates, Monterey, CA, United States

Giovanni Lapenta Departement Wiskunde, KULeuven, University of Leuven, Leuven, Belgium

Naomi Maruyama CIRES, Univ. of Colorado Boulder and NOAA Space Weather Prediction Center, Boulder, CO, United States

Massimo Materassi Institute for Complex Systems of the National Research Council (ISC-CNR), Florence, Italy

Michael Mendillo Department of Astronomy, Boston University, Boston, MA, United States

Paola De Michelis Istituto Nazionale di Geofisica e Vulcanologia, Roma, Italy

Bruno Nava The Abdus Salam International Centre for Theoretical Physics, Trieste, Italy

Mirko Piersanti National Institute of Nuclear Physics, University of Rome "Tor Vergata", Rome, Italy

Sandro M. Radicella The Abdus Salam International Centre for Theoretical Physics, Trieste, Italy

Joshua Semeter Department of Electrical and Computer Engineering and Center for Space Physics, Boston University, Boston, MA, United States

Andrew Silberfarb SRI International, Menlo Park, CA, United States

Susan Skone The University of Calgary | HBI Department of Geomatics Engineering, Calgary, AB, Canada

Luca Spogli Istituto Nazionale di Geofisica e Vulcanologia, Rome, Italy

Roberta Tozzi Istituto Nazionale di Geofisica e Vulcanologia, Roma, Italy

Preface

A new book on ionospheric physics can appear "not timely," if not "unnecessary," until one puts this enterprise in the contemporaneous "historical" and "cultural" context.

We live in a time characterized by a growing interest in what one may refer to as "large systems," i.e., portions of the universe, the state of which is described by large amounts of parameters. Remarkable examples of that may be found in several fields: the state of *human population* in terms of biological, medical, financial, or other personal data; *social networks*; *genetic mappings*; environmental, zoological, and botanical *large data*, *weather and climate* data; and data describing *markets*, the economy, and the wealth of nations.

This is historically determined by different factors.

First, in a world economically and politically "globalized," everything appears as in a "universal village" within which action-reaction chains propagate very rapidly and ostensibly, attracting the attention of stakeholders, politicians, and people in general.

Second, the unprecedented fast development of information technologies and supercomputers allows for lively processing of huge amounts of information, which renders it reasonably useful to collect very big data.

Last but not least, we have the development of *Complexity Science*, a "synergetic discipline" putting together many fields of investigation and thought, in which *the organizing effects of interactions* are investigated to understand the functioning of composite systems.

In a formula, we can refer to all this by stating that *we are in "the outbreak of Large Systems' Age"*

(LSA). In physical science, *the gifts* of LSA are new powerful data processing tools and machines; *the main request* is to develop (or rediscover and recover) the mathematical tools dealing with large systems: complex system dynamics, advanced statistics, and out-of-equilibrium thermodynamics.

Ionospheric physics will definitely benefit from the outbreak of LSA. On the one hand, we have been exploring the ionosphere from some decades now, collecting a huge amount of all kinds of possible data, and these could be processed by the new computing tools; on the other hand, the future ionospheric campaigns will be designed taking into account the new possibilities of the LSA technology (big data technology, neural networks, machine learning). It hence appears to be possible to refresh and strengthen ionospheric science with new external contributions, i.e., with what is taught theoretically by dynamical system and complexity science, and with the enormous opportunity of highly performing big data bases, particularly suited for large systems as population, healthcare, and geophysics. This new contribution to ionospheric physics is unprecedented, and as we are going to have new ways to interrogate ionospheric data, we must have a clear mind about what questions should be asked.

To focus on this, here an attempt is done *to revisit the Earth's ionosphere as a complex, dynamical system*, as complexity, i.e., "what comes out of the organizing role of interactions," appears to be the best way to understand the Earth's ionosphere dynamics. One must start from considering the ionosphere as a dynamical system, with a very high number of degrees of freedom, and governed by physical laws concentrated in few simple equations of motion, but leading to complex behaviors. According to this point of view, we list the phenomenology that points toward this vision, and collect some new experimental and theoretical results obtained thanks to it.

The structure of this book in some way tries to reflect a zoom-in approach to the ionospheric complex dynamics: one starts from the complexity of large time- and space-scale behavior of the Earth's ionosphere, regarded as part of the Sun-Earth dynamics. This global complexity is the effect of non-linear couplings among the different sub-systems forming the Heliosphere: the near-planet space, the interplanetary medium, the Sun itself. After treating some interesting aspects of global dynamics, the most challenging traits of local dynamics are discussed: the so-called ionospheric irregularities, basically due to plasma instabilities and turbulence.

Which consequences should be drawn, as the global and the local dynamics of the Earth's ionosphere are accepted to be "complex"? New theoretical data analysis and experimental tools should be introduced, and reviewing something in this field is the aim of our book.

This zoom-in approach, which consists of global complexity, local irregularity, and new ionospheric science, organizes the present book into four parts.

In Part I, entitled "The Earth's Ionosphere, An Overview," the ionosphere of our planet is described in a didactic way, yet addressing the subject with an emphasis on the aspects of phenomenology pointing toward a more dynamical, complexity science-oriented description of the ionosphere. In particular, after a general introduction (Chapter 1), Michal Mendillo focuses his attention on two challenging aspects of the ionospheric phenomenology: the "Day-to-Day Variability of the Ionosphere" (Chapter 2) and the "Ionospheric Conjugate Point Science: Hemispheric Coupling" (Chapter 3). He then

discusses the "Status" and the "Future Directions" of ionospheric science in Chapter 4. In Chapter 5, Phil Erickson gives an introduction to the "Natural Complexity in Action" describing the "Mid-Latitude Ionospheric Features," while Sandro Maria Radicella and Bruno Nava give a panoramic view of "Empirical Ionospheric Models" in Chapter 6. The first part of the book is closed by some recap about the need to adopt a complex-dynamical description of the Earth's ionosphere, given in Chapter 7 ("Wrap Up").

Part II of the book is entitled "Global Complexity," and describes the complexity of the Sun-Earth system on the large scale. Indeed, the helio-geospatial system involves complex dynamics, which are manifested on the planetary scale, or even larger scales.

Giovanni Lapenta gives a general and didactic view of the "Complex Dynamics of the Sun-Earth Interaction" in Chapter 8, while Chapter 9 is dedicated to the dynamics of geomagnetic storms and sub-storms, in Naomi Maruyama's "Storms and Sub-Storms." The part is closed by Chapter 10, in which Mirko Piersanti and Brett Carter describe the highly dynamical and technology-relevant phenomenon of "Geomagnetically Induced Currents."

In Part III, the small-scale effects of the ionospheric complex dynamics are discussed as "Local Irregularities."

First, David Lee Heysell guides us "From Instabilities to Irregularities" in his Chapter 11, in which the dynamic system approach is applied to ionospheric plasma instabilities; next, the "Equatorial F Region Irregularities" in particular are discussed by Mangalathayil Ali Abdu Sr. in Chapter 12. "Scintillation Theory," predicting the effects of local irregularities on trans-ionospheric signals, is described by Dennis L. Knepp (Chapter 13).

Part IV of the book, "The Future Era of Ionospheric Science," is dedicated to the consequence that we think one should draw on accepting the complex dynamical nature of the

Earth's ionospheric system: the idea is basically to reorient the experimental and the theoretical physics of the ionosphere borrowing ideas, languages, and techniques from dynamical system theory and complexity science. This part contains examples of present research results that we regard as already pointing toward this development.

Massimo Materassi discusses the origin of ionospheric complexity according to his point of view in Chapter 14 ("The Complex Ionosphere"), and describes how path integrals may be used for ionospheric turbulence; Joshua Semeter talks about "New High Resolution Techniques to Probe the Ionosphere" in Chapter 15, and what they can bring to ionospheric science. Ionospheric complex dynamics invokes the use of "Advanced Statistical Tools in Near-Earth Space Science," which Giuseppe Consolini and Massimo Materassi describe in Chapter 16, while the opportunity of interrogating big data bases for ionospheric studies is proposed by Asti Bhatta in Chapter 17, "Big Data Mining and Networking Applied to the Earth's Ionosphere." Massimo Materassi, Lucilla Alfonsi, Luca Spogli, and Biagio Forte describe some novel aspects of "Scintillation Modeling" in Chapter 18; still about small-scale structures, the discussion in Chapter 19 by Paola De Michelis and Roberta Tozzi sketches techniques for "Multiscale Analysis of the Turbulent Ionospheric Medium."

The final chapter, "The Future Ionospheric Physics" (Chapter 20), contains a synthesis of what the Editors consider to be the near future challenges of ionospheric physics, in view of what has been discussed throughout the book.

Massimo Materassi
Anthea J. Coster
Biagio Forte
Susan Skone

PART I

The earth's ionosphere, an overview

Introduction

Michael Mendillo

Department of Astronomy, Boston University, Boston, MA, United States

An ionosphere is that portion of a planet's upper atmosphere where solar photons impact neutral gases to yield a plasma of electrically charged ions and electrons. The basic physics and chemistry that govern terrestrial ionospheric structure and dynamics have been treated in a robust series of fundamental reference books (Ratcliffe, 1960; Rishbeth and Garriott, 1969; Banks and Kockarts, 1973; Rees, 1989; Hargreaves, 1995; Prölss, 2004; Kelley, 2009; Knipp, 2011). Extensions of ionospheric theory to other planets in our solar system are given in the monographs by Bauer (1973), Mendillo et al. (2002), Bauer and Lammer (2004), and Nagy et al. (2008). The most unified treatment of terrestrial and planetary ionospheric science is given in the comprehensive textbook by Schunk and Nagy (2009). Given these excellent sources of educational material, and a readership with experience in ionospheric research, this introductory chapter will not present a detailed repetition of theory. Rather, the goal is to set the stage for the innovative treatment of the Earth's ionosphere as a complex system, as presented in the chapters that follow.

Historically, ionospheric theory was approached as a unique solar-terrestrial phenomenon linking solar photons with the Earth's neutral gases to yield a plasma population

capable of being studied using ground-based instrumentation. Within this framework, the ionosphere is produced by a flux of solar photons versus wavelength (called "irradiance")—ranging from X-rays ($<\sim 10\,nm$) to extreme ultraviolet ($<\sim 120\,nm$). Collectively called XUV radiation, these photons penetrate to different heights in the upper atmosphere to ionize the primary gases N_2, O_2, and O. The strength of the XUV radiation varies over time scales ranging from minutes (solar flares) to decades (solar cycle). For a fixed daily value at the subsolar point, the ionizing radiation varies with latitude and local time (collectively described by solar zenith angle).

The fact that the solar irradiance components reach different altitudes resulted in each of the textbooks mentioned earlier describing the vertical structure of the ionosphere as a series of "layers" produced at different photon penetration heights. Similarly, the latitude structure was described as a series of zones ordered by solar zenith angles and magnetic field characteristics along north-south meridians. Thus, as shown in Fig. 1, the ionosphere was presented as having D, E, F1, and F2 layers in altitude, with each of these layers varying in latitude from the polar cap to the equator—subdividing the near-space environment into auroral, subauroral,

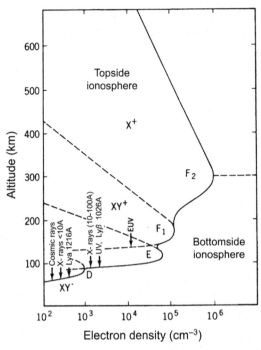

FIG. 1   Layers of the Earth's ionosphere. *From Bauer, S.J., Lammer, H., 2004. Planetary Aeronomy. Springer, New York.*

middle, low and equatorial ionospheric regimes. This view produced a high level of understanding of individual processes acting within the global ionosphere. Yet, such a compartmentalization of the ionosphere is purely historical and today seems as a somewhat limiting framework for progress. This ensemble-of-layers approach arose simply from sequential applications of the initial formulation of ionospheric theory within the context of photo-chemical-equilibrium conditions (Chapman, 1931). So powerful was the respect for Sydney Chapman's pioneering portrayal of the ionosphere that any observational departure from Chapman Theory was called an *anomaly* (e.g., seasonal anomaly, diurnal anomaly, equatorial anomaly). In reality, the shortfalls were in the physics used, and not with questionable diagnostic findings.

With the coming of the Space Age, new satellite instruments, much-improved radio and optical observing methods from the ground, and advanced computer modeling capabilities ushered in the modern era of ionospheric research. The quaint notion of individual electrified layers stacked on top of each other, with latitude zones isolated from each other, has been replaced by the new paradigm of *coupling*.

The historical foundation for coupling within the geospace domain had been set decades earlier via explanations of the causes of aurora. Thus, magnetosphere-ionosphere (M-I) coupling at high latitudes set the precedent for additional understanding of space physics *system-science*. Today, solar-terrestrial relationships involve far more than the photon source of the ionosphere. The chain of events leading to complexity starts with a coronal mass ejection producing modified solar wind plasma density, velocity, and magnetic field characteristics. These, in turn, cause solar wind-magnetosphere coupling—followed by M-I coupling. This classic scenario of Sun-Earth space physics is summarized schematically in Fig. 2. Yet, there is an additional component of coupling also shown in Fig. 2 that emerged from

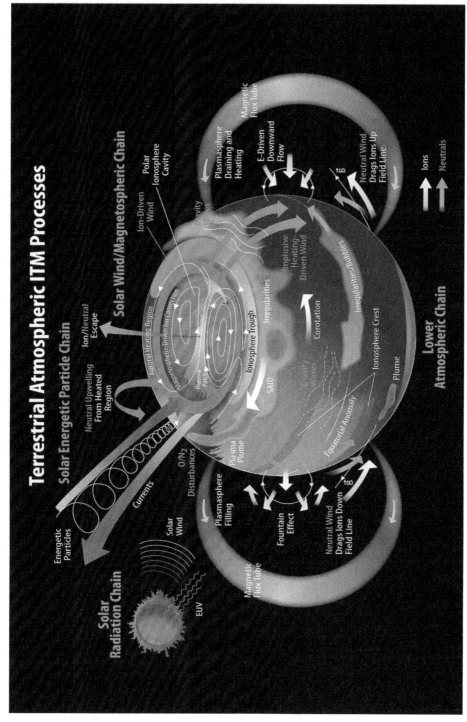

FIG. 2 Coupling components of upper atmosphere regions (NASA image).

I. The earth's ionosphere, an overview

more recent research. When ionospheric variability was found to be substantial during periods of very quiescent solar and magnetospheric conditions, a source of nondownward coupling was needed. This led to the concept of *coupling from below*—completing the paradigm of the ionosphere being a fully linked surface-to-Sun atmospheric-plasma system.

Altitude and latitude coupling on a global scale have introduced levels of complexity that are now the major foci of ionospheric research. This chapter is a prelude to the issues treated in subsequent chapters—where various types of complexity—a term still difficult to define (Charbonneau, 2017)—are introduced and described. Here, the agenda is set by describing a few processes that are not simply latitude or altitude dependent. Such topics illuminate core concepts of altitude and latitude coupling, but treat them as universal processes. The focus is on a difference in approach to problems formerly treated as issues confined by spatial and temporal boundaries. The goal is to continue fostering a transition of thinking about the ionosphere as depicted in Fig. 1 to the system depicted in Fig. 2.

References

Banks, P.M., Kockarts, G., 1973. Aeronomy (Parts A and B). Academic Press, New York.

Bauer, S., 1973. Physics of Planetary Ionospheres. Springer-Verlag, Berlin.

Bauer, S.J., Lammer, H., 2004. Planetary Aeronomy. Springer, New York.

Chapman, S., 1931. The absorption and dissociation or ionizing effect of monochromatic radiation in an atmosphere on a rotating earth. Proc. Phys. Soc. Lond. 43, 26–45.

Charbonneau, P., 2017. Natural Complexity: A Modeling Handbook. Princeton University Press, Princeton, NJ.

Hargreaves, J.K., 1995. The Solar-Terrestrial Environment. Cambridge University Press, Cambridge.

Kelley, M., 2009. The Earth's Ionosphere: Plasma Physics and Electrodynamics, second ed. Elsevier Academic Press, New York.

Knipp, D.J., 2011. Understanding Space Weather and the Physics Behind It. McGraw Hill, Boston, MA.

Mendillo, M., Nagy, A., Waite, J.H. (Eds.), 2002. Atmospheres in the Solar System: Comparative Aeronomy. In: Geophysical Monograph 130, American Geophysical Union, Washington, DC.

Nagy, A.F., Galogh, A., Cravens, T.E., Mendillo, M., Muller-Wodarg, I. (Eds.), 2008. Comparative Aeronomy. Springer, Berlin (also in Space Science Reviews, vol. 139, no. 1-4.).

Prölss, G.W., 2004. Physics of the Earth's Space Environment: An Introduction. Springer-Verlag, Berlin.

Ratcliffe, J.A. (Ed.), 1960. Physics of the Upper Atmosphere. Academic Press, New York.

Rees, M.H., 1989. Physics and Chemistry of the Upper Atmosphere. Cambridge University Press, Cambridge.

Rishbeth, H., Garriott, O.K., 1969. Introduction to Ionospheric Physics. Academic Press, New York.

Schunk, R.W., Nagy, A.F., 2009. Ionospheres: Physics, Plasma Physics and Chemistry. Cambridge University Press, Cambridge, UK.

2

Day-to-day variability of the ionosphere

Michael Mendillo

Department of Astronomy, Boston University, Boston, MA, United States

1 Overview

The defining characteristic of a planet's ionosphere is its profile of electron density versus height, $N_e(h)$. The ionosphere is not the same every day—and for reasons all ultimately linked to the Sun. As mentioned earlier, the observed $N_e(h)$ morphology depicted in Fig. 1 of Chapter 1 results from the differential heights reached by solar photons capable of ionizing the terrestrial neutral atmosphere. At its lowest altitudes, the ionosphere occurs in a dense neutral atmosphere where photo-chemical-equilibrium (PCE) dominates over plasma dynamics because its molecular ions (O_2^+, NO^+, N_2^+) combine so rapidly with electrons. Thus, plasma is lost at the same altitudes where it is produced. At greater heights, however, the recombination chemistry of the ionosphere's atomic ions (O^+) and electrons proceeds at a much slower rate, and thus plasma dynamics competes effectively with, and can dominate over, chemistry. The time constants for chemistry and dynamics are the key numbers that define complexity. For PCE-dominated plasmas in the low-altitude ionosphere, the chemistry can be complicated, but the overall system is not fundamentally complex. For dynamically dominated systems, e.g., at the "top" of the ionosphere, the interactions of diffusion, winds, and electrodynamics can again be complicated, but the overall system is still not complex merely due to a transition from photochemistry to plasma dynamics.

Given our basic understanding of the terrestrial ionosphere, present-day research deals with departures from "textbook" conditions. These fall primarily into categories associated with variable forms of solar output: (a) changes in the solar irradiance that occur with timescales ranging from flares ($\sim$minutes) to sunspot cycles (~ 11 years), and (b) solar wind variability due to coronal mass ejections ($\sim$days), rotating active regions ($\sim$monthly), and solar-cycle effects ($\sim$11 years). In addition to ionospheric variations due to direct solar photon and solar wind external drivers, ionospheres also vary as a result of processes driven by solar energy absorbed by the Earth's neutral atmosphere. These include (c) changes associated with upward coupling of waves and tides from the lower to upper atmosphere, and (d) composition changes due to thermal expansion and varying atmospheric dynamics and circulation. Atmospheric waves have time scales ranging from minutes to $\sim$hour—not all that different from solar flare effects. The time scales for tidal effects and global circulation are diurnal

The Dynamical Ionosphere
https://doi.org/10.1016/B978-0-12-814782-5.00002-9

(with subharmonics). Taken as an overall system, then, the ionosphere varies following known periodicities blended with statistically varying episodic patterns that range from minutes to decades. Is it possible to predict all of these cause-effect patterns? Is there some level of variability that is formally chaotic and thus an ever-present level of uncertainty that can never be predicted? The first example of complexity deals with this issue.

2 Quantifying ionospheric variability: Peak electron density and total electron content

Magnitudes of ionospheric variability vary with altitude. As an example, Fig. 1A shows electron density profiles at noon for 32 consecutive days as measured by the incoherent scatter radar at Millstone Hill. To the right, the standard deviations in percent are shown. For the maximum electron density (N_{max}), many studies have documented both magnitudes and sources of variability (e.g., Forbes et al., 2000; Rishbeth and Mendillo, 2001; Moore et al., 2006). The typical procedure has been to use radio reflection (ionosonde) observations of mid-day N_{max} values spanning a month at midlatitudes, and to characterize variability as the standard deviation in percent [σ(%)] about the monthly mean value. The selection of the midlatitude domain is made because it exhibits all of the coupling processes that are now the topics of intense study (Kintner et al., 2008).

The components of overall N_{max} variability [σ_{total}] can be attributed to changes in solar photon irradiance [σ_{sun}], solar-wind-induced geomagnetic activity [σ_{mag}], and meteorological coupling from below [σ_{met}]. The analysis method used by Rishbeth and Mendillo (2001) treated such contributions as independent functions. Following that approach, and guided by the numerical values used in Forbes et al. (2000) and Rishbeth and Mendillo (2001),

ionospheric variability about a monthly mean $<N_{max}>$ under mid-day conditions can be portrayed as follows:

$$[\sigma_{total}]^2 = [\sigma_{sun}]^2 + [\sigma_{mag}]^2 + [\sigma_{met}]^2$$
$$[20 - 25\%]^2 \approx [3 - 6\%]^2 + [14 - 17\%]^2 + [14 - 17\%]^2 \quad (1)$$

A dramatic variation of these results occurs when one of the sources of variability greatly exceeds the others. An example is shown in Fig. 1B where the same days used in panel (A) are selected from a high latitude location (the ISR in Svalbard). Due to auroral processes, the daytime variability at all altitudes is greatly above that found at midlatitudes.

An alternative way to monitor the terrestrial ionosphere is via observations utilizing transionosphere radio beacons (e.g., using geostationary or GPS satellites). This method provides the integral of the full electron density (N_e) profile—with total electron content defined as

$$TEC = \int N_e(h) \, dh.$$

Since most of TEC comes from the F-layer, TEC is highly correlated ($\approx 90\%$) with N_{max} (Fox et al., 1991), and thus σ_{TEC}(%) is also 20%–30% (Johanson et al., 1978). Fig. 2 shows examples of diurnal TEC variability throughout a year. No local time and no seasonal condition is free from variability. The extremes of high and low TEC patterns are due to geomagnetic storm effects (Mendillo, 2006). All of the other days conform to the implications of Eq. (1).

The overall message from Eq. (1) and, as portrayed in Figs. 1 and 2, is clear: for both N_{max} and TEC, the influence of solar irradiance is minimal in comparison to solar wind-magnetospheric sources of downward coupling and neutral atmosphere sources of upward coupling—with the latter two being comparable. This simply restates the fact that the bulk of the terrestrial ionosphere is not fully described by internal photochemical equilibrium (Chapman-esque)

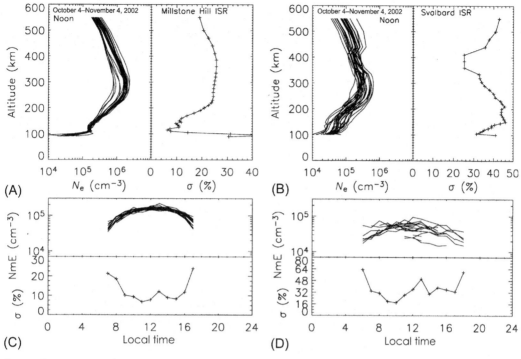

FIG. 1 Incoherent scatter radar measurements at Millstone Hill and Svalbard from October 4–November 4, 2002. Noontime electron density versus altitude for each of the 32 days: (A) Millstone, with standard deviations (%), and (B) Svalbard with standard deviations (%). Diurnal variation of the maximum electron density at 110 km (C) at Millstone Hill, with standard deviations (%), and (D) from Svalbard, with standard deviations in percent (%). *After Moore, L., Mendillo, M., Martinis, C., Bailey, S., 2006. Day-to-day variability of the E layer. J. Geophys. Res.: Space Physics 111 (A6), A06307, https://doi.org/10. 1029/2005JA011448.*

processes. Plasma dynamics (diffusion along magnetic field lines, neutral wind coupling, electrodynamics) compete with PCE, and changes in the thermosphere (waves and tides) affect the abundance of neutral gases that are ionized.

For the pure-PCE component of the terrestrial ionosphere, as found at ~110 km where molecular ions dominate, variability is much less because of the absence of significant contributions from plasma dynamical processes. In their study of sources of PCE variability, Moore et al. (2006) used observations and modeling of midday conditions at midlatitudes. An example is shown in Fig. 1C using ionosonde data from Millstone Hill. Their finding from this and other data sets led to the characterization of midlatitude, low-altitude observed variability to be

$$\sigma_{\text{total}}\,(\text{PCE}) = 7\% - 12\%. \qquad (2)$$

The Moore et al. (2006) modeling studies showed that the contribution from solar input (changes in flux and declination over a month) was 8%–9%. The remaining contributions came from small changes in the neutral atmosphere. Yet, at high latitudes affected by auroral processes, PCE variability can be as high as ~50%, as shown using ionosonde data from Svalbard (Fig. 1D). In summary, while the solar photon contribution to ionospheric variability dominates at low altitudes (Eq. 2), the opposite is true at high altitudes (Eq. 1). While downward and

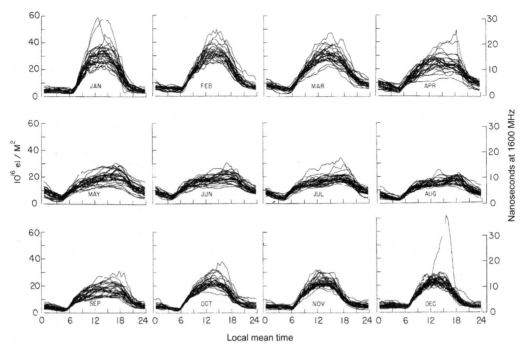

FIG. 2 A year of total electron content (TEC) data separated by month to portray ionospheric variability. The observations were made using Faraday rotation measurements of the 137-MHz plane polarized beacon on the geostationary satellite ATS-3 from the AFCRL observatory at midlatitudes (Hamilton, MA). The traditional MKS units for TEC (10^{16} e$^-$/m^2) are given on the left axis; the axis on the right gives their correspondingly imposed time delay (in nanoseconds, ns) for a GPS-type (L1=1.575GHz) frequency. A value of 20 TEC units results in a $\sim$10ns of delay, causing a range error of $\sim$3m. *From Mendillo, M., 2006. Storms in the ionosphere: patterns and processes for total electron content. Rev. Geophys. 47, RG4001. https://doi.org/10.1029/2005RG000193.*

upward coupling are comparable at middle latitudes (Eq. 1), downward coupling dominates all sources at high latitudes (Fig. 1D).

Understanding the complexity of ionospheric variability—to the point of achieving a predictive capability—defines a frontier topic of 21st century upper atmospheric physics. Initial steps, such as the separation of processes implied by Eqs. (1) and (2), are clearly an idealized approach. In reality, the *downward-propagating* changes due to solar irradiance and geomagnetic activity interact with independent *upward propagating* disturbances from the troposphere-stratosphere-mesosphere system that derive their energy from the Sun as well (but in very different ways). A self-consistent,

whole-atmosphere, global circulation model can, in principle, include all such processes. As will be seen, however, the complexity of the system-input and system-response functions currently must rely on parameterizations of basic processes ranging from secondary ionization by photoelectrons, to estimates of plasma convection and auroral precipitation patterns, to the spectrum of gravity wave and tidal patterns in the neutral atmosphere.

What should be the realistic goals for the simulation of such a complex system? Surely achieving a ±5% predictive capability would be a justifiable declaration of success. More probable is a ±20%–25% capability for basic morphologies. For geomagnetic storm effects and ionospheric instability disturbances, predictions within ±25%-50% of observational magnitudes, spatial locations, and time might be possible.

References

Forbes, J., Palo, S., Zhang, X., 2000. Variability of the ionosphere. J. Atmos. Solar Terr. Phys. 62, 685–693.

Fox, M., Mendillo, M., Klobuchar, J., 1991. Ionospheric equivalent slab thickness and its modeling applications. Radio Sci. 26, 429–438. https://doi.org/10.1029/90RS02624.

Johanson, J.M., Buonsanto, M.J., Klobuchar, J.A., 1978. The variability of ionospheric time delay. In: Proceedings of Ionospheric Effects Symposium. Naval Research Laboratory, U.S. Government Printing Office, Washington, DC, pp. 479–485.

Kintner Jr., P.M., Coster, A.J., Fuller-Rowell, T., Mannucci, A., Mendillo, M., Heelis, R. (Eds.), 2008. Midlatitude Ionospheric Dynamics and Disturbances, Geophysical Monograph 181. Amer. Geophys. Union, Washington, DC.

Mendillo, M., 2006. Storms in the ionosphere: patterns and processes for total electron content. Rev. Geophys. 47, RG4001. https://doi.org/10.1029/2005RG000193.

Moore, L., Mendillo, M., Martinis, C., Bailey, S., 2006. Day-to-day variability of the E layer. J. Geophys. Res. Space Physics 111 (A6), A06307. https://doi.org/10.1029/2005JA011448.

Rishbeth, H., Mendillo, M., 2001. Patterns of ionospheric variability. J. Atmos. Solar Terr. Phys. 63, 1661–1680.

3

Ionospheric conjugate point science: Hemispheric coupling

Michael Mendillo

Department of Astronomy, Boston University, Boston, MA, United States

The foregoing discussion of ionospheric variability focused on daytime conditions when the magnitudes of plasma densities are the largest. Yet, ionospheric variability does not cease at sunset. A second example of complexity deals with the plasma instabilities that can dominate nighttime conditions across the globe. Changes in low nighttime magnitudes of electron density can be quite large on a percentage basis, and thus the variability of the nighttime ionosphere might present the greatest quantitative challenge to complexity theory. Nighttime ionosphere topics were first studied at high latitudes during geomagnetic disturbances—where the optical manifestations of disturbance (aurora) are ordered by patterns of closed and open geomagnetic field lines. Given such control by **B**-field morphology, effects of auroral substorms and geomagnetic storms were found to be similar in each hemisphere if monitored at the north and south ends of the same geomagnetic field lines. Such interhemispheric sites are called *conjugate points*. Yet, as discussed later, significant cases of departures from conjugate consistency of the aurora have also been found.

For the regions from auroral to equatorial latitudes that span so much of the globe, the geomagnetic field lines remain "closed" and interhemispheric coupling occurs via electrodynamical processes. This vast ionospheric domain displays a systematic ordering of a multitude of processes controlled by geomagnetic fluxtubes—with each end having the same magnetic latitude (N and S) and longitude—the defining conditions for conjugate points. Yet, due to the tilt of the magnetic dipole axis, there can be pronounced differences in ionospheric storm effects at conjugate points (Mendillo and Narvaez, 2010). This occurs because of the differences between geographic latitude (the crucial factor for solar production of the ionospheric) and geomagnetic latitude (the controlling factor for M-I coupling). Thus, all conjugate points are not geophysically equivalent. Moreover, by its very definition, conjugate points have opposing seasonal conditions in the two hemispheres, and thus conjugate ionospheres present different "receptor conditions" to any single disturbance source ordered by (or affected by) season. There are several examples to consider.

1 Plasma instability conjugate science

The most intense and enigmatic disturbances in all of geospace plasma science are those associated with a gravitational Rayleigh-Taylor (GRT) plasma instability that occurs in the equatorial and low-latitude ionosphere. Following sunset, the maximum plasma density of the ionosphere (O^+, e^-) appears as a "dense-fluid" upon a "light fluid" (the remnant molecular ions and electrons in the bottomside ionosphere). A seed perturbation—intrusion of the low-density region into the denser region—can result in a classic GRT instability pattern. Low-density plasma along an entire geomagnetic meridian "percolates" upwards in explosive fashion via flux-tube interchange processes. Strong plasma gradients prompt a cascade of irregularities from large scale (100–1000s of kilometers) to small scales (centimeters) that cause serious effects upon radio signals (phase and amplitude scintillations). For historical reasons linked to signatures in early ionosonde measurements, the overall phenomenon is called Equatorial Spread-F (ESF). The theory and observational methods used to study ESF are well described in Kelley (2009), and reference therein.

Perhaps the most intriguing and challenging aspect of ESF from the perspectives of complex systems and applications areas is that ESF is a form of "Space Weather" that can occur *with and without* relationships to solar, solar wind, or magnetospheric disturbances. That is, ESF has an internal seasonal-longitude occurrence pattern of disturbances (Tsunoda, 1985; Aarons, 1993)—*as well as* instigations and suppressions that occur during geomagnetic storms (Martinis et al., 2005). Suppressions of ESF onset and growth unrelated to geomagnetic activity can also occur due the sudden appearance of plasma well below the ionospheric peak (Stephan et al., 2002). Achieving a functional understanding of ESF phenomena during both quiet and disturbed times is certainly one of the grand challenges for ionospheric complexity theory and practice.

An example of recent research on ESF deals with new insights gleamed from conjugate point observations. Given the electrodynamical nature of the fluxtube-integrated interchange process, ESF effects *must* affect the ionospheres at both ends of a geomagnetic meridian. This has been confirmed using interhemispheric radio diagnostics and optical imaging systems. For the latter, ESF is associated with *plasma depletions* that appear visually (e.g., in 630.0-nm images) as regions of *airglow depletions*. First observed as north-south-aligned 630.0-nm emission voids spanning the geomagnetic equator (Weber et al., 1978), they were soon shown to extend to geomagnetic latitudes of ±15–25 degrees (Mendillo and Baumgardner, 1982). This revealed ESF to be truly interhemispheric, magnetic fluxtube disturbances reaching lower midlatitudes—and thus **B**-field lines having apex heights of thousands of kilometers above the geomagnetic equator.

The first study of airglow depletions using all-sky-imagers (ASIs) at conjugate points (Otsuka et al., 2002) came from sites at Sata (Japan) and Darwin (Australia). With magnetic latitudes at zenith of 24°N and 22°S, respectively, and nearly identical magnetic longitudes, airglow depletions were observed in both hemispheres (Fig. 1). The 630.0-nm images showed depletion features having 40–100 km scale sizes in longitude, with some complex bifurcations in latitude. When mapped along geomagnetic field lines from one hemisphere to the other, Otsuka et al. (2002) and Shiokawa et al. (2004) found excellent morphology agreement between the two sites. Fukushima et al. (2015) used ASI systems in Kototabang (Indonesia) and Chiang Mai (Thailand) and also found excellent agreement for optical conjugate morphologies. Yet, Abdu et al. (2009) and Sobral et al. (2009) reporting on simultaneous ASI and ionosonde results from conjugate points in Brazil (Boa Vista and Campo Grande), found that while the airglow depletion characteristics were very similar, the ionosonde observations in each hemisphere showed somewhat different irregularity patterns.

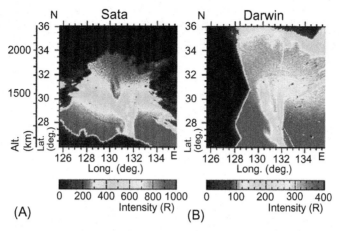

FIG. 1 Example of conjugate point ionospheric airglow depletions associated with equatorial spread-F (left). An all-sky-image of airglow patterns recorded in Sata (Japan) compared to an image taken at the same time in Darwin (Australia) that is mapped along geomagnetic field lines from the southern hemisphere to the northern hemisphere. Low values of airglow outline a plume of ionospheric irregularities that exhibits a bifurcation pattern—showing a coherence between hemispheres due to magnetic field control. The details agree to a scale size down to ~40km. *From Otsuka, Y., Shiokawa, K., Ogawa, T., Wilkinson, P., 2002. Geomagnetic conjugate observations of equatorial airglow depletions. Geophys. Res. Lett. 29, 15. https://doi.org/ 10.1029/2002GL015347.*

Studies using conjugate point ASIs from three longitude sectors have thus confirmed the basic electrodynamical mechanism of geomagnetic fluxtube-aligned control of the ESF signatures found from all-sky airglow imaging. Yet, when combining optical and radio results at conjugate points, Abdu et al. (2009) found coherence of effects in one hemisphere, but not in both. This suggests conjugate point science offers an important way to approach the complexity of large- versus small-scale irregularity signatures. If images mapped between hemispheres agree down to a scale size of ~40km, but radio soundings show differences, does that scale size define the onset of complexity?

2 Electrobuoyancy conjugate science

Incoherent scatter radar (ISR) observations at the Arecibo Observatory discovered unusual corrugations in ionospheric densities with horizontal scale size of 100s of kilometers (Behnke, 1979). The first optical studies of these midlatitude structures were carried out using an all-sky imager at the Arecibo Observatory (Mendillo et al., 1997; Miller et al., 1997). The wavelengths and speeds observed were typical of medium-scale traveling ionospheric disturbances (MSTIDs) associated with waves in the neutral atmosphere. Yet, the structures observed at Arecibo did not propagate over the full range of azimuths found with earlier studies of MSTIDs—rather, they had a narrow range of northeast-to-southwest propagation vectors. Those characteristics led to the suggestion that a specific type of plasma instability seemed to be involved (Perkins, 1973), but one not fully consistent with observations, and thus the phrase "Perkins-like" came into use. A more general name was suggested (electro-buoyancy-waves) by Kelley et al. (2000) in order to distinguish them from the ionospheric corrugations driven by waves in the neutral atmosphere (gravity waves) that propagate in all directions. Unfortunately, it has not been widely adopted. The MSTID designation is still used, as will be done here.

From a conjugate-point perspective, a phenomenon controlled by electrodynamics must map from one hemisphere to the other following magnetic field patterns (Otsuka et al., 2004). The airglow signatures of an MSTID are bright and dark bands moving from northeast to southwest in the northern hemisphere, and thus should be from southeast to northwest in the southern hemisphere. The first coordinate airglow and GPS observations of MSTID structures simultaneously seen at Arecibo and its conjugate point (Mercedes, Argentina) were described by Martinis et al. (2010, 2011). The NE-to-SW motion in the northern hemisphere and the SE-to-NW motion in the southern hemisphere thus showed them to be hemispherically coherent, electrodynamical phenomena.

An example of conjugate point MSTIDs is given in Fig. 2. Similarities and differences are noted—in this case, contrasting seasonal differences between northern winter and southern summer. In exploring such receptor condition differences, various components of a complex system are immediately apparent. It has been proposed that bottom-side ionospheric processes control MSTID occurrence patterns (Kelley et al., 2003). Specifically, their origin is in the summer hemisphere where low-altitude plasma can persist after sunset and/or appear in sudden sporadic layers—and thus structures in the nighttime summer hemisphere can be mapped to the opposite (winter) hemisphere. Such electrodynamical coupling from one hemisphere to the other must be affected by nonuniform B-field conditions versus longitude. With MSTID propagation vectors not perpendicular to magnetic meridians, and those meridians having different orientations (declinations) with respect to geographic meridians, a complex longitude-dependent propagation pattern can emerge (Martinis et al., 2018, 2019). Clearly, new approaches from the ionospheric complexity perspective are needed to advance our understanding of the electrobuoyancy/MSTID wave phenomena.

3 Transitions from conjugate to nonconjugate science: Coherence of ionospheres not connected by B-field lines

Solar-wind-induced modifications to a planet's global magnetic field topology define the field of heliosphere-magnetosphere interactions throughout the solar system. At Earth, the result is a dramatic departure from dipole geometry driven by merging of magnetic fields from the solar wind and those of the magnetosphere. The result is an overall geospace domain defined magnetically by a dayside boundary (magnetopause) and a long geomagnetic tail on the nightside. The basic physics of magnetosphere formation is covered in many of the reference books listed in the Introduction; more highly focused books on the auroral and polar ionosphere are those by Akasofu (1968, 2003), Eather (1980), and Hunsucher and Hargreaves (2003).

From an ionospheric perspective, there are three dramatic consequences that result from the highly modified dipole geometry: (a) magnetic field-line merging results in energetic particle precipitation upon the high-latitude neutral atmosphere. As with photons, there is a range of heights reached by these energetic ions and electrons. This leads to the production of plasma by non-solar-photon means to create the auroral ionosphere—and the auroral emissions associated with it; (b) plasma motions driven by the electrodynamics of the solar wind – magnetosphere interaction transport solar-produced plasma from the dayside to the nightside, as well as high-speed return flows that move nightside plasma toward the dayside. Thus, a high-latitude circulation pattern arises that governs virtually all of the ionospheric patterns at high latitudes; and (c) given that aurora are associated with the last closed B-line, classic conjugate point science is no longer a meaningful concept poleward of the aurora. All of these processes are indicated schematically in Fig. 2 of Chapter 1.

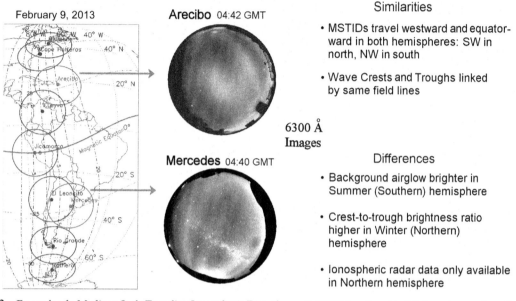

Boston University all-sky-imagers

Geomagnetic conjugate science feature: Medium Scale Travelling Ionospheric Disturbances

February 9, 2013

Arecibo 04:42 GMT

Mercedes 04:40 GMT

6300 Å
Images

Similarities

- MSTIDs travel westward and equatorward in both hemispheres: SW in north, NW in south
- Wave Crests and Troughs linked by same field lines

Differences

- Background airglow brighter in Summer (Southern) hemisphere
- Crest-to-trough brightness ratio higher in Winter (Northern) hemisphere
- Ionospheric radar data only available in Northern hemisphere

FIG. 2 Example of a Medium-Scale Traveling Ionospheric Disturbance (MSTIDs) captured in 630.0-nm images from Boston University all-sky imagers at the Arecibo Observatory (Puerto Rico) and the Mercedes Observatory (Argentina). Similarities and difference occur at these conjugate points. *From Martinis, C., Baumgardner, J., Wroten, J., Mendillo, M., 2018. All-sky-imaging capabilities for ionospheric space weather research using geomagnetic conjugate point observing sites. Adv. Space Res. 61 (7), 1636–1651. https://doi.org/10.1016/j.asr.2017.07.021.*

The complexity of the auroral and polar ionosphere arises from the fact that while specific processes are not mapped from one hemisphere to the other, there is a remarkable degree of coherence in high-latitude ionospheric morphologies. The features shown for the northern hemisphere in Fig. 2 of Chapter 1 also exist in the southern hemisphere—auroral ovals, plasma convection patterns, and polar cap phenomena. It is the solar wind-magnetosphere interaction that creates this symmetry on the largest of scales. During periods of enhanced geomagnetic activity, disturbances launched from tail regions due to magnetic field reconnection produce signatures in both hemispheres that should be magnetically conjugate. Indeed, this has been observed for many years (e.g., Frey et al., 1999). Yet, departures from conjugacy have also been noted (Watanabe et al., 2007; Reistad et al., 2013; Østgaard et al., 2015), and thus the specifics (and complexity) of auroral conjugate effects are an active area of investigation. In addition to conjugate topics, current research focuses on the ever-decreasing scales sizes of plasma structures and optical emissions capable of being observed with new high-resolution (spatial and temporal) radars and optical systems.

A recent example of the multisensor data fusion approach of auroral effects deals with studies of GPS phase and amplitude scintillations using an array of state-of-the-art instruments in Alaska (Semeter et al., 2017). Radio propagation disruptions have long been known to occur in ionospheric regions experiencing aurora (see textbook references mentioned earlier). The vast use of Global Navigation Satellite Systems (GNSS)—such as GPS for geolocation needs—has resulted in two coupled issues: (1) When will a GPS-to-ground-receiver raypath experience loss of lock (LL) and thus removing it from service? and (2) How can studies of the spatial-temporal details of the LL phenomenon advance our knowledge of the geospace system at high latitudes? Using an array of nine GPS receivers within the field-of-view (FOV) of a single wide-field imaging system, Semeter et al. (2017) advanced such

studies by increasing the time resolution of optical and radar instruments to 1 s. For reference, the images shown in Figs. 5 and 6 of Chapter 1 required 2 min of integration times, and ISR data shown in Fig. 1 of Chapter 2 and the TEC data in Fig. 2 of Chapter 2 typically have resolutions of several minutes.

Fig. 3 gives a set of high-resolution results spanning 4 min on the night of October 7, 2015. The optical images show aurora at 557.7 nm and the red circles show all of the GPS raypath intersection points recorded by the nine receivers. The yellow star marks the imaging system's location. In frame (a) all of the GPS signals are received unaffected by the aurora. As time progresses, a loss of lock is shown where

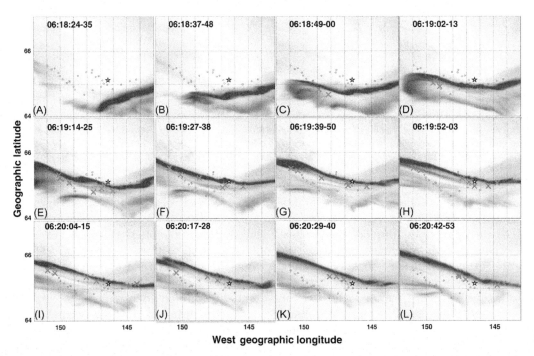

FIG. 3 An example of auroral research using new levels of spatial and temporal resolution for optical and radio observations. The images document the auroral green emission during a geomagnetic storm. Inserted within each frame are small red circles that show where GPS raypaths to nine stations occur. When loss of lock (LL) happens, the red circles are turned into bold X-notations. With data available every second, unprecedented definition occurs revealing that GPS degradation occurs along the edges of auroral forms at E-region altitudes. *From Semeter, J., Mrak, S., Hirsch, M., Swoboda, J., Akbari, H., Starr, G., et al., 2017. GPS signal corruption by the discrete aurora: precise measurements from the Mahali experiment, Geophys. Res. Lett. 44, 9539–9546. https://doi.org/10.1002/2017GL073570; see text.*

a thin red circle becomes a bold red cross. As the auroral forms move poleward a storm-time pattern emerges. The LL events consistently occur at edges of the auroral forms where precipitation causes ionospheric irregularities that results in phase and amplitude scintillations. The fact that the optical emissions are known to occur at heights of ~120 km (within the "auroral E-layer"), coupled to the fact that the LL points are consistently located at the edge of the aurora, show that the cause of the LL is within the E-layer. This, in turn, sets limits upon instability mechanisms, introducing levels of complexity not previously envisioned.

References

Aarons, J., 1993. The longitudinal morphology of equatorial F layer irregularities relevant to their occurrence. Space Sci. Rev. 63, 209.

Abdu, M., Batista, I., Reinisch, B., de Souza, J., Sobral, J., Pedersen, T., Medeiros, A., Schuch, N., de Paula, E., Groves, K., 2009. Conjugate point equatorial experiment (COPEX) campaign in Brazil: electrodynamics highlights on spread-F development conditions and day-to-day variability. J. Geophys. Res. 114 (A4), A04308. https://doi.org/10.1029/2008JA013749.

Akasofu, S.-I., 1968. Polar and Magnetospheric Substorms. Springer-Verlag, New York.

Akasofu, S.-I., 2003. Exploring the Secrets of the Aurora. Kluwer Academic Publishers, Dordrecht.

Behnke, R.A., 1979. F layer height bands in the nocturnal ionosphere over Arecibo. J. Geophys. Res. 84, 974–978.

Eather, R., 1980. Majestic Lights: The Aurora in Science. History and the Arts. AGU, Washington, DC.

Frey, H., Mende, S., Vo, H., Parks, G., 1999. Conjugate observations of optical aurora with polar satellite and ground-based camera. Adv. Space Res. 23 (10), 1647–1652.

Fukushima, D., Shiokawa, K., Otsuda, Y., Nishioka, M., Kubota, M., Tsugawa, T., Nagatsuma, T., Komonjinda, S., Yatini, C.Y., 2015. Geomagnetically conjugate observations of plasma bubbles and thermospheric neutral winds at low latitudes. J. Geophys. Res. Space Physics 120, 2222–2231. https://doi.org/10.1002/2014JA020398.

Hunsucher, R.D., Hargreaves, J.K., 2003. The High-Latitude Ionosphere and Its Effects on Radio Propagation. Cambridge University Press, Cambridge.

Kelley, M., 2009. The Earth's Ionosphere: Plasma Physics and Electrodynamics, second ed. Elsevier Academic Press, New York.

Kelley, M., Makela, J., Saito, A., Aponte, N., Sulzer, M., Gonzalez, S., 2000. On the electrical structure of airglow depletion/height layer bands over Arecibo. Geophys. Res. Lett. 27 (18), 2837–2840.

Kelley, M., Haldoupis, C., Nichols, M., Makela, J., Belehadi, A., Shalimov, S., Wong, V., 2003. Case studies of coupling between the E and F regions during unstable sporadic-E conditions. J. Geophys. Res. 108 (A12), 1447. https://doi.org/10.1029/2003JA009933.

Martinis, C., Mendillo, M., Aarons, J., 2005. Toward a synthesis of equatorial spread F onset and suppression during geomagnetic storms. J. Geophys. Res. 110, A07306. https://doi.org/10.1029/2003JA010362.

Martinis, C., Baumgardner, J., Wroten, J., Mendillo, M., 2010. Seasonal dependence of MSTIDs obtained from 630.0 nm airglow imaging at Arecibo. Geophys. Res. Lett. 37. https://doi.org/10.1029/2010GL043569.

Martinis, C., Baumgardner, J., Wroten, J., Mendillo, M., 2011. All-sky imaging observations of conjugate medium scale traveling ionospheric disturbances in the American sector. J. Geophys. Res. 116. https://doi.org/10.1029/2010JA016264.

Martinis, C., Baumgardner, J., Wroten, J., Mendillo, M., 2018. All-sky-imaging capabilities for ionospheric space weather research using geomagnetic conjugate point observing sites. Adv. Space Res. 61 (7), 1636–1651. https://doi.org/10.1016/j.asr.2017.07.021.

Martinis, C., Baumgardner, J., Mendillo, M., Wroten, J., MacDonald, T., Kosch, M., Lazzarin, M., Umbriaco, G., 2019. First conjugate observations of medium scale travelling Ionospheric disturbances (MSTIDs) in the Europe-Africa longitude sector. J. Geophys. Res. Space Physics 124. https://doi.org/10.1029/2018JA026018.

Mendillo, M., Baumgardner, J., 1982. Airglow characteristics of equatorial plasma depletions. J. Geophys. Res. 87, 7641–7652. https://doi.org/10.1029/JA087iA09p07641.

Mendillo, M., Narvaez, C., 2010. Ionospheric storms at geophysically-equivalent sites—part 2: local time storm patterns for sub-auroral ionospheres. Ann. Geophys. 28 (7), 1449–1462. https://doi.org/10.5194/angeo-28-1449-2010.

Mendillo, M., Baumgardner, J., Nottingham, D., Aarons, J., Reinisch, B., Scali, J., Kelley, M., 1997. Investigations of thermospheric-ionospheric dynamics with 6300-Å images from the Arecibo Observatory. J. Geophys. Res. 102, 7331–7343. https://doi.org/10.1029/96JA02786.

Miller, C., Swartz, W., Kelley, M., Mendillo, M., Nottingham, D., Scali, J., Reinisch, B., 1997. Electrodynamics of midlatitude spread F, 1. Observations of unstable gravity wave-induced ionospheric electric fields at tropical latitudes. J. Geophys. Res. 102 (A6), 11521–11532.

Østgaard, N., Reistad, J., Tenfjord, P., Laundal, K., Snekvik, K., Milan, S., Haaland, S., 2015. Mechanisms that produce auroral asymmetries in conjugate hemispheres. In: Zhang, Y., Paxton, L.J. (Eds.), Auroral Dynamics and

Space Weather. In: Geophysical Monograph Series, vol. 215. American Geophysical Union (Chapter 11) https://doi.org/10.1002/9781118978719.

Otsuka, Y., Shiokawa, K., Ogawa, T., Wilkinson, P., 2002. Geomagnetic conjugate observations of equatorial airglow depletions. Geophys. Res. Lett. 29, 15. https://doi.org/10.1029/2002GL015347.

Otsuka, Y., Shiokawa, K., Ogawa, T., Wilkinson, P., 2004. Geomagnetic conjugate observations of medium-scale traveling ionospheric disturbances at midlatitude using all-sky airglow imagers. Geophys. Res. Lett. 31. https://doi.org/10.1029/2004GL020262.

Perkins, F., 1973. Spread F and ionospheric currents. J. Geophys. Res. 78, 218.

Reistad, J., Østgaard, N., Laundal, K., Oksavik, K., 2013. On the non-conjugacy of nightside aurora and their generator mechanisms. J. Geophys. Res. Space Physics 118, 3394–3406.

Semeter, J., Mrak, S., Hirsch, M., Swoboda, J., Akbari, H., Starr, G., Hampton, D., Erickson, P., Lind, F., Coster, A., Pankratius, V., 2017. GPS signal corruption by the discrete aurora: precise measurements from the Mahali experiment. Geophys. Res. Lett. 44, 9539–9546. https://doi.org/10.1002/2017GL073570.

Shiokawa, K., Otsuka, Y., Ogawa, T., Wilkinson, P., 2004. Time evolution of high-altitude plasma bubbles imaged at

geomagnetic conjugate points. Ann. Geophys. 22, 3137–3143. https://doi.org/10.5194/angeo-22=3137-2004.

Sobral, J.H.A., Abdu, M.A., Pedersen, T.R., Castilho, V.M., Arruda, D.C.S., Muella, M.T.A.H., …Bertoni, F.C.P., 2009. Ionospheric zonal velocities at conjugate points over Brazil during the COPEX campaign: experimental observations and theoretical validations. J. Geophys. Res. 114 (A4), A04309. https://doi.org/10.1029/2008JA013896.

Stephan, A., Colerico, M., Mendillo, M., Reinisch, B., Anderson, D., 2002. Suppression of equatorial spread F by sporadic E. J. Geophys. Res. 107 (A2), 1021. https://doi.org/10.1029/2001JA000162.

Tsunoda, R., 1985. Control of the seasonal and longitudinal occurrence of equatorial scintillations by the longitudinal gradient in integrated E region Pederson conductivity. J. Geophys. Res. 90, 447.

Watanabe, M., Kadokura, A., Sato, N., Saemundsson, T., 2007. Absence of geomagnetic conjugacy in pulsating auroras. Geophys. Res. Lett. 34. https://doi.org/10.1029/2007GL030469.

Weber, E., Buchau, J., Eather, R., Mende, S., 1978. North-south aligned equatorial airglow depletions. J. Geophys. Res. 83, 712–716. https://doi.org/10.1029/JA083iA02p00712.

Status and future directions

Michael Mendillo

Department of Astronomy, Boston University, Boston, MA, United States

In this brief overview of ionospheric morphologies and processes, a number of well-understood mechanisms have been identified. These include production and loss of ionospheric plasma, dynamical influences upon photochemical systems, and instabilities that occur within those systems. The concept that coupling occurs between individual processes within the ionosphere, as well as between altitude and latitude regimes above and below the ionosphere, is not a new idea. Considerable progress has been made in the understanding of individual coupling scenarios. These include energetic particles in the magnetosphere and the aurora they produce in the high-latitude atmosphere; waves and tides in the mesosphere as a source of dynamics and energy deposit in the thermosphere; couplings between gravity waves in the thermosphere and ionospheric perturbations within the same system; the electro-buoyance wave phenomena generated entirely within the ionosphere; relationships between large- and small-scale spatial effects and long- and short-term temporal effects in the auroral ionosphere. The need for new insights from complexity theory arises from the fact that so many mechanisms are in progress within the

same system that, in turn, is formed by many coupled subsystems. The foremost lesson from the foregoing examples is that ionospheric variability, ESF onset/growth/suppression, and MSTID dynamics occur when the Sun and solar wind are *either* quiet or disturbed. That does not mean that solar and heliospheric conditions are uninvolved, but that the interplay between local effects and external drivers is poorly understood due to the complexity of the system. Equally complex is the auroral and polar ionosphere where the response is regulated entirely by the degree of disturbance of the solar wind. That fundamental paradigms exist that are mutually inclusive and exclusive surely encourage the search for meaningful analyses from complexity theory.

Finally, ionospheric physics was and remains a data-driven science. The tools of complexity research can thus be tuned to every-increasing data sets and to varying modes of data usage. Observations of the ionosphere's total electron content are now made from over 6000 ground stations observing over 30 GPS satellites. Nearly 100 all-sky-imagers are distributed worldwide, and there are over 30 SuperDarn radars at polar, high, and middle latitudes. At the onset of 2019, current satellites making ionospheric observations include DMSP and SWARM, with GOLD and ICON in service. An unprecedented impact upon auroral research from "citizen science" teams is now a reality (MacDonald et al., 2018). Such a data-driven science requires innovative uses of scientific observations. These include data assimilation by models running in real time and the application of complexity protocols to created science yield. As shown in Fig. 1, McGranaghan et al. (2017) have identified four aspects of data termed Volume, Variety, Veracity, and Velocity. These refer to the number of observing instruments, the diversity of such platforms, the reliability of data obtained, and speed at which data can be analyzed. When the message of Fig. 2 of Chapter 1 (showing the diversity of science topics relevant to ionospheric physics) is merged with the Geospace data sources in Fig. 1, there is little doubt that the theories and tools being developed for complex systems have a worthy topic in the Earth's ionosphere.

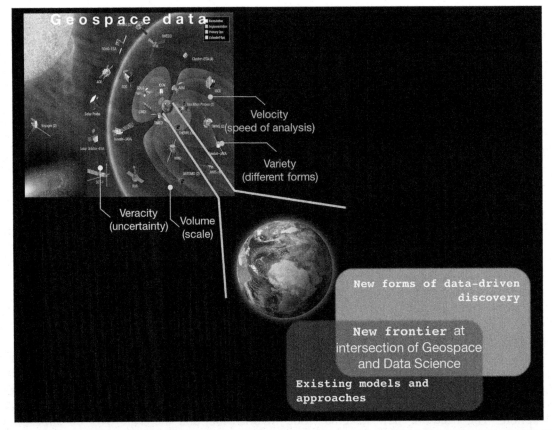

FIG. 1 An illustration of the relationship between new sources of data and their impact upon the data-driven science of the geospace environment. *From McGranaghan, R., Bhatt, A., Matsuo, T., Mannucci, A., Semeter, J., Datta-Barus, S. 2017. Ushering in a new frontier in geospace through data science. J. Geophys. Res. Space Phys. 122, 12,586–12,590. https://doi.org/10.1002/2017JA024835.*

References

MacDonald, E., Donovan, E., Nishimura, Y., Case, N.A., Gillies, D.M., Gallardo-Lacourt, B., Archer, W.E., Spanswick, E.L., Bourassa, N., Connors, M., Heavner, M., Jacket, B., Kosar, B., Knudsen, D.J., Ratzlaff, C., Schofield, I., 2018. New science in plain sight: citizen scientists lead to the discovery of optical structure in the upper atmosphere. Sci. Adv. 4(3). eaaq0030, https://doi.org/10.1126/sciadv.aaq0030.

McGranaghan, R., Bhatt, A., Matsuo, T., Mannucci, A., Semeter, J., Datta-Barus, S., 2017. Ushering in a new frontier in geospace through data science. J. Geophys. Res. Space Phys. 122, 12,586–12,590. https://doi.org/10.1002/2017JA024835.

Mid-latitude ionospheric features: Natural complexity in action

Philip J. Erickson

Atmospheric and Geospace Sciences Group, MIT Haystack Observatory, Westford, MA, United States

1 Introduction

Mendillo (this volume) has summarized the benefits of complexity approaches that treat Earth's ionosphere as one element of a multifaceted system with many interlocking parts. This approach, involving the information-rich paradigm of coupling, has many benefits beyond a purely reductionist framework that has dominated earlier treatments of the ionosphere as a largely self-contained system with static upper and lower boundaries. We amplify Mendillo's discussion with additional examples taken from magnetosphere-ionosphere dynamic system effects in the mid-latitude and subauroral ionosphere, defined here as threaded by field lines which cross the equatorial plane at $L \sim 2$–5 in classic McIlwain coordinates (McIlwain, 1961). We emphasize that this treatment is primarily illustrative, and for more detail we direct the reader to more extensive references such as Kelley (2009), Schunk and Nagy (2009), and Heelis (2004).

2 Plasmasphere boundary layer

2.1 Overview and definition

Historically, study of the cold, dense plasma characterizing mid-latitudes used the physical organizing principle of diffusive processes strongly controlled by the background magnetic field, in a manner similar to solar XUV-driven altitude-dependent ionosphere descriptions from Chapman production and loss theory. Here, the strong preference for charged particles to move along magnetic field lines in the strongly magnetized terrestrial ionosphere is a guiding principle. Work at the Stanford radio group beginning in the 1950s concentrated on observations of naturally generated electromagnetic plasma waves ("whistlers") at low frequencies, typically ~ 1–20 kHz, launched by terrestrial lightning and propagating from one hemisphere to the other along the strong background magnetic field (see Helliwell, 1965 and the excellent history of the Stanford VLF group by Carpenter, 2015). Dispersive properties of the VLF waves as received at ground stations allowed derivation of the total electron density encountered as the wave propagated along the field line. Using a relatively small number of spatially distributed southern latitude monitoring stations, VLF-based studies quickly found that a large change in field-aligned electron content existed as one proceeded magnetically north within mid and subauroral magnetic latitudes. This was thought to be characteristic of a solar XUV created and diffusively dominated mid-latitude ionosphere, bound to corotating field lines, transitioning to dynamic high-latitude regions, where diffusion processes were overtaken by strong electrodynamic and transport forcing. A relatively simple organizing principle emerged, consisting of a sharp, abrupt, and single magnetically aligned plasma density transition in each hemisphere, termed the plasmapause (Angerami and Carpenter, 1966; Carpenter, 1966). The plasmapause thus delineated more dense, corotating plasma regions from less dense but more transport-dominated regions.

However, within a decade, the relatively fixed view of a static inner plasmasphere was quickly observed to have significant and complex time-dependent deviations from this principle. Statistical studies of electron density spatial structure using ionosonde and VLF observation networks along with fixed satellite beacon TEC observations (e.g., Mendillo and Klobuchar, 1975) concluded that the simple, static picture of a sharp and stable plasmapause almost never occurred. In fact, at an aggregate level, the coupled inner magnetosphere-ionosphere system was found to be nearly always in a state of dynamic recovery from geomagnetic storm-driven energy inputs (Carpenter and Park, 1973). This was partially due to the long postdisturbance times (hours to days) required to diffusively refill a long (1000 s of km), hemisphere-spanning magnetic mid-latitude flux tube with plasma through

neutral atmosphere ionization from solar sources (Singh and Horwitz, 1992; Lemaire and Gringauz, 1998; Denton et al., 2012). The time span dictated by classical production and loss theory is in practice long enough that the system does not remain stationary to subsequent storm-driven reconfigurations.

In the modern era, global navigation satellite systems (GNSS)-based TEC global measurements (Rideout and Coster, 2006) provide a far greater spatial and temporal sampling cadence that has been profitably applied to mid-latitude studies. When combined with mid-latitude incoherent scatter radar altitude-resolved measurements of plasma density and temperature (e.g., Foster et al., 2002), observations have provided further and even more compelling evidence of mid-latitude plasma structure and variation that is characterized by not a single boundary, but an interchange region whose characteristics are constantly in dynamic flux. Accordingly, Carpenter and Lemaire (2004) and Darrouzet et al. (2009) assimilated these and other observational ionosphere and plasmasphere findings and replaced the notion of a single plasmapause with a plasmasphere boundary layer (PBL). This more comprehensive paradigm is well suited to complexity approaches, since variability and instability naturally arise from the combined overlap of cold, dense solar produced plasma with hot, tenuous plasma having origins in the equatorial plane inner magnetosphere's plasma sheet.

2.2 Frontier questions on PBL-region ionospheric variability

Since PBL regions are often observed in a state of flux, research questions for PBL dynamics are similar to those of ionospheric variability and provide ample motivation for complexity approaches. For instance, what drivers of PBL complexity are quasistable and can be numerically determined to the point of predictability? Alternately, which are seemingly random agents of change? How many states exist and what are the triggers of transition between one point and another? Fortunately, the community does not need to start from a blank page, as multiple mechanisms are known to exist which link the mid-latitude ionosphere with other regions, and some of these mechanisms have been examined extensively for parameterization purposes.

For instance, field aligned refilling of F-region O^+ dominated plasma from the vast H^+ reservoir at higher topside altitudes has been known for some decades to be a potentially important factor at night, through the efficiency of resonant charge exchange reactions between H^+, O^+, H, and O (Hanson and Ortenburger, 1961; Hanson and Patterson, 1964; MacPherson et al., 1998). These questions of production and loss versus transport reach back to the earliest days of ionospheric study, including the puzzle of why the F-region ionosphere does not disappear at night due to recombination (Hanson and Ortenburger, 1961). Recently, the community has realized that global knowledge of these time-dependent coupling processes is not yet available with sufficient sophistication to properly quantify variations, for example, in how H^+ refilling competes numerically with in-place F-region XUV production and horizontal cross-field transport. Furthermore, in the PBL, the frequent storm-time reconfiguration of formerly closed field lines to a relatively open state provides a dynamic situation unique to those regions since evacuation and refilling of flux tubes is highly dependent on ionospheric and plasmaspheric preconditioning in both space and time.

Similarly, magnetic conjugacy can be a dominant factor when combined with the altitude-dependent strength of the background terrestrial magnetic field, as conservation of magnetic flux and field line isopotentiality implies spatial scale "filtering" of cross-field structures (Farley, 1960). However, accurate knowledge of coupling between ionosphere, atmosphere, and magnetosphere is needed well beyond a static picture in order to quantify when local changes in E-region Pedersen and Hall conductivity

provide a preconditioning means of electrically "shorting out" the magnetic conjugacy filtering principle between source and sink, with corresponding impacts on ionospheric spatial electron density structure. Multiple studies have shown in fact that straightforward application of conjugacy is not always correct. Mendillo (this volume) has discussed a number of these conjugacy processes. As an additional example, Foster and Rideout (2007) unexpectedly found that strong TEC enhancements at the base of a storm-enhanced density (SED) plume in the North American sector were considerably more extensive than in its magnetic conjugate region. A complexity approach could focus on quantifying those cases when conductivity variations might dominate and contrast them with situations, where changes in neutral atmosphere sources (such as O to N_2 ratios) are more important. Such a class of studies has a large potential to greatly clarify understanding of the generation of important spatial ionospheric structuring, with effects on prediction of ionospheric variations.

3 Ionospheric consequences of mid-latitude M-I electrodynamics

Consideration of mid-latitude ionospheric electric field/cross-field ion velocity variations, and their relation to SED structures, provides another set of processes with electrodynamic origin whose understanding demands the use of complexity and system-scale approaches. PBL structuring in the form of defined electron density plumes originates primarily on inner magnetospheric field lines in the premidnight and dusk sector. During geomagnetic disturbances, the overlying ring current in the inner magnetosphere (Daglis et al., 1999) becomes asymmetric, with a resulting pressure imbalance that injects particles within the PBL earthward of those locations magnetically tied to the electron plasma sheet edge. This is the region of Birkeland Region 2 currents within the subauroral ionosphere, manifesting itself as a primarily poleward electric field driven by Pedersen current closure.

3.1 SED and SAPS/SAID

During storm times, the afternoon sector ionosphere can have greatly elevated electron density in spatially restricted regions, with features variously known as the positive phase storm effect (Jones, 1971; Mendillo and Klobuchar, 1975; Rodger et al., 1989; Huang et al., 2005) or SED (Foster et al., 2002; Coster and Foster, 2007). Closely coupled to the creation of these electron density features is the existence of a large storm-time mid-latitude poleward ionospheric electric field, originally known as the polarization jet (Galperin, 2002) and now as the subauroral polarization stream (SAPS) (Foster and Burke, 2002). SAPS drives strong cross-field flows at levels of hundreds of m s^{-1} up to 2+km s^{-1} in a structure extending over a few degrees in magnetic latitude. The appearance of SAPS can be prompt, within $\sim$30 min of a substorm onset (Anderson et al., 1993). SAPS flows cause considerable ionospheric spatial variability. In particular, at their equatorward edge within subauroral regions, SAPS overlaps the normal SED-associated ionospheric electron density gradient within the PBL in the dusk sector. This overlap and fast flow, especially compared to normal quiescent ionospheric motions <100 m s^{-1}, creates considerable ionospheric electron density variations and provides significant sunward ionospheric mass flux at values exceeding 10^{14}m^{-2}s^{-1} (e.g., Foster et al., 2004b). A prime associated mechanism for electron density structuring is the large acceleration of recombination processes over normal levels (Schunk et al., 1975), leading not only to SED production but also to dramatic plasma density reductions poleward of SED structures such as seen at the edge of the PBL and in the mid-latitude trough. An excellent review of trough features and processes can be found in Moffett and Quegan (1983).

Figs. 1 and 2 show example observations of SAPS and SED features in the storm-time mid-latitude ionosphere. The top panel of Fig. 1 plots electron density as a function of geodetic latitude and longitude in the American sector dusk ionosphere at 22:40 UTC during the large March 17, 2013 storm event as observed by the Millstone Hill incoherent scatter radar (42.6 N geodetic latitude, 288.5 E geodetic latitude; $\sim$53 degrees invariant magnetic latitude) as it scanned to the northwest. The low elevation nature of the radar scan means that altitude increases as the radar observes further away in latitude and longitude, and locations corresponding to the ionospheric F-region at 310 km altitude are marked as magenta stars on the scan. A clear SED plume of spatially confined and enhanced density with sharp density gradients on its poleward edge was encountered near 46 degrees geodetic latitude, extending beyond the F-region peak into the near topside (far ranges of the radar scan). The second panel provides a line plot as a function of invariant magnetic latitude of F-region electron density and magnetic west velocity, extracted from scan locations data along the line of magenta stars. The demarcation of high-latitude processes, defined as the equatorward extent of electron precipitation, is provided as a magenta line on the density and velocity plots from particle data (not shown) on the DMSP F16A spacecraft at 840 km altitude, passing northbound through the sector slightly earlier at 22:03 UTC. The SED plume is seen as a "shoulder" of two times density enhancement at the edge of the plasmasphere near 56 Λ, with the mid-latitude trough poleward of this feature at approximately 60 Λ ($\sim L = 4$). A gradually increasing several degree wide SAPS velocity feature approached 600–700 m s^{-1} magnetic west velocity, overlapping the SED plume and providing a significant sunward mass flux channel. Fig. 2 plots a polar projection two-dimensional global TEC map, using the worldwide GNSS receiver network and employing data at the same UTC

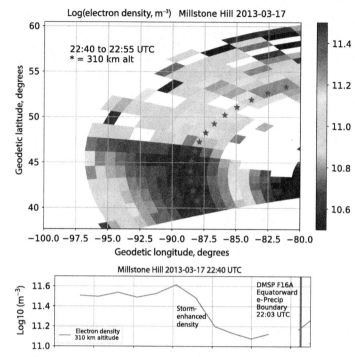

FIG. 1 Mid-latitude ionospheric dynamic structure in the dusk sector at subauroral latitudes during the March 17, 2013 large geomagnetic storm, measured using the Millstone Hill incoherent scatter radar. The radar scan (*top panel*) shows a snapshot of mid-latitude electron density distribution as a function of latitude and longitude. Electron density and magnetic westward velocity in the F-region at 310 km altitude (*magenta stars* in the scan) is plotted as a function of invariant latitude Λ in the *middle and bottom panels*. SED and subauroral polarization stream (SAPS) are marked.

as the Millstone Hill ionospheric measurements. TEC is an integrated quantity out to 20,000+km altitude, but is dominated by ionospheric electron density (typically more than ∼50%). A clear ionospheric SED plume, associated with fast SAPS flow, can be seen stretching over North America from the dusk sector toward the noontime high-latitude ionosphere.

3.2 Mid-latitude fine-scale structure associated with SAPS/SAID

The presence of mid-latitude ion flow through the low-velocity background ionosphere is inherently unstable to the Farley-Buneman two stream instability (Farley, 1960). This sets up a highly coherent plasma wave structure which can interact and shape fine-scale electron density and velocity features within the larger SAPS envelope through electrodynamic action driven by a highly variable electric field (Foster et al., 2004a; Mishin and Burke, 2005). Fig. 3 provides an example, adapted from Erickson et al. (2002), of dynamic and spatially diverse poleward electric field structure observed with coherent scatter radar techniques (Foster and Erickson, 2000). The plot demonstrates that even within a broad SAPS envelope (not shown), the observed ionospheric poleward electric field or equivalently magnetic westward velocity can have superimposed

GNSS TEC 2013-03-17 22:20 UTC

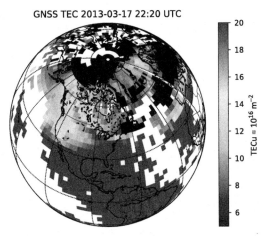

FIG. 2 North American sector total columnar electron content in TEC units ($10^{16}m^{-2}$) as measured by GNSS during March 17, 2013 at 22:20 UTC during the Millstone Hill ionospheric measurements in Fig. 1. A clear SED plume can be seen stretching over North America toward the noontime cusp.

multiple instances of very narrow (~0.1 degree magnetic latitude; ~10-km spatial scale), intense electric fields having lifetimes as short as 1.5 min and gradients on their equatorward edge as large as 4 mV m^{-1} km^{-1}. A number of subauroral studies using DMSP satellite passes through the SAPS region have found similar narrow and intense spatial scales, primarily at substorm times, known as subauroral ion drifts (SAIDs) (Spiro et al., 1979). Some SAID structures in fact provide ionospheric sunward velocities of greater than 3 km s^{-1} imposed on the SAPS envelope of 500 +m s^{-1} (Anderson et al., 1993, 2001).

The March 17, 2013 storm presented already in Figs. 1 and 2 provides two other complementary examples of highly structured and dynamic ionospheric velocity and density structure, using in situ observations by the DMSP satellite

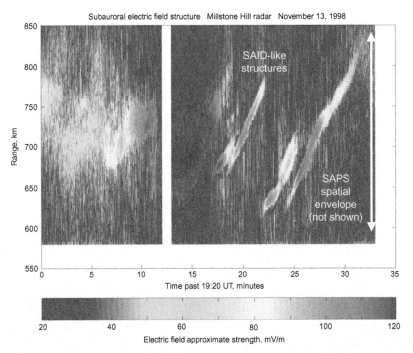

FIG. 3 SAPS fine-scale electric field structure. During this intense geomagnetic storm event, the broad mid-latitude SAPS envelope is seen to break up into multiple narrow features with short lifetimes, providing highly structured velocity fields within the mid-latitude ionosphere. *Modified from Erickson, P.J., Foster, J.C., Holt, J.M., 2002. Inferred electric field variability in the polarization jet from Millstone Hill E region coherent scatter observations. Radio Sci. 37 (2), 1014–1027.*

cluster in the topside ionosphere. The first example comes from the initial main phase onset of the storm, and Fig. 4 presents data just after 08 UTC from two different DMSP satellite passes at 840 km altitude that both pass through the same 16 MLT afternoon sector with approximately 19 min separation. The top panel plots the DMSP F16A electron particle environment, showing differential electron flux as a function of energy and magnetic latitude. The equatorward edge of electron precipitation, marking the high-latitude boundary, occurs at ~67Λ. The second panel plots the DMSP F16A ion differential flux as a function of energy and magnetic latitude. Below the high-latitude cutoff at 67 Λ, a position-dependent separatrix in ion particles at 10^3-10^4 eV marks the ionospheric

footprint of the magnetospheric zero energy Alfvén boundary closely associated with the ring current boundary. This latter feature associates the ~63−67Λ region with structured inner magnetospheric electrodynamic drivers. The third and fourth panels plot electron density and cross-track horizontal ion velocity from both the F16A and F17A platforms. At ~63−67Λ, the relatively smooth ionospheric velocity field at 08:16 UTC abruptly becomes highly structured on ~10-km spatial scales just 19 min later at 08:34 UTC. Furthermore, electron density in the same region is associated with SED effects and increases by more than three times over the same ~19-min interval. In general, this dynamic ionospheric response provides a hallmark of fast mid-latitude storm-time

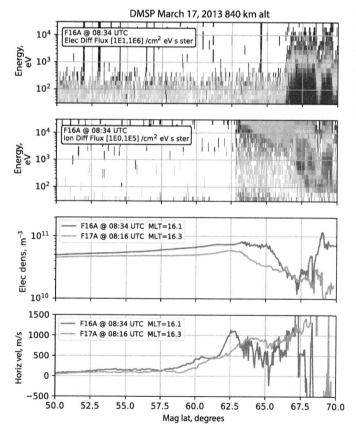

FIG. 4 Subauroral precipitation and severe electrodynamic electric field structure in the dusk sector topside ionosphere measured using the DMSP F16A and F17A satellites at ~08:30 UTC on March 17, 2013. Abrupt onset of highly structured subauroral ion velocities occurs at 08:34 UTC associated with ring current ion precipitation.

ionospheric structuring that can at times rival that seen at equatorial latitudes.

A second example from DMSP observations later in the event at 18:42 UTC, near the peak of the magnetic storm, is plotted in Fig. 5, with data products shown in each panel in a similar manner as in Fig. 4. Here, the equatorial extent of electron precipitation is seen at much lower magnetic latitudes of 54Λ but without the clear ring current precipitation effects of the earlier storm period. The main dynamic feature at subauroral latitudes in this example is the motion and larger-scale structuring of the SAPS field over ~19-min time separation at the same afternoon sector MLT (curves in lower panel) despite a nearly time-stable electron density latitude profile within the PBL (third panel). Since

sunward mass flux is a product of both available electron density and cross-field ion velocity, this change shows that large storm-time mid-latitude horizontal mass fluxes in the ionosphere can abruptly shift by hundreds of kilometers in spatial location in response to rapid changes in electrodynamic forcing and response.

3.3 Resolving theories of SAPS/SAID and SED structure

Relevant for this chapter, comprehensive understanding of the dynamics of these features, and especially their associated ionospheric fine-scale temporal and spatial

FIG. 5 Subauroral precipitation and severe electrodynamic electric field structure in the dusk sector topside ionosphere measured using the DMSP F16A and F17A satellites at ~18:42 UTC on March 17, 2013. Sudden equatorward movement of the broad SAPS velocity envelope implies a dynamic change in cross-field ionospheric mass flux delivery to the high-latitude cusp.

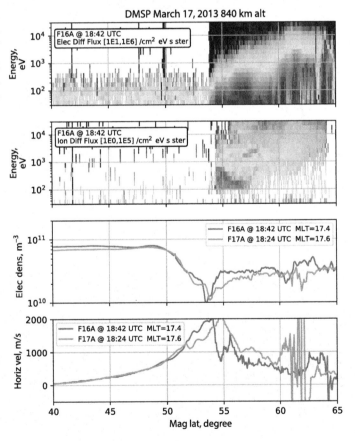

structure, once again requires a system-level analysis approach. In particular, two-way coupling paradigms are needed to move well beyond the community's original attempts to apply simple, one-way mechanisms for explanation of SAPS and SAID features. A detailed summary of initial mechanistic possible drivers of subauroral structuring is provided in the first part of the discussion by Makarevich et al. (2011). In brief, the most important prior isolated mechanisms used a framework of primary magnetospheric control and one-way ionospheric response in either a voltage generator mode driven by an electron precipitation created potential drop (variable field aligned current) (Southwood and Wolf, 1978) or a current generator mode (variable potential drop) (Anderson et al., 2001; Landry and Anderson, 2018). In contrast, however, a number of observations concluded that many SAPS and SAID features are insufficiently predicted with these approaches, particularly the appearance of SAPS electric fields within 10–15 min after the onset of asymmetric ring current pressure (e.g., Fig. 4) as opposed to the expected delay of hours to build up the driving radial electric field if pressure is the sole driver (Galperin, 2002; Mishin and Mishin, 2007; Mishin et al., 2017). Once again, system-level two-way coupled analysis, using more than one energy source and sink, is essential to resolve the observational and theoretical discrepancy. In fact, this approach is amenable as well to novel "out of the box" ideas which explain mid-latitude ionospheric dynamic effects using known processes from other plasma physics realms. For example, Mishin and Mishin (2007) and Mishin (2016) use particle accelerator findings to assert that "plasmoid" injections within the plasma sheet can act as a key dynamic mechanism explaining the fast temporal response of SAPS and SAID ionospheric electric fields to substorm onsets.

4 Geospace system impacts of PBL electron density structuring

Mid-latitude ionospheric plasma major restructuring and dynamic complexity during geomagnetic storm events is now known to have truly global consequences across the entire geospace system, well beyond ionospheric dynamics itself. This greatly increases the impact and reach of mid-latitude ionospheric understanding employing more sophisticated, system-level complexity approaches.

In particular, the mid-latitude ionosphere provides a strong source for heavy, cold O^+ ion deposition far out into the magnetospheric ring current, plasma sheet, and indeed throughout geospace, through subauroral ionospheric mass flux feeding the high-latitude cusp (Freeman et al., 1977; Elphic et al., 1997; Williams, 1981; Daglis et al., 1999; Keika et al., 2013; Burke et al., 2016). This latter finding displaced an earlier, more isolated theory that inner magnetospheric O^+ ions observed in situ could only originate from solar wind particles. Further evidence of the ionosphere's crucial role in inner magnetospheric dynamics is available from multiple sources. Chappell (1974) observed regions of detached cold plasma in the inner magnetosphere of clear ionospheric origin. Heavy cold O^+ plasma has been identified in ionospheric plumes (Sandel et al., 2001) and in inner magnetosphere plasmaspheric tails (Goldstein and Sandel, 2005). The strong influence of the background magnetic field means that ionospheric and magnetospheric structuring of cold, dense plasma have tightly linked morphologies (Foster et al., 2002), implying that both regions play an active role in controlling ionospheric mass flow through electrodynamic coupling and must be studied jointly.

System-level theory provides arguably the only way forward in joint study of these effects, since no one region or process can be considered in isolation. For ionospheric research, a coupling-based approach also has a unifying property, adapting into the larger framework many previous findings on storm-time electron density structuring. These include positive phase dusk sector ionospheric density enhancements in limited longitude regions (Mendillo and Klobuchar, 1975), SED at F-region heights (e.g., Foster, 1993), ionospheric source connections to width and density of plasmaspheric plumes (Foster et al., 2002; Goldstein and Sandel, 2005), and large cross-field transport of plasma against corotation in the dusk sector (Foster and Erickson, 2013) where it competes with large positive mid-latitude storm-time electron density production (Heelis et al., 2009).

A dynamic complexity approach also informs predictions and refinements in understanding for ionospheric and PBL-connected effects on other geospace system processes known to be important. Recent successes of this methodology include the identification of ionospheric origin plasmaspheric plume material as having direct impact on storm-time Sun-Earth system energy input, as cold ionospheric plasma flow outwards through the inner magnetosphere participates in negative feedback regulation to large coronal mass ejection and geomagnetic inputs by slowing dayside reconnection through mass-loading field lines at the magnetopause (Walsh et al., 2014; Borovsky, 2014). The presence of cold dense ionospheric origin plasma in the inner magnetosphere also has profound impacts for understanding outer radiation belt production mechanisms through wave-particle interactions which create the relativistic energy outer radiation belt. In particular, elevated cold ionospheric electron density emanating from the highly structured PBL is now understood in superstorm cases to have a large preconditioning impact on electron cyclotron resonance efficiency (e.g., Abel and Thorne, 1998), shifting resonance to be highly efficient at producing ultrarelativistic electrons from 100 s of keV substorm-injected "seed" particles (Jaynes et al., 2015) in combination with strong VLF frequency waves provided by whistlers and human origin VLF transmissions (Foster et al., 2016).

5 Mid-latitude ionospheric irregularities

As a final mid-latitude ionospheric complexity example, PBL-region plasma interchange and overlap provides ample energy for the development of sharp, embedded temporal, and spatial gradients, along with cascade processes that can transfer energy from large scales to small ones. Steadily improving observational capabilities within magnetosphere-ionosphere coupling regions led to the realization that a number of embedded ionospheric irregularities (Fejer and Kelley, 1980), considered not to occur at mid- and subauroral latitudes, were in fact ubiquitous. In particular, decameter irregularities were unexpectedly discovered at magnetically quiet times within the PBL. These came from observations with the mid-latitude Super-DARN HF radar network (Greenwald et al., 2006), originally constructed for the purpose of detecting high-latitude convection patterns during intense storms with greatly expanded auroral ionospheric flows. The low-velocity irregularities, acting as tracers for quiet time ionospheric motions, unexpectedly occurred ~70% of the time (Ribeiro et al., 2012). However, more than one plausible instability generation mechanism exists, including gradient drift instabilities (GDI) fed by opposed electron density and ionospheric velocity gradients (Simon, 1963) and nonlinear evolution of temperature gradient instabilities (TGI) with opposed electron density and electron temperature gradients at the plasmasphere edge (Hudson and Kelley, 1976; Greenwald et al., 2006; de Larquier et al., 2014; Eltrass and Scales, 2014; Eltrass et al., 2014). Both conditions can exist in the PBL, and nonlinear effects also become important (e.g., Keskinen, 1984). The exact quantitative balance of these two mechanisms in creating PBL irregularities and ionospheric structure remains an unresolved frontier research topic, requiring both theoretical and observational joint work. Complexity-driven approaches to these problems are essential. For instance, Eltrass et al. (2016) combine theoretical calculations and observations to conclude that the most likely mechanism for mid-latitude irregularities arises not from single pathways but from cascades where TGI and/or GDI processes work together to move energy from kilometer-scale source regions to observed decameter irregularity scales. Furthermore, primacy of TGI and GDI drivers can switch back and forth depending on preconditioning and electrodynamic drivers. Such analysis by its nature cannot be done separately on each instability scale in isolation.

6 Conclusion

We have presented several examples of mid-latitude ionospheric dynamic reconfigurations connected to the coupled magnetosphere-ionosphere geospace system, including PBL electron density structuring, fine-scale SAPS electric field variability, and mid-latitude ionospheric irregularities. All these processes are prime examples of ionospheric dynamic features that must be considered within a multipart system, rather than by examining each area in isolation. Complexity methods are attractive for a more sophisticated determination of which mechanisms dominate in which regions, when they gradually or abruptly change form, and what implications they have for system-scale ionospheric configuration. Future work on mid-latitude ionospheric structure, and its accompanying advancement in knowledge is not only of the ionosphere but also all of geospace, will benefit considerably from a multifaceted, coupling-based approach to study of this fascinating part of the near-Earth space environment.

Acknowledgments

The author wishes to thank J.C. Foster and colleagues at MIT Haystack Observatory for many illuminating discussions on mid-latitude ionospheric features. Millstone Hill Geospace Facility's ionospheric radar experiments, GNSS TEC data products, Madrigal distributed data system, and scientific analysis activities are supported by NSF Grant AGS-1762141 to the Massachusetts Institute of Technology.

References

Abel, R., Thorne, R.M., 1998. Electron scattering loss in Earth's inner magnetosphere 1. Dominant physical processes. J. Geophys. Res. 103 (A), 2385–2396.

Anderson, P.C., Hanson, W.B., Heelis, R.A., Craven, J.D., Baker, D.N., Frank, L.A., 1993. A proposed production model of rapid subauroral ion drifts and their relationship to substorm evolution. J. Geophys. Res. Space Phys. 98 (A4), 6069–6078. https://doi.org/10.1029/92JA01975.

Anderson, P.C., Carpenter, D.L., Tsuruda, K., Mukai, T., Rich, F.J., 2001. Multisatellite observations of rapid subauroral ion drifts (SAID). J. Geophys. Res. Space Phys. 106 (A12), 29585–29599. https://doi.org/10.1029/2001JA000128.

Angerami, J.J., Carpenter, D.L., 1966. Whistler studies of the plasmapause in the magnetosphere: 2. Electron density and total tube electron content near the knee in magnetospheric ionization. J. Geophys. Res. 71 (3), 711–725. https://doi.org/10.1029/JZ071i003p00711.

Borovsky, J.E., 2014. Feedback of the magnetosphere. Science 343 (6175), 1086–1087. https://doi.org/10.1126/science.1250590.

Burke, W.J., Erickson, P.J., Yang, J., Foster, J., Wygant, J., Reeves, G., Kletzing, C., 2016. O⁺ ion conic and plasma sheet dynamics observed by Van Allen probe satellites during the 1 June 2013 magnetic storm. J. Geophys. Res. Space Phys. https://doi.org/10.1002/2015JA021795.

Carpenter, D.L., 1966. Whistler studies of the plasmapause in the magnetosphere: 1. Temporal variations in the position of the knee and some evidence on plasma motions near the knee. J. Geophys. Res. 71 (3), 693–709. https://doi.org/10.1029/JZ071i003p00693.

Carpenter, D.L., 2015. Very Low Frequency Space Radio Research at Stanford, 1950–1990, first ed. Lulu.com. http://www.lulu.com/shop/donald-carpenter/vlf-history-hardcover/hardcover/product-22554687.html.

Carpenter, D., Lemaire, J., 2004. The plasmasphere boundary layer. Ann. Geophys. 22, 4291–4298.

Carpenter, D.L., Park, C.G., 1973. On what ionospheric workers should know about the plasmapause-plasmasphere. Rev. Geophys. 11 (1), 133. https://doi.org/10.1029/RG011i001p00133.

Chappell, C.R., 1974. Detached plasma regions in the magnetosphere. J. Geophys. Res. 79 (13), 1861–1870. https://doi.org/10.1029/JA079i013p01861.

Coster, A.J., Foster, J.C., 2007. Space-weather impacts of the sub-auroral polarization stream. Radio Sci. Bull. 321, 28–36.

Daglis, I.A., Thorne, R.M., Baumjohann, W., Orsini, S., 1999. The terrestrial ring current: origin, formation, and decay. Rev. Geophys. 37 (4), 407–438. https://doi.org/10.1029/1999RG900009.

Darrouzet, F., Gallagher, D.L., André, N., Carpenter, D.L., Dandouras, I., Décréau, P.M.E., De Keyser, J., Denton, R.E., Foster, J.C., Goldstein, J., Moldwin, M.B., Reinisch, B.W., Sandel, B.R., Tu, J., 2009. Plasmaspheric density structures and dynamics: properties observed by the CLUSTER and IMAGE missions. Space Sci. Rev. 145 (1–2), 55–106. https://doi.org/10.1007/s11214-008-9438-9.

de Larquier, S., Eltrass, A., Mahmoudian, A., Ruohoniemi, J.M., Baker, J., Scales, W.A., Erickson, P.J., Greenwald, R.A., 2014. Investigation of the temperature gradient instability as the source of midlatitude quiet time decameter-scale ionospheric irregularities: 1. Observations. J. Geophys. Res. Space Phys. 119 (6), 4872–4881. https://doi.org/10.1002/2013JA019643.

Denton, R.E., Wang, Y., Webb, P.A., Tengdin, P.M., Goldstein, J., Redfern, J.A., Reinisch, B.W., 2012. Magnetospheric electron density long-term (>1 day) refilling rates inferred from passive radio emissions measured by IMAGE RPI during geomagnetically quiet times. J. Geophys. Res. Space Phys. 117(A3). https://doi.org/10.1029/2011JA017274.

Elphic, R.C., Thomsen, M.F., Borovsky, J.E., 1997. The fate of the outer plasmasphere. Geophys. Res. Lett. 24 (4), 365–368. https://doi.org/10.1029/97GL00141.

Eltrass, A., Scales, W.A., 2014. Nonlinear evolution of the temperature gradient instability in the midlatitude ionosphere. J. Geophys. Res. Space Phys. 119 (9), 7889–7901. https://doi.org/10.1002/2014JA020314.

Eltrass, A., Mahmoudian, A., Scales, W.A., de Larquier, S., Ruohoniemi, J.M., Baker, J.B.H., Greenwald, R.A., Erickson, P.J., 2014. Investigation of the temperature gradient instability as the source of midlatitude quiet time decameter-scale ionospheric irregularities: 2. Linear analysis. J. Geophys. Res. Space Phys. 119 (6), 4882–4893. https://doi.org/10.1002/2013JA019644.

Eltrass, A., Scales, W.A., Erickson, P.J., Ruohoniemi, J.M., Baker, J.B.H., 2016. Investigation of the role of plasma wave cascading processes in the formation of midlatitude irregularities utilizing GPS and radar observations. Radio Sci. 51 (6), 836–851. https://doi.org/10.1002/2015RS005790.

Erickson, P.J., Foster, J.C., Holt, J.M., 2002. Inferred electric field variability in the polarization jet from Millstone Hill E region coherent scatter observations. Radio Sci. 37 (2), 1014–1027.

Farley, D.T., 1960. A theory of electrostatic fields in the ionosphere at nonpolar geomagnetic latitudes. J. Geophys. Res. 65 (3), 869–877. https://doi.org/10.1029/JZ065i003p00869.

Fejer, B.G., Kelley, M.C., 1980. Ionospheric irregularities. Rev. Geophys. 18 (2), 401. https://doi.org/10.1029/RG018i002p00401.

Foster, J.C., 1993. Storm time plasma transport at middle and high latitudes. J. Geophys. Res. Space Phys. 98 (A2), 1675–1689. https://doi.org/10.1029/92JA02032.

Foster, J., Burke, W., 2002. {SAPS}: a new characterization for sub-auroral electric fields. EOS Trans. AGU 83, 393–394.

Foster, J.C., Erickson, P.J., 2000. Simultaneous observations of E-region coherent backscatter and electric field amplitude at F-region heights with the Millstone Hill UHF Radar. Geophys. Res. Lett. 27 (19), 3177–3180. https://doi.org/10.1029/2000GL000042.

Foster, J.C., Erickson, P.J., 2013. Ionospheric superstorms: polarization terminator effects in the Atlantic sector. J. Atmos. Sol. Terr. Phys. 103, 147–156.

Foster, J.C., Rideout, W., 2007. Storm enhanced density: magnetic conjugacy effects. Ann. Geophys. 25 (8), 1791–1799. https://doi.org/10.5194/angeo-25-1791-2007.

Foster, J.C., Erickson, P.J., Coster, A.J., Goldstein, J., Rich, F.J., 2002. Ionospheric signatures of plasmaspheric tails. Geophys. Res. Lett. 29 (1), 1623.

Foster, J.C., Coster, A.J., Erickson, P.J., Rich, F.J., Sandel, B.R., 2004a. Stormtime observations of the flux of plasmaspheric ions to the dayside cusp/magnetopause. Geophys. Res. Lett. 31 (8), 8809.

Foster, J.C., Erickson, P.J., Lind, F.D., Rideout, W., 2004b. Millstone Hill coherent-scatter radar observations of electric field variability in the sub-auroral polarization stream. Geophys. Res. Lett. 31(21). https://doi.org/10.1029/2004GL021271.

Foster, J.C., Erickson, P.J., Baker, D.N., Jaynes, A.N., Mishin, E.V., Fennel, J.F., Li, X., Henderson, M.G., Kanekal, S.G., 2016. Observations of the impenetrable barrier, the plasmapause, and the VLF bubble during the 17 March 2015 storm. J. Geophys. Res. Space Phys. 121 (6), 5537–5548. https://doi.org/10.1002/2016JA022509.

Freeman, J.W., Hills, H.K., Hill, T.W., Reiff, P.H., Hardy, D.A., 1977. Heavy ion circulation in the Earth's magnetosphere. Geophys. Res. Lett. 4 (5), 195–197. https://doi.org/10.1029/GL004i005p00195.

Galperin, Y.I., 2002. Polarization jet: characteristics and a model. Ann. Geophys. 20, 391.

Goldstein, J., Sandel, B.R., 2005. The global pattern of evolution of plasmaspheric drainage plumes. In: Inner Magnetosphere Interactions: New Perspectives From Imaging, American Geophysical Union (AGU), pp. 1–22. https://doi.org/10.1029/159GM02.

Greenwald, R.A., Oksavik, K., Erickson, P.J., Lind, F.D., Ruohoniemi, J.M., Baker, J.B.H., Gjerloev, J.W., 2006. Identification of the temperature gradient instability as the source of decameter-scale ionospheric irregularities on plasmapause field lines. Geophys. Res. Lett. 33(18). https://doi.org/10.1029/2006GL026581.

Hanson, W.B., Ortenburger, I.B., 1961. The coupling between the protonosphere and the normal F region. J. Geophys. Res. 66 (5), 1425–1435. https://doi.org/10.1029/JZ066i005p01425.

Hanson, W.B., Patterson, T.N.L., 1964. The maintenance of the night-time F-layer. Planet. Space Sci. 12 (10), 979–997. https://doi.org/10.1016/0032-0633(64)90112-6.

Heelis, R.A., 2004. Electrodynamics in the low and middle latitude ionosphere: a tutorial. J. Atmos. Sol. Terr. Phys. 66 (10), 825–838. https://doi.org/10.1016/J.JASTP.2004.01.034.

Heelis, R.A., Sojka, J.J., David, M., Schunk, R.W., 2009. Storm time density enhancements in the middle-latitude dayside ionosphere. J. Geophys. Res. Space Phys. 114(A3). https://doi.org/10.1029/2008JA013690.

Helliwell, R.A., 1965. Whistlers and Related Ionospheric Phenomena. Stanford University Press, Stanford, CA, p. 349. http://adsabs.harvard.edu/abs/1965wrip.book...H.

Huang, C.-S., Foster, J.C., Goncharenko, L.P., Erickson, P.J., Rideout, W., Coster, A.J., 2005. A strong positive phase of ionospheric storms observed by the Millstone Hill incoherent scatter radar and global GPS network. J. Geophys. Res. 110 (A6), A06303. https://doi.org/10.1029/2004JA010865.

Hudson, M.K., Kelley, M.C., 1976. The temperature gradient drift instability at the equatorward edge of the ionospheric plasma trough. J. Geophys. Res. 81 (22), 3913–3918. https://doi.org/10.1029/JA081i022p03913.

Jaynes, A.N., Baker, D.N., Singer, H.J., Rodriguez, J.V., Loto'aniu, T.M., Ali, A.F., Elkington, S.R., Li, X., Kanekal, S.G., Claudepierre, S.G., Fennell, J.F., Li, W., Thorne, R.M., Kletzing, C.A., Spence, H.E., Reeves, G.D., 2015. Source and seed populations for relativistic electrons: their roles in radiation belt changes. J. Geophys. Res. Space Phys. 120 (9), 7240–7254. https://doi.org/10.1002/2015JA021234.

Jones, K., 1971. Storm time variation of F2-layer electron concentration. J. Atmos. Terr. Phys. 33 (3), 379–389. https://doi.org/10.1016/0021-9169(71)90143-7.

Keika, K., Kistler, L.M., Brandt, P.C., 2013. Energization of O^+ ions in the Earth's inner magnetosphere and the effects on ring current buildup: a review of previous observations and possible mechanisms. J. Geophys. Res. Space Phys. 118 (7), 4441–4464.

Kelley, M.C., 2009. The Earth's Ionosphere: Plasma Physics and Electrodynamics. Academic Press, ISBN: 9780120884254, p. 556.

Keskinen, M.J., 1984. Nonlinear theory of the $\mathbf{E} \times \mathbf{B}$ instability with an inhomogeneous electric field. J. Geophys. Res. 89 (A6), 3913. https://doi.org/10.1029/JA089iA06p03913.

Landry, R.G., Anderson, P.C., 2018. An auroral boundary-oriented model of subauroral polarization streams (SAPS). J. Geophys. Res. Space Phys. 123 (4), 3154–3169. https://doi.org/10.1002/2017JA024921.

Lemaire, J., Gringauz, K.I., 1998. The Earth's Plasmasphere. Cambridge University Press, ISBN: 9780521430913, p. 350.

MacPherson, B., González, S.A., Bailey, G.J., Moffett, R.J., Sulzer, M.P., 1998. The effects of meridional neutral winds on the O^+-H^+ transition altitude over Arecibo. J. Geophys. Res. Space Phys. 103 (A12), 29183–29198. https://doi.org/10.1029/98JA02660.

Makarevich, R.A., Kellerman, A.C., Devlin, J.C., Ye, H., Lyons, L.R., Nishimura, Y., 2011. SAPS intensification during substorm recovery: a multi-instrument case study. J. Geophys. Res. Space Phys. 116(A11). https://doi.org/10.1029/2011JA016916.

McIlwain, C.E., 1961. Coordinates for mapping the distribution of magnetically trapped particles. J. Geophys. Res. 66 (11), 3681–3691. https://doi.org/10.1029/JZ066i011p03681.

Mendillo, M., Klobuchar, J., 1975. Investigations of the ionospheric F region using multistation total electron content observations. J. Geophys. Res. Space Phys. 80 (4), 643–649.

Mishin, E.V., 2016. SAPS onset timing during substorms and the westward traveling surge. Geophys. Res. Lett. 43 (13), 6687–6693. https://doi.org/10.1002/2016GL069693.

Mishin, E.V., Burke, W.J., 2005. Stormtime coupling of the ring current, plasmasphere, and topside ionosphere: electromagnetic and plasma disturbances. J. Geophys. Res. 110 (A7), A07209. https://doi.org/10.1029/2005JA011021.

Mishin, E.V., Mishin, V.M., 2007. Prompt response of SAPS to stormtime substorms. J. Atmos. Sol. Terr. Phys. 69 (10–11), 1233–1240. https://doi.org/10.1016/j.jastp.2006.09.009.

Mishin, E., Nishimura, Y., Foster, J., 2017. SAPS/SAID revisited: a causal relation to the substorm current wedge. J. Geophys. Res. Space Phys. 122 (8), 8516–8535. https://doi.org/10.1002/2017JA024263.

Moffett, R.J., Quegan, S., 1983. The mid-latitude trough in the electron concentration of the ionospheric F-layer: a review of observations and modelling. J. Atmos. Terr. Phys. 45 (5), 315–343. https://doi.org/10.1016/S0021-9169(83)80038-5.

Ribeiro, A.J., Ruohoniemi, J.M., Baker, J.B.H., Clausen, L.B.N., Greenwald, R.A., Lester, M., 2012. A survey of plasma irregularities as seen by the midlatitude Blackstone SuperDARN radar. J. Geophys. Res. Space Phys. 117(A2). https://doi.org/10.1029/2011JA017207.

Rideout, W., Coster, A., 2006. Automated GPS processing for global total electron content data. GPS Solut. 10 (3), 219–228. https://doi.org/10.1007/s10291-006-0029-5.

Rodger, A.S., Wrenn, G.L., Rishbeth, H., 1989. Geomagnetic storms in the Antarctic F-region. II. Physical interpretation. J. Atmos. Terr. Phys. 51 (11–12), 851–866. https://doi.org/10.1016/0021-9169(89)90002-0.

Sandel, B.R., King, R.A., Forrester, W.T., Gallagher, D.L., Broadfoot, A.L., Curtis, C.C., 2001. Initial results from the IMAGE Extreme Ultraviolet Imager. Geophys. Res.

Lett. 28 (8), 1439–1442. https://doi.org/10.1029/2001GL012885@10.1002/(ISSN)1944-8007.IMAGMAG1.

Schunk, R.W., Nagy, A.F., 2009. Ionospheres: Physics, Plasma Physics, and Chemistry, second ed. Cambridge University Press, ISBN: 978-0-521-87706-0, p. 628.

Schunk, R.W., Raitt, W.J., Banks, P.M., 1975. Effect of electric fields on the daytime high-latitude E and F regions. J. Geophys. Res. 80 (22), 3121–3130. https://doi.org/10.1029/JA080i022p03121.

Simon, A., 1963. Instability of a partially ionized plasma in crossed electric and magnetic fields. Phys. Fluids 6 (3), 382. https://doi.org/10.1063/1.1706743.

Singh, N., Horwitz, J.L., 1992. Plasmasphere refilling: recent observations and modeling. J. Geophys. Res. 97 (A2), 1049. https://doi.org/10.1029/91JA02602.

Southwood, D.J., Wolf, R.A., 1978. An assessment of the role of precipitation in magnetospheric convection. J. Geophys. Res. 83 (A11), 5227. https://doi.org/10.1029/JA083iA11p05227.

Spiro, R.W., Heelis, R.A., Hanson, W.B., 1979. Rapid subauroral ion drifts observed by atmosphere explorer C. Geophys. Res. Lett. 6 (8), 657–660. https://doi.org/10.1029/GL006i008p00657.

Walsh, B.M., Foster, J.C., Erickson, P.J., Sibeck, D.G., 2014. Simultaneous ground- and space-based observations of the plasmaspheric plume and reconnection. Science 343 (6175), 1122–1125.

Williams, D.J., 1981. Ring current composition and sources: an update. Planet. Space Sci. 29 (11), 1195–1203. https://doi.org/10.1016/0032-0633(81)90124-0.

6

Empirical ionospheric models

Sandro M. Radicella, Bruno Nava

The Abdus Salam International Centre for Theoretical Physics, Trieste, Italy

1 Conceptual introduction

Before entering into the specific subject of this chapter, it is convenient to address some basic concepts about "models" in geophysics and, in particular, "models" related to ionospheric studies.

A "model" in geophysical sciences, including ionospheric research, represents essentially a "theory." This is also true for statistical or empirical models. Any model provides a basis for making predictions of the outcome of new measurements, but it has to be understood that models are always simpler than the real natural phenomena. There are two main types of geophysical models: empirical and physics-based. Empirical models are descriptive and based on data, thus not relying on physical first principles. Physics-based models are deterministic and explain and predict natural phenomena using mathematical representations of physical laws. However, it cannot be said that each of these categories of models ignores the other. The construction of empirical models is guided by the physics that determines the variables and the data sets to be analyzed. On the other hand, physics-based models rely on observations to validate their results and to estimate key quantities like initial and boundary conditions.

Any geophysical model, including ionospheric models, has intrinsic limits that can be summarized as follows:

1. It is impossible to perform controlled experiments to study geophysical phenomena;
2. The development of a geophysical model has to rely entirely upon the observation of a given geophysical phenomenon;
3. Although past data are a "guide" to predict future behavior, this is always an "uncertain guide" because of the intrinsic limited database used;
4. It is not always possible to draw exact physical conclusions from the observations because it is difficult to separate the physical processes controlling the observation;
5. The physical conclusions are uncontrolled synthesis of many mutually dependent physical processes.

Having mentioned these limits, it must be said that models able to predict parameter values in geophysical sciences are becoming more and more "success stories." This is due to the rapid growth of the number and quality of the observations, the continuous improvement of existing models, and the constant increase of computing power since the 1970s of last century.

As far as ionospheric models are concerned, they are needed essentially to "predict" in time and space the behavior of ionospheric characteristics and to specify the conditions of the ionosphere as required by the advanced technological systems that depend on radio signal propagation. Of course they are needed also to improve our understanding about the ionosphere and the near space environment. Indeed numerical experiments done with the models to study different processes can be a convenient alternative to direct measurements.

Like all geophysical models, the ionospheric models may need different types of input data. Well-established databases are also needed to validate the models and to test systematically the modifications leading to model improvements and evolution.

2 Ionospheric models

In this chapter, the use of "ionospheric model" denotes a description of an ionospheric parameter, like the electron density, in three space dimensions and time. Global or regional specifications of ionospheric parameters like total electron content (TEC) derived from GNSS observations are not considered here.

Ionospheric models are based on the need to specify the behavior in space and time of the electron density in the ionosphere, but several of them are able to describe other parameters like ion and electron temperatures, ion densities, and ionospheric drifts. An important aspect to take into account is the fact that "all-purpose" ionospheric models do not yet exist.

Empirical models are based on the description of ionospheric parameters with mathematical functions derived from historic experimental data. Sources of data are ground ionosondes, topside sounders, incoherent scatter radars, rockets, and satellites. They are easy to use for assessment and prediction purposes. In their standard way of operation, they describe the

"climate" or average conditions and regular variations of ionospheric parameters. Their representation of the ionosphere is realistic in geographical areas sufficiently covered by observations.

Being the scope of this chapter, these types of models will be described in detail in the next sections.

In the *physics-based* or theoretical models, conservation equations (continuity, momentum, energy, etc.) are solved numerically for electrons and ions as a function of spatial and time coordinates to calculate electron and ions densities, temperatures, and flow velocities of species. Data sources are magnetospheric (convection electric field and particle precipitation) and atmospheric parameters (neutral densities, temperatures, and winds). These models are powerful tools to understand the physical and chemical processes of the upper atmosphere. Their accuracy depends on the quality and quantity of input data (including their error estimates) but also on possible "missing physics," when not enough knowledge of all physical processes involved is available. Examples of these models are the Global Ionosphere Thermosphere Model (GITM) described by Ridley et al. (2006) and the SAMI 3 three-dimensional global ionospheric model described by Huba et al. (2008).

The *parameterized models* are based on orthogonal function fits to data output of physics-based models. These data are obtained from model runs performed for various heliogeophysical conditions and the parameterization is usually done in terms of solar, geomagnetic activity, and season. These models describe the climatology of the ionosphere. They are computationally fast and able to retain the physics of the theoretical models, mainly for well-specified geophysical conditions. One example of this type of model is the Parameterized Ionospheric Model (PIM) built combining physics-based models output of GITM for low and middle latitudes, Utah State University Time-Dependent

Ionospheric Model (TDIM) for high latitudes, and the empirical model for the plasmasphere by Gallagher et al. (1988). The PIM model is described by Daniell et al. (1995).

In particular, following the needs of ionospheric specifications for advanced technological applications like telecommunications and satellite navigation, the trend of ionospheric modeling is to shift from "climate" to "weather-like" representations of the ionized medium. Two approaches have been used for this purpose. One of these is the systemic approach that takes into account the earth atmosphere as a single system. Following this approach, it is necessary to consider taking into account both the forcing from the sun and the magnetosphere and the forcing of the lower atmosphere to model the observed day-to-day variability of ionospheric parameters.

Using this method, Liu et al. (2013) have shown that the Thermospheric General Circulation Model constrained in the stratosphere and mesosphere by the Whole Atmosphere Community Climate Model (Garcia et al., 2007) is capable of reproducing observed features of day-to-day variability in the thermosphere-ionosphere that cannot be attributed to solar-geomagnetic forcing.

The other approach to obtain "weather-like" specification of the ionosphere is the data assimilation or ingestion in both physics-based and empirical models. During the last 20 years, several techniques have been developed to assimilate data of different types in such models. This has been possible because of the increased availability of solar and geomagnetic data in addition to the ionospheric ground and satellite-based data.

An important example of data assimilation in a physics-based model is the Utah State University (USU) Global Assimilation of Ionospheric Measurements (GAIM) models (Scherliess et al., 2004). It uses a physics-based model of the ionosphere and a Kalman filter as a basis for assimilating real-time (or near real-time) bottom-side N_e profiles, slant TEC from ground GPS stations, in situ N_e from four DMSP satellites, and line-of-sight solar UV emissions measured by satellites. Upon request, it is possible to get specific runs of this model at https://ccmc.gsfc.nasa.gov/requests/IT/USUGAIM/usugaim_user_registration.php.

Two examples of data assimilation/data ingestion in empirical models are those introduced by Nava et al. (2011) and Galkin et al. (2012). Nava et al. (2011) ingest global TEC maps into NeQuick 2 model and Galkin et al. (2012) in their IRI Real-Time Assimilative Modeling (IRTAM) system assimilate digisonde data from the Global Ionospheric Radio Observatory (GIRO) network into the IRI model. Both the NeQuick 2 and the IRI models are going to be described in detail in other sections of this chapter.

3 Brief history of the empirical ionospheric models

The historical evolution of the empirical models of the ionosphere is linked to the applications of such models. Applications are related essentially to the propagation of radio waves into the ionospheric environment that needs to be specified in relation to the relevant use. The discovery itself of the ionosphere was related to the need for an explanation of Marconi's demonstration of the crossing of the Atlantic with a radio wave transmission from England to Canada in 1901. Kennelly (1902) and Heaviside (1902) independently suggested that the long-distance propagation radio wave experiment by Marconi could be explained by the reflection of the waves by an ionized layer in the atmosphere. It was another experiment with radio waves done by Appleton and Barnett (1925) that proved the existence of the ionized layer, in addition to determining its height.

For decades, the use of radio waves in applications particularly related to ground-to-ground communications at long distance was the driver for the development of empirical models of the electron density in the ionosphere. The emphasis on communications via the ionosphere, and consequently on electron density modeling, started decreasing with the advent of satellite communications in the 1960s. The era of satellite navigation and positioning initiated in the last decades of the 20th century shifted the focus more toward electron density models used for transionospheric propagation applications. The interest about the structure of the ionosphere below the peak of the F2 layer was then extended to the region above that peak, the ionospheric topside.

The empirical models that will be mentioned in this chapter are considered as global representation of a specific ionospheric parameter: the electron density, usually represented in terms of vertical profiles. These models use, generally, layer peak characteristics maps as anchor points in the mathematical description of the profile. The main source of experimental data for these maps is the ionosonde global network, but they use also data from other sources like topside sounders, backscatter radars, and in situ satellite measurements. Some of the peak characteristics like the height of the F2 layer are often derived from empirical formulae together with the very important layer thickness parameters that play a fundamental role in describing the shape of the vertical profile given by the models and therefore have also impact on total electron content (TEC) calculation.

It is possible to trace the first attempt to define a vertical profile of the electron density up to the peak of the F2 layer, to the paper by Appleton and Beynon (1947). These authors were studying the use of ionospheric data obtained from ionosondes to concrete communications applications between distant locations over spherical Earth using radio waves in the HF band. They introduced the concept of one "parabolic" layer of electron density determined by the critical frequency of the layer (related to its maximum

electron density) scaled from the ionogram and computed values of the height of the ledge and the semithickness parameter of the layer (related to its shape).

A model that introduced a second parabolic layer to consider the presence of the E layer in the electron density profile was the ITS 78 model by Barghausen et al. (1969). The main limitation of this model was the absence of the F1 layer.

Later, Bradley and Dudeney (1973) introduced a new model of electron density that takes into account the existence of both the F2 and the E layer but includes also a crude solution for the presence of the F1 layer. The model gives the plasma frequency as a function of height and is driven by the ionospheric characteristics foE, foF2, M(3000)F2 and either the virtual height h'F or h'F2 as appropriate, whichever is the minimum virtual height of reflection from the F2 layer. It considers a half parabola for the bottom side of the E layer and a linear increase from the peak of the E layer to the point where a parabolic F2 layer reaches 1.7 times foE, to take into account the presence of electron density between the E and F2 layers. The E layer peak height hmE and semithickness parameter ymE are taken constant at 110 and 20 km, respectively. Both hmF2 and the semithickness parameter ymF2 are obtained from empirical formulae. It has to be noted that the F1 layer is ignored in this model and is replaced by a simple linear increase. The International Radio Consultative Committee (CCIR) of the International Telecommunication Union (CCIR, 1967) has adopted Bradley and Dudeney (1973) model for the description of the electron density distribution in the E and F layers of the ionosphere. The main limitations of the model are the electron density gradient discontinuities at the F1 and E layers that are relevant particularly for raytracing and the absence of the D region and the valley above the E layer and the F1 layer.

Dudeney (1978) introduced improvements in the Bradley and Dudeney (1973) model that overcame two main limitations mentioned here.

The gradient discontinuities were eliminated in the case of the absence of the F1 layer by changing the linear increase between the parabolic E and F2 layers into a combination of secant and cosine functions from the peak of the E layer to the peak of the F2 layer. The presence of the F1 layer, represented in ionograms by a value of foF1, or what used to be called the "L condition," has been introduced by the author in such a way that again no discontinuities were found in the electron density vertical profile gradient. The Dudeney (1978) model has been used successfully in radio propagation studies using raytracing techniques thanks to the elimination of those discontinuities.

Another description of the electron density profile, up to the peak of the F2 layer and based on parabolic approximations, was introduced later by the IONCAP model (Teters et al., 1983). This model, in addition to parabolic approximations for the E and F2 layers, introduces a linear valley above the E layer peak and an exponential tail in correspondence of the D layer.

The same decade of the 1970s saw the appearance of the first empirical model of ionospheric electron density profile particularly designed for transionospheric radio propagation, although it was still based essentially on a parabolic approximation. This model, by Bent et al. (1972) and Llewellyn and Bent (1973), considers only the F2 layer as a biparabola and the topside of the profile as a parabola and as exponential segments. The topside of the model has been designed on the basis of about 50,000 ionograms, obtained by the Alouette satellite topside ionosonde. Klobuchar (Klobuchar, 1982, 1987) has used the Bent model as the basis for his GPS ionospheric delay correction algorithm, which in turn is a model representation of the vertical TEC over the earth.

Ching and Chiu (1973) proposed a model that, instead of a parabolic approximation of the ionospheric layers, uses a more physically oriented description of the electron density profile in the ionosphere. They introduced modified Chapman shape mathematical expressions for each of the three main layers E, F1, and F2. In addition, the model describes electron density not only to the F2 peak but also up to 600 km to make possible comparisons to the experimental results obtained with backscatter radar observations. The model was later improved by Chiu (1975). The main limitation of these models is that their use is restricted essentially to middle latitudes. However, they extended their description of the electron density profile above the peak of the F2 layer, introducing a new era for ionospheric modeling.

In the framework of the European Cooperation in Science and Technology (COST) actions devoted to ionospheric research and applications, electron density models have been developed. Di Giovanni and Radicella (1990) introduced a model (DGR from now on) that describes the electron density vertical profile, up to the F2 peak, by the combination of three Epstein layers. The model uses as anchor points the peak characteristics of the E, F1, and F2 layers with the addition of another anchor point related to the semithickness of the F2 layer. The mathematical formulation used in the model is such that the function and its first derivative are always continuous. Radicella and Zhang (1995) modified the DGR model introducing improvements in the description of the profile between the E and F1 layers and adding a topside profile. This last addition is based on experimental values of TEC obtained from satellite data that allow introducing an empirical "shape factor" in the Epstein description of the electron density topside profile. The output of the model adds to the profile the corresponding TEC value up to the height of a satellite.

Based on the same principles of the DGR model, a new "family" of three models of ionospheric electron density has been developed. They have different degrees of complexity but related areas of transionospheric radio-propagation applications. These models are

NeQuick, COSTProf, and NeUoG-plas (Hochegger et al., 2000; Radicella and Leitinger, 2001).

NeQuick model: a quick-run model that will be described in detail later, taking into account its evolution and international use.

COSTProf: used for ionospheric and plasmaspheric satellite to ground radiopropagation studies and applications and adopted in the output of COST 251 Action, one of the European COST actions related to ionospheric research and applications.

NeUoG-plas: used particularly for assessment studies involving satellite to satellite radiopropagation containing a magnetic field aligned formulation for the plasmasphere above 2000 km of height.

The ionospheric electron density empirical models mentioned in this section have been developed driven essentially by application purposes related to the ionospheric and transionospheric radiopropagation. In parallel to them, a more ambitious (and oriented to both geophysics research and applications) empirical model has been developed and continuously improved: the International Reference Ionosphere (IRI). This model is the result of the joint work of a large number of researchers from all over the world and is recognized as a climatological standard model by the International Standardization Organization (ISO) since 2009 (ISO/TS 16457:2009 replaced by the ISO 16457:2014).The model in its present development stage gives a climatological description of electron density, electron temperature, ion temperature, and ion composition in the ionosphere from 50 to 2000 km of height. The model includes the vertical TEC calculation, the occurrence probability for Spread-F and also of the F1-region, and the equatorial vertical ion drift. The latest version provides a suggestion for the way to convert the model from a climatological model to a "weather-like" one, driven by data in quasi real time.

In this section, only a brief history of the IRI evolution will be given. Considering its relevance, a more extended description of the model will be provided in a separate section of this chapter.

The Committee on Space Research (COSPAR) established by the International Council of Scientific Unions (ICSU), now the International Council for Science under the same acronym, initiated the International Reference Ionosphere (IRI) project in 1968 inspired by the first Chairperson, Prof. Karl Rawer. He was able to join the efforts of ionospheric experts from both the ground-based and the newly developed space-based scientific communities from many different countries. He also was able to get the International Union of Radio Science (URSI) involved in the project. Initially starting with 20 members of the IRI Working Group recognized by both COSPAR and URSI organizations, the IRI Working Group has now grown to more than 60 members from 26 countries. The evolution of the IRI model was and is presented and discussed during yearly IRI Workshops held in many locations, in addition to official COSPAR or URSI meetings. A particular contribution to the more accurate representation of the bottomside of this model electron density profile based on experimental data from low-latitude locations was given during the special IRI Task Force Activities held at the Abdus Salam International Centre for Theoretical Physics (ICTP) yearly from 1994 to 2003 (Bilitza, 2002). The evolution of the IRI model can be followed through a series of papers: Rawer et al. (1978), Bilitza (1986), (1990), (1997), (2001), Bilitza and Reinisch (2008), Bilitza et al. (2014), and Bilitza et al. (2017).

4 IRI model

For given location and epoch, IRI describes the electron density, electron temperature, ion temperature, and ion composition (O+, H+, He+, N+, NO+, O2+, cluster ions) in the altitude range from about 50 km to about 2000 km and also the corresponding electron content. As mentioned earlier, it gives the occurrence probability for Spread-F and also of the F1-region, and the equatorial vertical ion drift.

The IRI electron density profile is divided in six subregions: the topside, the F2 bottomside, the F1 layer, the intermediate region, the E region valley, the bottomside E and D regions. The boundaries are defined by the presence of characteristic points that include the F2, F1, and E peaks. The shape of the IRI topside electron density profile was based on the descriptive compilation of Alouette topside sounder data and Epstein functions. Two parameters, B0 and B1, determine the bottomside thickness and shape, respectively (Bilitza, 2001). IRI includes a model to describe ionospheric storm-time conditions (Araujo-Pradere et al., 2002) and has adopted as one of the options the NeQuick 2 model topside formulation (Bilitza and Reinisch, 2008). IRI has also the capability to describe real-time ionospheric weather conditions based on the ingestion of real-time measurements (Bilitza et al., 2017) like ionosonde-derived peak parameter values (Galkin et al., 2012).

The main evolution of the electron density vertical profile given by the IRI model from the version of the year 2001 (IRI 2001) to the one of 2007 (IRI 2007) deals with the description of the topside profile above the peak of the F2 Layer and the bottomside at the height of the D and E layers (Bilitza and Reinisch, 2008).

Two new options were given for the topside. A correction factor for the 2001 model based on over 150,000 topside profiles from Alouette 1, 2, and ISIS 1, 2 was the first option (Bilitza, 2004). With this correction term, the IRI model represents better the experimental topside sounder data improving also the TEC estimate. The last version of the NeQuick topside model (Nava et al., 2008) is the second option for the topside of IRI 2007 and has been considered the most mature of the different proposals for the IRI topside and has been taken as default option (Bilitza and Reinisch, 2008).

Fig. 1 gives examples of the three options of the IRI topside electron density profiles: the one that corresponds to the IRI 2001 version and the two new options introduced in IRI 2007.

The FIRI model for the D and E layers (Friedrich and Torkar, 2001) based on a compilation of rocket data with simultaneous radio propagation and in situ measurements, was introduced in the IRI 2007 as a standalone model of the electron density. The reason being that it could not be normalized to the E layer peak of the IRI model was because of its formalism.

The improvements introduced in the electron density description of its IRI 2012 version refer particularly to the introduction of new models for the thickness and shape parameters Bo and B1 (Bilitza et al., 2014). These models by Altadill et al. (2009) are obtained from data of 27 ionosondes globally distributed after applying spherical harmonics analysis to describe the parameters in modified dip latitude (modip), LT, month, and sunspot number. Fig. 2 shows examples of the bottom-side profile of IRI model electron density when the three options of Bo and B1 given by IRI 2012 are used.

A further step ahead of IRI description of the ionospheric electron density has been the introduction of new models for the F2 layer, peak height hmF2 in the IRI 2016 latest version of the model. They replace the older formulation of IRI based on the CCIR (1967) models of foF2 and M(3000)F2. The hmF2 models adopted by the IRI 2016 version (Bilitza et al., 2017) are based on ionosonde data (Magdaleno et al., 2011; Altadill et al., 2013) and on radio occultation data (Shubin et al., 2013)

Fig. 3 shows IRI 2016 model profiles from 100 to 500 km of height when the old and the new models of hmF2 are used.

Fig. 4 displays an example of IRI 2016 global TEC representation.

IRI 2016 introduces also the transition from a climatological model to the development of an IRI Real-time model. The most advanced option of this new approach is the IRI Real-time

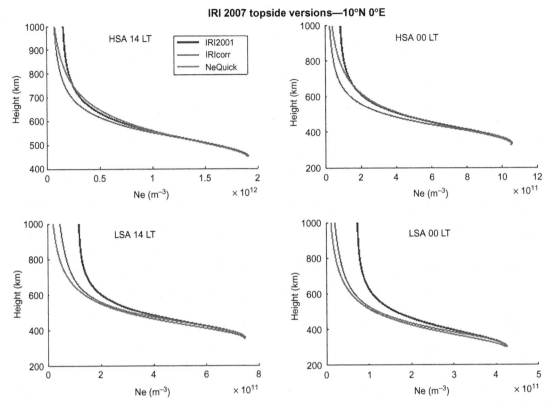

FIG. 1 Topside electron density profile from the IRI 2007 when the three options given by the model are used for a location at 10°N and 0°E. The profiles correspond to 14:00 and 00:00 LT for 15th March 2000 for High Solar Activity (HAS) and 15th March 2008 for Low Solar Activity (LSA). *Data obtained from https://ccmc.gsfc.nasa.gov/modelweb/models/iri_vitmo.php.*

Assimilative Modeling (IRTAM) by Galkin et al. (2012) that assimilates digisonde data from the Global Ionospheric Radio Observatory (GIRO).

IRI has performed very well in a series of model evaluations and assessments that were initiated by NSF's Coupling, Energetics, and Dynamics of Atmospheric Regions (CEDAR) program and were undertaken by NASA's Coordinated Community Modeling Center (CCMC) (Shim et al., 2011, 2012, 2017). In the group of up to 10 models, IRI was the only empirical model, and all others were theoretical or assimilative (GAIM) models. It is important to note that the assessment was done by a team that was not involved in any of the tested models. The assessment involved multiple different data sources and several different assessment parameters. In most cases, IRI outperformed the theoretical models.

5 NeQuick model

The NeQuick model is an ionospheric electron density model developed at the T-ICT4D (former ARPL) of The Abdus Salam International Centre for Theoretical Physics (ICTP), Trieste, Italy, and at the Institute for Geophysics, Astrophysics and Meteorology (IGAM) of the University of Graz, Austria. It is a quick-run model particularly designed for transionospheric propagation applications that has been

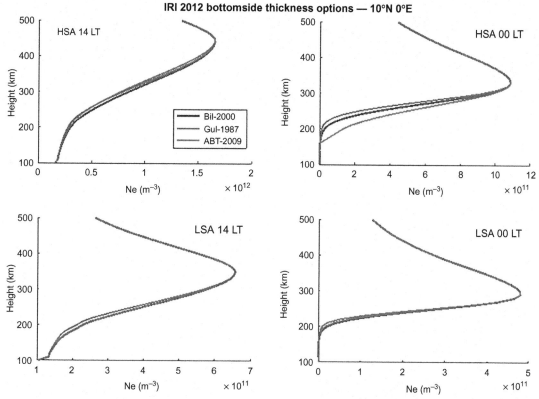

FIG. 2 IRI 2012 electron density profile when the three options of the bottomside thickness and shape parameters Bo and B1 are used for a location at 10°N and 0°E. The profiles correspond to 14:00 and 00:00 LT for 15th March 2000 for High Solar Activity (HAS) and 15th March 2008 for Low Solar Activity (LSA). *Data obtained from https://ccmc.gsfc.nasa.gov/modelweb/ models/iri2012_vitmo.php.*

conceived as a climatic model to reproduce the median behavior of the ionosphere.

As mentioned earlier, NeQuick is based on the DGR model, subsequently modified by Radicella and Zhang (1995). A first version of the model has been reported by Hochegger et al. (2000) and further described in Radicella and Leitinger (2001). A modified and simplified bottomside has been introduced by Leitinger et al. (2005) and modified topside has been introduced by Coïsson et al. (2006). All these efforts, directed toward the developments of a new version of the model, have led to the implementation of the NeQuick 2 (Nava et al., 2008).

The first version of NeQuick has been adopted by the ITU-R Recommendation P.531-6 of 2001 as a procedure for estimating TEC. The second version of the model, NeQuick 2, has been adopted by the ITU-R Recommendation P.531-12/13 of 2013–2016, therefore superseding the previous version of the model. It has to be noted that the version adopted by ITU-R is considered at the moment the standard version of the NeQuick 2 model. This version contains the values of modip that correspond to the International Geomagnetic Reference Field (IGRF) of 1960. This is done to be consistent with the same geomagnetic field description

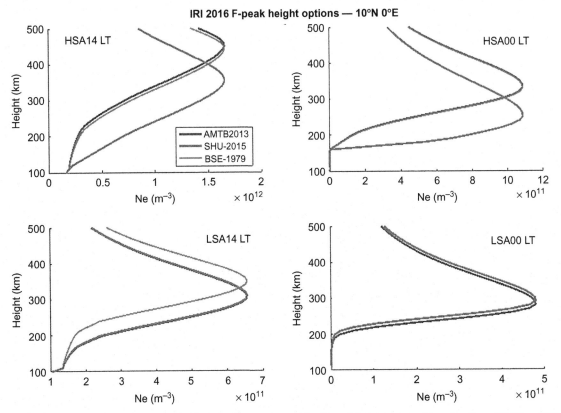

FIG. 3 IRI 2016 electron density profile when the three options of the F2 peak height are used for a location at 10°N and 0°E. The profiles correspond to 14:00 and 00:00 LT for 15th March 2000 for High Solar Activity (HAS) and 15th March 2008 for Low Solar Activity (LSA). *Data obtained from https://ccmc.gsfc.nasa.gov/modelweb/models/iri2016_vitmo.php.*

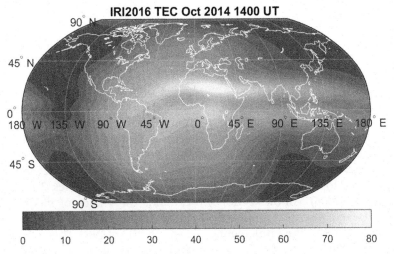

FIG. 4 IRI 2016 TEC global representation for October 2014 at 14:00 UT.

I. The earth's ionosphere, an overview

adopted by the Recommendation ITU-R P.1239 that gives the coefficients for the determination of the global foF2 and M(3000) maps used as drivers for the NeQuick models. Outside the standard version adopted by ITU-R, a version with modip values derived from the IGRF 2005 has been circulated for research purposes.

Fig. 5 shows examples of the vertical electron density profiles obtained from three different versions of NeQuick model: the first version is that as adopted by ITU-R Recommendation P.531-6 of 2001 (indicated as NeQuick1-ITUR), the second standard version is that as adopted by ITU-R Recommendation P.531-12/13 (indicated as NeQuick2-ITU). The figure shows also the profile obtained from NeQuick 2 formulation but with modip derived from IGRF 2005 (indicated as NeQuick2).

Fig. 6 shows an example of NeQuick2 TEC global representation.

A specific version of NeQuick (NeQuick G, implemented by ESA) has been adopted as Galileo Single-Frequency Ionospheric Correction algorithm and its performance has been confirmed during In-Orbit Validation (Prieto-Cerdeira et al., 2014). In a paper by Orus Perez et al. (2018), it was shown that NeQuick G performance is very good globally.

To describe the electron density of the ionosphere above ~100 km and up to the peak of

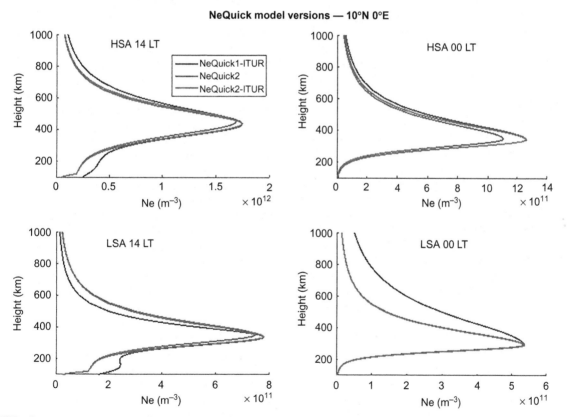

FIG. 5 Vertical electron density profiles obtained from three different versions of NeQuick model (see text). The profiles correspond to 14:00 and 00:00 LT for 15th March 2000 for High Solar Activity (HAS) and 15th March 2008 for Low Solar Activity (LSA).

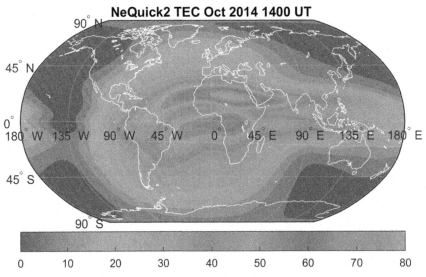

FIG. 6   NeQuick2 TEC global representation for October 2014 at 14:00 UT.

the F2 layer, NeQuick uses a profile formulation, which includes five semi-Epstein layers with empirically modeled thickness parameters. Three profile anchor points are used: the E layer peak, the F1 peak, and the F2 peak, that are modeled in terms of the ionosonde parameters foE, foF1, foF2, and M(3000)F2. These values can be modeled or experimentally derived. A semi-Epstein layer represents the model topside with a height-dependent thickness parameter empirically determined. The basic inputs of the NeQuick model are position, time, and solar flux (or sunspot number); the output is the electron concentration at the given location and time. NeQuick package includes routines to evaluate the electron density along any ground-to-satellite ray-path and the corresponding TEC by numerical integration.

To provide 3-D specification of the ionosphere electron density for current conditions, different ionosphere electron density retrieval techniques based on the NeQuick adaptation to GPS-derived TEC data and ionosonde measured peak parameters values have been developed (e.g., Nava et al., 2006, 2011).

As far as empirical models are concerned, the IRI and the NeQuick 2 are widely used in ionospheric studies, as a priori information (Lin et al., 2015; Minkwitz et al., 2016) when data assimilation methods are implemented.

6 The future

Empirical and first-principles based ionospheric models are evolving from climatological to "weather-like." To make this transition possible, the basic technique is the assimilation or ingestion of experimental data of different types in the models as it is shown in Galkin et al. (2012) and by Nava et al. (2006), Nava et al. (2011), and Aa et al. (2018). It is expected that this trend will continue with the objective of obtaining models able to forecast or predict the behavior of the ionosphere. An essential need is to determine the "missing physics" in the formulation of the first-principle models. In the case of empirical models, it will be necessary to continue the efforts for the improvement of the mathematical expressions that describe parameters like, for example, the ionospheric layer thickness. Attempts like those of Altadill et al. (2009) in the case of the IRI and Alazo-Cuartas and Radicella (2017) in the case of NeQuick indicate this direction. Such developments would allow improving the effectiveness of data assimilation or ingestion techniques to specify and predict ionospheric conditions.

Acknowledgment

The authors thank Yenca Migoya-Oruè for her contribution in producing the figures that appear in this chapter.

References

Aa, E., Ridley, A.J., Huang, W., Zou, S., Liu, S., Coster, A.J., Zhang, S., 2018. An ionosphere specification technique based on data ingestion algorithm and empirical orthogonal function analysis method. Space Weather 16 https://doi.org/10.1029/2018SW001987.

Alazo-Cuartas, K., Radicella, S.M., 2017. An improved empirical formulation of an ionospherebottomside electron density profile thickness parameter. Adv. Space Res. 60 (8), 1725–1731. https://doi.org/10.1016/j.asr.2017.06.030.

Altadill, D., Torta, J.M., Blanch, E., 2009. Proposal of new models of the bottom-side B0 and B1 parameters for IRI. Adv. Space Res. 43, 1825–1834. https://doi.org/10.1016/j.asr.2008.08.014.

Altadill, D., Magdaleno, S., Torta, J.M., Blanch, E., 2013. Global empirical models of the density peak height and of the equivalent scale height for quiet conditions. Adv. Space Res. 52, 1756–1769. https://doi.org/10.1016/j.asr.2012.11.018.

Appleton, E.V., Barnett, M.A.F., 1925. Local reflection of wireless waves from the upper atmosphere. Nature 115, 333–334.

Appleton, E.V., Beynon, W.J.G., 1947. The application of ionospheric data to radio communications problems. Part II. Proc. Phys. Soc. 59, 58–76.

Araujo-Pradere, E.A., Fuller-Rowell, T.J., Bilitza, D., 2002. Comprehensive validation of the STORM response in IRI2000. J. Geophys. Res., submitted.

Barghausen, A.L., Finney, J.W., Proctor, L.L., Schultz, L.D., 1969. Predicting longterm operational parameters of high-frequency sky-wave telecommunication systems, ESSA Tech. Report, ERL 110-ITS 78.

Bent, R.B., Llewellyn, S.K., Schmid, P.E., 1972. A Highly Successful Empirical Model for the Worldwide Ionospheric Electron Density Profile. DBA Systems, Melbourne, FL.

Bilitza, D., 1986. International reference ionosphere: recent developments. Radio Sci. 21, 343–346. https://doi.org/10.1029/RS021i003p0034.

Bilitza, D.: International Reference Ionosphere 1990, National Space Science Data Center, Report 90-22, Greenbelt, MA, 1990.

Bilitza, D., 1997. International reference ionosphere—status 1995/96. Adv. Space Res. 20, 1751–1754.

Bilitza, D., 2001. International reference ionosphere 2000. Radio Sci. 36 (2), 261–275. https://doi.org/10.1029/2000RS002432.

Bilitza, D., 2002. Ionospheric models for radio propagation studies. In: Stone, W.R. (Ed.), The Review of Radio Science 1999–2002. URSI.

Bilitza, D., 2004. A correction for the IRI topside electron density model based on Alouette/ISIS topside sounder data. Adv. Space Res. 33 (6), 838–843.

Bilitza, D., Reinisch, B.W., 2008. International reference ionosphere 2007: improvements and new parameters. Adv. Space Res. 42, 599–609. https://doi.org/10.1016/j.asr.2007.07.048.

Bilitza, D., Altadill, D., Zhang, Y., Mertens, C., Truhlik, V., Richards, P., McKinnell, L.-A., Reinisch, B., 2014. The international reference ionosphere 2012—a model of international collaboration. J. Space Weather Space Clim. 4, 1–12. https://doi.org/10.1051/swsc/2014004.

Bilitza, D., Altadill, D., Truhlik, V., Shubin, V., Galkin, I., Reinisch, B., Huang, X., 2017. International reference ionosphere 2016: from ionospheric climate to real-time weather predictions. Space Weather 15, 418–429. https://doi.org/10.1002/2016SW001593.

Bradley, P.A., Dudeney, J.R., 1973. Vertical distribution of electron concentration in the ionosphere. J. Atmos. Terr. Phys. 35, 2131–214.

CCIR, 1967. Atlas of Ionospheric Characteristics, Comité Consultatif International des Radiocommunications. Report 340-4. International Telecommunications Union, Geneva.

Ching, B.K., Chiu, Y.T., 1973. A phenomenological model of global ionospheric electron density in the E-, F1-, and F2-regions. J. Atmos. Terr. Phys. 35, 1615.

Chiu, Y.T., 1975. An improved phenomenological model of ionospheric density. J. Atmos. Terr. Phys. 37, 1563.

Coïsson, P., Radicella, S.M., Leitinger, R., Nava, B., 2006. Topside electron density in IRI and NeQuick: features and limitations. Adv. Space Res. 37, 937–942.

Daniell, R.E., Brown, L.D., Anderson, D.N., Fox, M.W., Doherty, P.H., Decker, D.T., Sojka, J.J., Schunk, R.W., 1995. Parameterized ionospheric model: a global ionospheric parameterization based on first principle models. Radio Sci. 30 (5), 1499–1510.

Di Giovanni, G., Radicella, S.M., 1990. An analytical model of the electron density profile in the ionosphere. Adv. Space Res. 10 (11), 27–30.

Dudeney, J.R., 1978. An improved model of the variation of electron concentration with height in the ionosphere. J. Atmos. Terr. Phys. 40 (2), 195–203. https://doi.org/10.1016/0021-9169(78)90024-7.

Friedrich, M., Torkar, K., 2001. FIRI: a semiempirical model of the lower ionosphere. J. Geophys. Res. 106 (A10), 21409–21418.

Galkin, I.A., Reinisch, B.W., Huang, X., Bilitza, D., 2012. Assimilation of GIRO data into a real-time IRI. Radio Sci. 47, RS0L07. https://doi.org/10.1029/2011RS004952.

Gallagher, D.L., Craven, P.D., Comfort, R.H., 1988. An empirical model of the Earth's plasmasphere. Adv. Space Res. 8 (8), 15–24.

Garcia, R.R., Marsh, D.R., Kinnison, D.E., Boville, B.A., Sassi, F., 2007. Simulation of secular trends in the middle atmosphere, 1950–2003. J. Geophys. Res. 112, D09301. https://doi.org/10.1029/2006JD007485.

Heaviside, O., 1902. Telegraphy in Encyclopedia Britannica IX, vol. 33, p. 215.

Hochegger, G., Nava, B., Radicella, S.M., Leitinger, R., 2000. A family of ionospheric models for different uses. Phys. Chem. Earth Part C 25 (4), 307–310.

Huba, J.D., Joyce, G., Krall, J., 2008. Three-dimensional equatorial spread F modeling. Geophys. Res. Lett. 35, https://doi.org/10.1029/2008GL033509.

Kennelly, A.E., 1902. On the elevation of the electrically conducting strata of the earth's atmosphere. Elec. World Eng. 39, 473.

Klobuchar, J.A., 1982. Ionospheric corrections for the single frequency user of the global positioning system. In: National Telesystems Conference, NTC'82. Systems for the Eighties. IEEE, Galveston, Texas, USA, New York.

Klobuchar, J.A., 1987. Ionospheric time-delay algorithm for single-frequency GPS users. IEEE Trans. Aerosp. Electron. Syst. AES-23 (3), 325–331.

Leitinger, R., Zhang, M.L., Radicella, S.M., 2005. An improved bottomside for the ionospheric electron density model NeQuick. Ann. Geophys. 48 (3), 525–534.

Lin, C.Y., Matsuo, T., Liu, J.Y., Lin, C.H., Tsai, H.F., Araujo-Pradere, E.A., 2015. Ionospheric assimilation of radio occultation and ground-based GPS data using nonstationary background model error covariance. Atmos. Meas. Tech. 8, 171–182.

Liu, H.-L., Yudin, V.A., Roble, R.G., 2013. Day-to-day ionospheric variability due to lower atmosphere perturbations. Geophys. Res. Lett. 40, 665–670. https://doi.org/10.1002/GRL.50125.

Llewellyn, S. K. and R. B. Bent, Documentation and Description of the Bent Ionospheric Model, Air Force Geophysics Laboratory, Report AFCRL-TR-73-0657, Hanscom AFB, Massachusetts, 1973.

Magdaleno, S., Altadill, D., Herraiz, M., Blanch, E., de la Morena, B., 2011. Ionospheric peak height behavior for low middle and high latitudes: a potential empirical model for quiet conditions—comparison with the IRI-2007 model. J. Atmos. Solar Terr. Phys. 73, 1810–1817. https://doi.org/10.1016/j.jastp.2011.04.019.

Minkwitz, D., van den Boogaart, K.G., Gerzen, T., Hoque, M., Hernández-Pajares, M., 2016. Ionospheric tomography by gradient-enhanced kriging with STEC measurements and ionosonde characteristics. Ann. Geophys. 34, 999–1010.

Nava, B., Radicella, S.M., Leitinger, R., Coïsson, P., 2006. A near-real-time model-assisted ionosphere electron density retrieval method. Radio Sci. 41, RS6S16. https://doi.org/10.1029/2005RS003386.

Nava, B., Coïsson, P., Radicella, S.M., 2008. A new version of the NeQuick ionosphere electron density model. J. Atmos. Sol. Terr. Phys. https://doi.org/10.1016/j.jastp.2008.01.015.

Nava, B., Radicella, S.M., Azpilicueta, F., 2011. Data ingestion into NeQuick 2. Radio Sci. 46, RS0D17. https://doi.org/10.1029/2010RS004635.

Orus Perez, R., Parro-Jimenez, J.M., Prieto-Cerdeira, R., 2018. Status of NeQuick G after the solar maximum of cycle 24. Radio Sci. 53, 257–268. https://doi.org/10.1002/2017RS006373.

Prieto-Cerdeira, R., Orús-Perez, R., Breeuwer, E., Lucas-Rodriguez, R., Falcone, M., 2014. Performance of the Galileo Single-Frequency Ionospheric Correction During In-Orbit Validation. GPS World.

Radicella, S.M., Leitinger, R., 2001. The evolution of the DGR approach to model electron density profiles. Adv. Space Res. 27 (1), 35–40.

Radicella, S.M., Zhang, M.L., 1995. The improved DGR analytical model of electron density height profile and total electron content in the ionosphere. Ann. Geofis. XXXVIII (1), 35–41.

Rawer, K., Bilitza, D., Ramakrishnan, S., 1978. Goals and status of the international reference ionosphere. Rev. Geophys. Space Phys. 16 (2), 177–182. https://doi.org/10.1029/RG016i002p00177.

Ridley, A.J., Deng, Y., Toth, G., 2006. The global ionosphere-thermosphere model (GITM). J. Atmos. Solar Terr. Phys. 68, 839–864.

Scherliess, L., Schunk, R.W., Sojka, J.J., Thompson, D., 2004. Development of a physics-based reduced state Kalman filter for the ionosphere. Radio Sci. 39, RS1S04. https://doi.org/10.1029/2002RS002797.

Shim, J.S., et al., 2011. CEDAR Electrodynamics Thermosphere Ionosphere 1 (ETI) challenge for systematic assessment of ionosphere/thermosphere models 1: NmF2, hmF2, and vertical drift using ground based observations. Space Weather 9, S12003. https://doi.org/10.1029/2011SW000727.

Shim, J.S., et al., 2012. CEDAR Electrodynamics Thermosphere Ionosphere (ETI) challenge for systematic assessment of ionosphere/thermosphere models: electron density, neutral density, NmF2, and hmF2 using space based observations. Space Weather 10, S10004. https://doi.org/10.1029/2012SW000851.

Shim, J.S., et al., 2017. CEDAR-GEM challenge for systematic assessment of ionosphere/thermosphere models in predicting TEC during the 2006 December storm event. Space Weather 15, 1238–1256. https://doi.org/10.1002/2017SW001649.

Shubin, V.N., Karpachev, A.T., Tsybulya, K.G., 2013. Global model of the F2 layer peak height for low solar activity based on GPS radiooccultation data. J. Atmos. Solar Terr. Phys. 104, 106–115. https://doi.org/10.1016/j.jastp.2013.08.024.

Teters, L. R. J. L. Lloyd, G. W. Haydon, and D. L. Lucas, Estimating the Performance of Telecommunication Systems Using the Ionospheric Transmission Channel-Ionospheric Communications Analysis and Prediction Program User's Manual, National Telecommunication and Information Administration, Report NTIA 83–127, Boulder, 1983.

7

Wrap up

Massimo Materassi[a], Anthea J. Coster[b], Biagio Forte[c], Susan Skone[d]

[a]Institute for Complex Systems of the National Research Council (ISC-CNR), Florence, Italy
[b]MIT Haystack Observatory, Westford, MA, United States
[c]Department of Electronic and Electrical Engineering, University of Bath, Bath, United Kingdom
[d]The University of Calgary | HBI Department of Geomatics Engineering, Calgary, AB, Canada

The Earth's ionosphere (EI) configuration and evolution is determined by several physical factors, each of which tends to shape it differently: the three most effective and "regular" ones are mimicked in Figs. 1–3.

The *gravitational field* of the planet makes the EI get stratified perpendicular to its (grossly Earth-radial) field lines, as depicted in Fig. 1; the *solar irradiation*, that is the factor making the ionosphere exist *tout court* ionizing the near-Earth neutrals, shapes the EI according to the interplay between Sun's radiation propagation and the planetary geometry and "orientation," see Fig. 2.

Last but not least, as reported in Fig. 3, the geomagnetic field tends to constrain the charged particles along its field lines, packaging the plasma in *flux tubes* tangent to the dipole L-shells (Kelley, 1989).

The interplay among these different contributions would be able by itself to produce *complicated, highly organized space-time patterns*: not only because they are characterized by *different symmetries*, but also because they have *different characteristic time paces* and *different degrees of irregularity*. This "basic structure of the EI," let us refer to its ionization density as $N_{basic}(x,t)$

(where x represents the position at which the ionospheric density is evaluated and t the time), shaped according to the competition among these three forcing factors, is further enriched, and rendered more dynamical, by *the interactions among the different regions* of the ionosphere of ionization field $N_{basic}(x,t)$. The "interaction paradigm" introduces a first level of *complexity*, as interactions among the subparts of the system (such as transport) will structure the system as a whole: one may imagine this resulting into some richer, but still rather space and time regular in time and space, density $N_0(x,t)$: this should depend on daytime, season and on the phase of the Solar Cycle, in a complicated but rather predictable way.

The day-to-day variability of the real EI is not exactly what would be referred to as "predictable" (Mendillo, 2019; Materassi, 2019): indeed, it suggests one investigate about the real degree of predictability of the ionosphere; the "real" ionospheric density must be represented as some superposition $N_{real} = N_0 + \delta N$, in which local irregularities δN come into the play as an "erratic" contribution highly rough in space and time (see Materassi, 2019; Knepp, 2019;

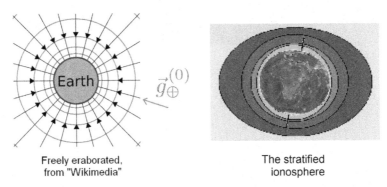

Freely eraborated,
from "Wikimedia"

The stratified
ionosphere

FIG. 1   A cartoon mimicking the stratifying effect on the Earth's ionosphere due to the action of the gravitational field of the planet.

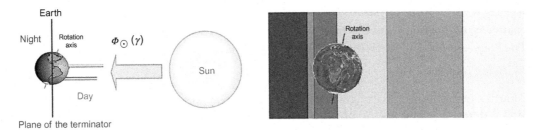

FIG. 2   A cartoon mimicking the effect on the Earth's ionosphere due to the Sun's irradiation, with the interplay between irradiation and planet's geometry. The symbol $\Phi_{\odot}(\gamma)$ represents the electromagnetic radiation of the Sun. In the right-hand panel, the darkening from right to left mimics the attenuation of Sun's radiation due to the distance from the star.

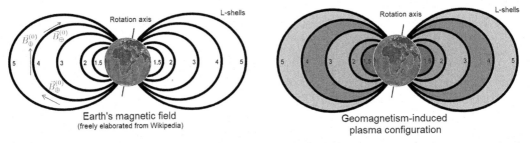

FIG. 3   A cartoon mimicking how the geomagnetic field $B^{(0)}$ tends to shape the near-Earth plasma.

Materassi et al., 2019 and references therein). These irregularities δN are generated by complicated dynamical processes, namely, plasma instabilities (Heysell, 2019), but the physical factors controlling them have to do with the very complicated interplay between the local ionospheric turbulence and aspects of the global Sun-Earth interaction.

All in all, the challenge of ionospheric physics is to understand the dynamics producing all the different aspects of the density $N_0(x,t) + \delta N(x,t)$, possibly tracing the cause-effect relationships so to be able to make *predictions* about ionospheric variability.

Empirical modeling has been for sure, and is, an important contribution to this attempt, showing the important limits discussed in Radicella and Nava (2019): no controllable experiments can be performed in this field; most of all, one has to rely only on geophysical observations; past data may be of help, but the geophysical phenomena appear to be *intrinsically unique*; physical conclusions are very difficult to be drawn after observations, as any attempt to synthesize the competition of many interdependent processes. On the other hand, *physical models*, trying to portray the ionospheric configuration predicting it "from first principles," have the advantage to track in extreme detail the logical causal relationships between processes, and how these depend on physical hypotheses, but miss the "improvisation freedom" the ionosphere appears to permit herself, through the irregular fluctuations $\delta N(x,t)$.

Lastly, let us observe that the fruitful compromise between empirical and physical modeling, namely, *data-driven physical models*, that are so effective, e.g., in meteorology, have a big difficulty in the ionospheric physics: the EI is so large that sampling it with a meteorological detail would be impossible.

The following Parts of this book deal with the attempts to go beyond these aforementioned difficulties, learning how to study the EI *as a complex dynamical system*.

References

Heysell, D.L. 2019. From Instabilities to Irregularities. (Chapter 11 of this book).

Kelley, M.C., 1989. The Earth's Ionosphere. Academic Press Inc.

Knepp, D.L., 2019. Scintillation Theory. (Chapter 13 of this book).

Materassi, M., 2019. The Complex Ionosphere. (Chapter 15 of this book).

Materassi, M., Alfonsi, L., Spogli, L., Forte, B., 2019. Scintillation Modelling. (Chapter 19 of this book).

Mendillo, M., 2019. The Earth's Ionosphere: An Overview. Introduction, (Chapter 1 of this book).

Radicella, S.M., Nava, B., 2019. Empirical Ionospheric Models. (Chapter 6 of this book).

PART II

Global complexity

8

Space weather: Variability in the Sun-Earth connection

Giovanni Lapenta

Department of Mathematics, KULeuven, University of Leuven, Leuven, Belgium

1 The Sun

The Sun is often quoted as being an unimpressive star at the center of our solar system. There are very many stars like the Sun and many are much bigger and exotic, but the Sun is neither unimpressive nor at the center of the solar system.

The Sun is a G-type star (G2V), and in the local galactic neighborhood only 7% of the stars are G-type. In fact, 90% of the stars in the local galactic neighborhood have a smaller mass than the Sun.

The center of the solar system is the center of mass of the solar system, not the center of the Sun. Taking just as an example the Sun-Jupiter two-body problem, the center of mass is about 0.005 AU away from the Sun, which is about 1 solar radius. When all planets are considered, the motion of the Sun around the center of mass of the solar system is very complex and spans up to a diameter of 0.02 AU (Charvátová, 2000), or 4.4 $R_\odot$ ($R_\odot$ is the solar radius, approximately 6.957×10^5 km). Similarly the orbit of the Earth is not a simple Keplerian ellipse, and its detailed position needs to be computed numerically. The general heliospheric data website OMNIWeb provides the position of all planets, while more accurate estimates can be obtained from the JPL solar system Dynamics website.

We will begin our tour of the solar system by studying its central star and its variability; we will then follow the effects of the Sun on the rest of the system. The focus will be on the processes that represent the so-called *space weather*: the study of the dynamical conditions in space, around the Sun, the Earth, and the solar system.

1.1 Structure of the interior of the Sun

Like all stars, the Sun is composed of matter that is at such a high temperature that the nuclei and the electrons of its atoms are free to move with respect to each other, a state called *plasma*. Plasma is the fourth state of matter, after solid, liquid, and gas, and it is reached when the energy content is sufficiently large to energize the electrons so much that they are freed from the orbital motion around their nuclei. A great achievement of the 20th century was the discovery of what gives this great energy to the stars: *nuclear fusion*. This discovery led to Hans Bethe being awarded the Nobel Prize in Physics 1967. In the core of the Sun, nuclei of light

The Dynamical Ionosphere
https://doi.org/10.1016/B978-0-12-814782-5.00008-X

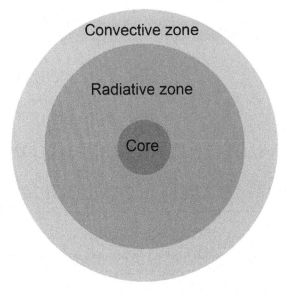

FIG. 1   Structure of the Sun.

elements such as helium and hydrogen fuse to form heavier nuclei, releasing part of their mass in energy. This is the same process happening in hydrogen bombs and in experimental nuclear fusion reactors for energy production, such as ITER, an international experiment being built in France.

Fig. 1 shows the internal structure of the Sun. The process of nuclear fusion is active in the core (up to about 0.2 $R_\odot$). The energy produced in the *core* travels outward, encountering first a *radiative zone* and then the *convective zone*, respectively, named after their main mechanism for energy transport. The convective zone is also host to motions in the plasma that composes the Sun. Often an analogy with boiling water in a pan is used. The bubbling of energy in this region arrives all the way to the surface. But motions in a plasma, unlike those in a water pan, lead also to electric currents and magnetic fields in a process similar to that active in the dynamo of a bicycle, and for this reason it is called a *dynamo process*. All the dynamical interactions of the Sun with the Earth can be tracked back to two originating agents: the energy released from the core and the action of the magnetic field produced in the convective zone.

1.2 Surface of the Sun: Photosphere

The hot plasma of the Sun is held together by one force: gravity. As with planets, this force confines the mass of the Sun ($M_\odot = 1.99 \times 10^{31}$ kg) into a roughly spherical shape of radius $R_\odot = 6.957 \times 10^5$ km. This plasma is surrounded by a well-defined surface called the *photosphere*, where most of the visible light is emitted (its name comes from the Greek word for light).

The photosphere has been observed for centuries. The reader should be well aware that looking at the Sun is extremely dangerous and can lead to irreversible damage to the retina and to blindness. Instruments need to be used to look at the photosphere. Fig. 2 shows images taken during a period of more intense solar activity by the current most highly resolved space-based mission, the SDO. The left image was taken by the HMI instrument in the continuum (white light) channel and the right image by the AIA instrument in the ultraviolet

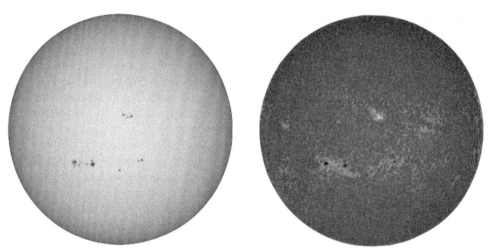

FIG. 2 Photosphere and chromosphere observed by the Solar Dynamic Observatory (SDO) on September 25, 2014 at 09:14:49 UT. *Left*: HMI continuum (*white light*) image. *Right*: AIA 1600 Å.

1600 Å. The *convective cells* carrying energy from the surface are clearly visible, but more prominent are the darker regions called *Sunspots*. The two bottom left sunspots are expanded in Fig. 3.

The locations of sunspots are the loci of active regions where some of the most interesting processes developing on the Sun start producing effects in the whole solar system, creating the domain of research of space weather. As shown in Figs. 2 and 3, sunspots tend to form groups. The internal structure of sunspots is complex, presenting clear striations emerging from the center. Fig. 4 shows a detailed image taken by

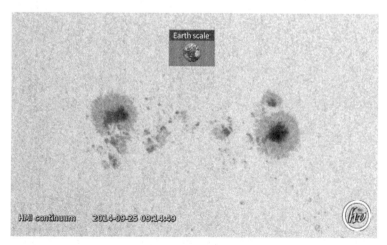

FIG. 3 Sunspots observed by the Solar Dynamic Observatory (SDO) on September 25, 2014 at 09:14:49 UT with the instrument HMI, continuum channel. The Earth is reported for size comparison. Image (as other images below carrying the same *hv* logo) obtained using helioviewer.

II. Global complexity

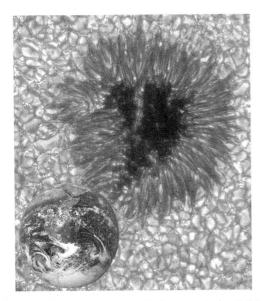

FIG. 4   A colorized photograph of a sunspot taken in May 2010, with Earth shown to scale. The image has been colorized for esthetic reasons. This image with 0.1 arcsecond resolution from the Swedish 1-m Solar Telescope represents the limit of what is currently possible in terms of spatial resolution. *Courtesy: The Royal Swedish Academy of Sciences, V.M.-J. Henriques (sunspot image taken from the Swedish 1-m Solar Telescope), NASA Apollo 17 (Earth).*

the Swedish ground solar observatory showing a highly detailed view.

These striations betray the origin of sunspots. These regions are loci of very intense magnetic field and the striations are aligned with magnetic field lines. Fig. 5 reports the intensity of the magnetic field in the exact same location reported in Fig. 3. Obviously, the sunspots are regions of very intense fields. A well-known property of plasmas is that of being diamagnetic: a plasma wants to expel the magnetic field from its interior and in Nature (as well as in the laboratory) plasmas and magnetic fields tend to separate (Park et al., 2019). A sunspot is a form of magnetic condensation: the magnetic field congregates in the sunspot where a lower density of plasma is present. A simple mathematical model can be formulated assuming constant pressure and imposing that the magnetic pressure (proportional to the square of the magnetic field, $B^2/2\mu_0$) inside the sunspot is balanced by the plasma pressure outside. In reality of course, the condensation is not perfect and there is a transition layer around the sunspot, visible in the striated region around the

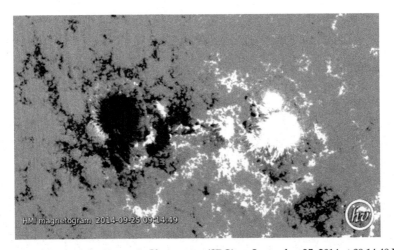

FIG. 5   Sunspots observed by the Solar Dynamic Observatory (SDO) on September 25, 2014 at 09:14:49 UT with the instrument HMI, magnetogram. *White* is positive polarity and *black* in negative polarity, where polarity is meant along the line of sight.

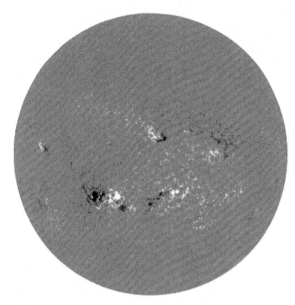

FIG. 6 Magnetogram observed by the Solar Dynamic Observatory (SDO) on September 25, 2014 at 09:14:49 UT with the instrument HMI.

sunspot where magnetic field lines from the sunspot exit into the plasma around.

Fig. 5 shows another typical feature of sunspots: they come in pairs of opposite polarity: in essence, the magnetic field coming out of one sunspot enters the other. Magnetic field lines need to close since there are no magnetic monopoles in Nature and the divergence of the magnetic field is zero. Any magnetic field line exiting the Sun must be compensated by one entering.

At any given time, we can see only half of the Sun. Fig. 6 reports the Sun observed at the same moment used in Figs. 2, 3, and 5. The back side of the Sun is not visible from Earth, but it can be observed by the space mission Stereo which launched two spacecraft, both staying on the same orbit the Earth has around the Sun, but one progressing ahead and the other behind. At various times, the spacecraft have been progressively moving away from the Earth providing for part of the mission a complete view of the

Sun around 360 degrees. Except for Stereo, the state-of-the-art approach to model the total photospheric field is to form a *synoptic map* made by recording each day a central meridian band of magnetograms such as the one in Fig. 6. By putting side by side each day's measurement, a map of the whole magnetic field can be made; see Fig. 7. The map looks like a typical wallmap of the Earth, but it represents the magnetic field. It should be kept in mind that the magnetic field is not static while the synoptic map is not simultaneous and covers several days of data. One day we may have a whole fleet of spacecraft looking at the Sun from many angles and we may be able to do a simultaneous synoptic map, but that would require a multibillion mission. Preferably that fleet should be in the orbit of Venus or even closer to give us sufficient advanced warning. The day such a mission will be launched will be the day space weather becomes a mature science.

The magnetic field is rather complex; the sunspots are prominent but the field extends over the whole photosphere, with typically one polarity dominating one hemisphere and the opposite the other. The magnetic field at the polar regions is inaccurate because of the angle of view from the Earth. The view in Fig. 7 provides quantitative information about the solar magnetic field and it is the basis of most space weather modeling efforts: it is the number one source of information about the conditions near the Sun and how those impact the rest of the solar system. Many models use this information as their primary input.

1.3 Atmosphere of the Sun: Chromosphere, transition region, and corona

The surface of the Sun is surrounded by a stratified atmosphere. The atmosphere of the Sun becomes visible during a solar eclipse because the Moon blocks the bright photosphere, allowing (with proper protection because even

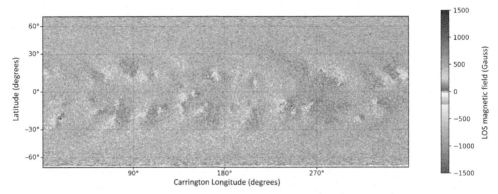

FIG. 7 Synoptic map for Carrington rotation number 2155 based on SDO HMI data. Note that CR 2155 includes September 25, 2014 used in Figs. 2, 3, 5, and 6. The line of sight (LOS) magnetic field in Gauss is plotted using SunPy.

watching an eclipse is very dangerous, and only certified viewing glasses should be used) one to see the solar atmosphere. Closest to the photosphere, there is the chromosphere (so-called from the Greek word for color because it shows color; it is not white when observed during an eclipse), followed by the transition region and by the corona.

Fig. 8 shows the strata of the solar atmosphere above the photosphere. This graph includes one of the greatest puzzles in modern science: while as one would expect the density drops with altitude, the temperature increases at a tremendous pace. The photosphere of the Sun is on average at 5780 K, while the sunspots are significantly cooler, at 4200 K. While these temperatures are high, they pale in comparison to those in the core of the Sun: 1.5×10^7 K. Inside the Sun the energy is carried outward and the surface is much colder than the core. But just above the chromosphere, the temperature starts increasing exponentially, reaching 2×10^6 K in the corona. Where is this heating coming from? This is a fascinating area of research, but we focus here on the dynamical changes of the Sun impacting the solar system and will not address this topic (Klimchuk, 2006).

The magnetic field emerging from the interior of the Sun, through the photosphere and into the corona, also expands out. Fig. 9 shows the

corona above the same pair of sunspots observed by SDO on September 25, 2014 discussed in Figs. 2, 3, 5–7. The image reports UV light, but the eye can track easily the magnetic field lines. What we see is the light emitted by the plasma particles captured by the magnetic field lines. These lines link the two sunspots, emerging from the right (positive polarity) sunspot and entering back under the photosphere in the left sunspot (negative polarity). Above the sunspots in the corona, the magnetic field lines form loops and groups of loops called *magnetic arcades*.

As we move away from the photosphere, the magnetic field expands and its features become more macroscopic. The field on the photosphere can be represented in a multipole expansion, using a polynomial expansion. As one moves away from the photosphere the high-order polynomials decay more rapidly, giving the field a progressively smoother structure. As we will see later, this effect can be misleading because there are new processes developing in the corona and producing new structures. But it is nevertheless true that the overall magnetic field structure becomes simpler at larger radii. An often used and very simple model to represent the field at different heights is the potential field source surface (PFSS) (Altschuler and Newkirk, 1969).

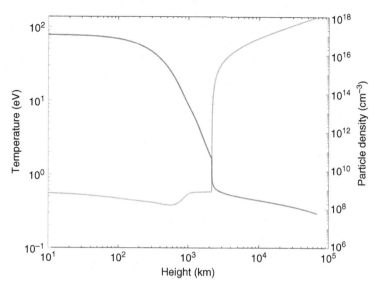

FIG. 8 The density and temperature are reported on the two distinct vertical axes as a function of the height above the photosphere, using there C7 model (Avrett and Loeser, 2008)

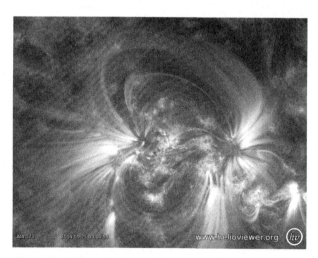

FIG. 9 View from SDO on September 25, 2014 at 09:14:49 UT of the same region of the two sunspots observed above but over a wider domain. The image is taken by the AIA instrument, channel at 171 Å.

The PFSS model extrapolates the magnetic field of the photosphere with a potential assumption for the magnetic field: that is, it assumes there are no currents in the corona so that $\nabla \times \mathbf{B} = 0$. The PFSS model produces the magnetic field anywhere in the corona, allowing one to track field lines. A very useful aspect of the model is to produce the magnetic field on the so-called *source surface*, a surface located at an arbitrary distance above the Sun. Typically the radius of the source surface is 2.5 $R_\odot$. If we apply the PFSS model to the conditions in the photosphere reported in Fig. 7, we obtain the magnetic field map at the source surface in Fig. 10. The magnetic field at 2.5 $R_\odot$ is far simpler and is often used as the basis of models for the conditions of the solar system.

More advanced models based on less restrictive assumptions have been developed to improve the description of the solar corona magnetic field, especially during active times.

While magnetic fields in the corona are not directly visible, images of the plasma can, as in Fig. 9, identify typical magnetic structures indirectly by the typical striations observed. The same approach can be used in global coronal images. Such images can be obtained during solar eclipses or using a coronagraph: these instruments produce an artificial eclipse by putting an occulting disk in front of the camera. The distance between the camera and the occultation disk determines how close to the surface the images are accurate (due to light diffraction at the edge of the occulting disk).

These instruments are on board space missions and especially important in the history of solar exploration is the LASCO coronagraph on board the SOHO spacecraft, still operating since 1995. LASCO has three different fields of view; C1 was the closest to the Sun but broke down early in the mission, while C2 and C3 are still operating at the time of writing. Fig. 11 shows a combined picture of the corona taken using ground-based images of the solar eclipse of March 9, 2016 and the LASCO C2 image at the same time. The striations due to the magnetic field are evident.

Especially evident are two features: closed field lines regions forming spiked pointy features called traditionally *helmet streamers* and regions instead of open field lines, called *coronal holes*. Inside the helmet streamers, the field lines are closed (i.e., with both ends tied to the photosphere), while outside, in the coronal holes, field lines extend out into the solar system.

A mission is in preparation, PROBA-3, to use two spacecraft: one takes the picture and the other carries the occultation disk to cause the artificial eclipse. By increasing tremendously the distance between camera and occultation

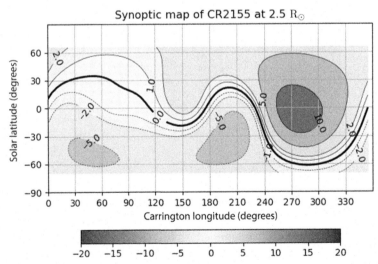

FIG. 10 Synoptic map for Carrington rotation number 2155 extrapolated to the source surface 2.5 $R_\odot$ using the AIDApy PFSS implementation. The colormap indicates the magnetic field in μT.

Eclipse Solar Corona of March 9, 2016 0.52 UT
Composite of W-L coronal images Space/Ground-based
LASCO C2 (SoHO); J. Vilinga/ R. Wittich/ S. Koutchmy Obs. & Processing

FIG. 11 Combined view of ground photo of a solar eclipse and a LASCO image taken on March 9, 2016. *Image processing by J. Vilinga, R. Wittich, and S. Koutchmy.*

disk, compared with that available within a standard coronagraph on a single spacecraft, the new mission will be able to make coronagraphs much closer to the photosphere and monitor regions of the corona that we cannot see now except during eclipses. At the time of writing, the expected launch is late 2020.

2 Changes in the Sun

2.1 Rotation of the Sun and Carrington solar coordinates

We noted earlier how the Sun is not located in the center of the Solar System but actually revolves around the center of mass of the solar system in a complicated orbit. The Sun is also not stationary; it rotates with an axis of rotation inclined by 7 degrees with respect to the ecliptic plane. But the most important aspect of the Sun's rotation is that our star does not rotate as a solid body: the equator rotates faster than the poles, creating a stirring action that is central to the dynamical evolution of the Sun. At its equator, the rotation period is 24.47 days, increasing with latitude to reach a value of 34.3 days at the poles.

As can be imagined, it is thus not obvious how to refer to observable features on the photosphere and their relative position, since this changes in time. Our method of referring to the positions on the Sun dates back to the 1850s when Richard C. Carrington determined the solar rotation rate by tracking low-latitude sunspots. He defined a fixed solar coordinate system that rotates in a sidereal frame exactly once every 25.38 days. But we are not sitting in a sidereal frame, we are sitting on Earth, which itself revolves around the Sun. As the Sun rotates, the Earth moves, so we will see the same feature on the Sun after one rotation from a different position around the Sun.

The heliographic coordinate system going back to Carrington uses a canonical prime solar meridian that was the central meridian visible from Earth at noon (GMT) on January 1, 1854. Since then, every time the prime meridian passes the central meridian of the Sun, we count a new *Carrington rotation*. As an example, September 25, 2014 used in the present chapter happens during Carrington rotation number 2155. Web tools and solar software (e.g., SunPy) give the equivalence for any given date.

The Carrington longitude of the central meridian is 360 degrees at the beginning of the Carrington rotation, and decreases to zero at the end. The rotation period of the Carrington heliographic coordinate system varies throughout the year because the Earth's revolution is not circular. The average value is 27.2753 days (called the mean synodic period). The latitude is simply measured from the equator to the poles in the usual way.

With this convention, the Carrington longitude of a feature on the Sun remains approximately fixed in time. However, the differential rotation at different latitudes tends to make features at different latitudes drift relative to each other.

2.2 Evolution of the photospheric magnetic field

The main driver of the dynamical evolution on the Sun is the magnetic field. The magnetic field originates from the dynamo processes in the convective regions under the photosphere. The domain of solar seismology is devoted to the study of waves on the Sun to obtain information on this changing field under the photosphere. However, direct measurements are not possible and the main source of information we have on the evolving solar magnetic field is the magnetogram data, obtained by measuring directly the magnetic field on the photosphere. The magnetic field of the corona can also be deduced by monitoring the evolution of the characteristic striations and arcades observed. The field of corona magnetic field and subphotosperic magnetic field observation is actively growing, with new detectors giving ever better data. For example, the DKI Solar Telescope under construction in Maui will provide the first ongoing measurements of the magnetic fields in the Sun's corona.

The changes of the solar magnetic field are both long term and short term. In the long term, the Sun has a 22-year cycle. The magnetic field on the Sun is far from a simple dipole, but still a very clear dipole component is present, with one pole predominantly of one polarity and the other of the opposite. Every 11 years this polarity flips. In 22 years, therefore, the same polarity returns. During the period of polarity change, the magnetic field of the Sun is very far from being dipolar and is most active and changing: for this reason it is called *solar maximum*. At solar maximum, the helmet streamers can be far displaced from the equatorial regions and the coronal holes can descend closer to the equator. Conversely, the periods of stable polarity when the Sun field looks more dipolar are less active and called *solar minimum*. In those times, the helmet streamers tend to be more equatorial and the coronal holes more positioned over the higher latitudes. Nevertheless, even during solar minimum, the magnetic field can be rather nondipolar and its features might be nontextbook.

The maximum and minimum are best characterized by the count of sunspots visible on the Sun. During maximum, the sunspots are most numerous. Fig. 12 reports the variation of the sunspot number over several centuries. As the reader might suspect, the observations of sunspots evolved in time as technology developed.

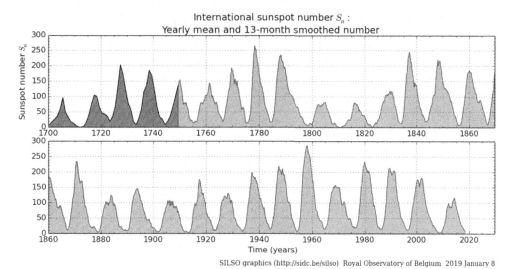

International sunspot number S_n :
Yearly mean and 13-month smoothed number

SILSO graphics (http://sidc.be/silso) Royal Observatory of Belgium 2019 January 8

FIG. 12 Yearly mean sunspot number (*black*) up to 1749 and monthly 13-month smoothed sunspot number (*blue*) (*light gray in print version*) from 1749 up to the present. *Data from The Royal Observatory of Belgium, Uccle, Belgium.*

However, for consistency, sunspots are still measured by eye and hand using the same techniques since the 18th century. In essence, the photosphere image from a telescope is projected on a paper where an astronomer circles each sunspot and counts the number. This operation is done in many observatories and by amateurs around the world. These data are collected and aggregated by the Royal Observatory of Belgium, which produces the official international sunspot number. These data are very important for their relevance to the climate debate.

Some features of the sunspot number are especially important to mention. The data seems to identify longer trends that modulate the height of the peaks with several solar cycles when the maxima are weaker and the minima longer. Among them, especially famous is the period around 1645–1715 (not reported in the figure, because the data is more sketchy and less established), which had extremely few sunspots, with essentially an extended minimum lasting for decades. Astronomers already had the telescope and were monitoring sunspots, but very few existed to be seen. In 1894, Maunder

published a study about this peculiar period, now called the Maunder minimum. The years of the Maunder minimum went down in history as years of exceedingly cold temperatures in northern Europe and as a period of low agricultural yield, causing famine, misery, and migrations. Conversely, the maxima since the dawn of the space age, started by the launch of the Sputnik1 on October 4, 1957, have been relatively stronger. With the exception of the last maximum that was preceded by an unusually extended minimum. This situation caused a significant debate both on the possible effect on global warming of stronger solar maxima in the last several decades and as to the circumstances and implications of the last weaker maximum. Predictions for the future vary so much that we must wait and see as illustrated in Fig. 13. The gray band reports the range of predictions, which go all the way from something close to a new Maunder minimum scenario back to strong space-age normality.

The study of the long-term changes in solar activity are the domain of *space climate*. *Space weather* focuses on much shorter scales, from

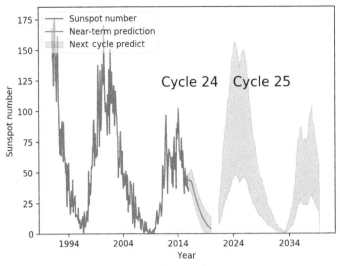

FIG. 13   Number of sunspots in recent solar cycles and predictions based on a suite of methods. *Data provided by NOAA via SunPy and based on predictions made in 2016.*

minutes to days. The global photospheric magnetic field changes little on those scales, but local features around the active regions surrounding the sunspots can change dramatically even in a matter of seconds. On those scales, the solar activity produces *flares*, intense flashes of energy emission, and *coronal mass ejections* (CMEs), vast eruptions of matter and energy, the subject of Section 4.

The sunspot cycle has another important feature: when a solar minimum begins to end and the activity tends to pick up again, the sunspots tend to appear first at higher latitudes and then migrate progressively equator-ward. Fig. 14 shows this trend over several solar cycles. The characteristic shape of the plot gives it the name of a butterfly diagram.

2.3 Variability of the solar energetic output

The variability of the solar magnetic field is accompanied by a corresponding variability of the total energy emission of the Sun, in all spectra of radiation. The total energy density

arriving at the Earth orbit is the so-called *solar constant* (an old name that is still used even though we now know it varies in time): on average 1380 W/cm^2. The orbit of the Earth changes its distance slightly from the Sun and this number fluctuates during the year. This change should not be confused with the seasonal change of the energy density arriving on a surface on Earth. That energy density depends on the angle of the Sun to that surface and is why the northern hemisphere experiences a winter while the southern hemisphere enjoys a summer, and vice versa. The most important change in the solar energetic output comes from the solar 11-year cycle: at solar maximum there are more sunspots (darker regions), but the presence of more intense radiation emitted around them overcompensates and results in a higher total energy output. Measuring the total output of the Sun over a long period is a complex task because the measurements are taken by different instruments over different times with different calibrations. The task, however, is of critical importance as our life depends on the solar output and how the Earth's climate responds to it: it

Daily sunspot area averaged over individual solar rotations

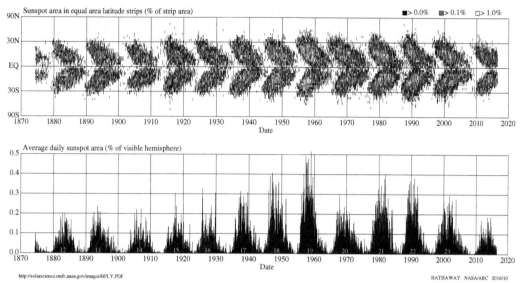

FIG. 14 Number of sunspots in recent solar cycles and their latitudinal location. *Data provided by NASA Marshall Space Flight Center.*

is then of clear and present importance to know this in detail. For this reason, detailed studies have been carried out, showing that over the period of observation with the first mission in the 1970s up to now, the variation is well correlated with the sunspot number, but overall the solar constant varies by less than 0.1%. However, if the total energy output varies so little, the variability in different parts of the radiation spectrum can change much more, as reported in Fig. 15.

Following the frequency bands of the emitted radiation, in order of increasing wave-length:

- *X-ray and extreme UV*: 10^{-3} of the total energy output, it varies with the solar cycle.
- *UV*: 9% of the total energy output, it varies with the solar cycle.
- *Visible*: 40% of the total energy output, almost no variation with the solar cycle.
- *Infra-red*: 51% of the total energy output, almost no variation with the solar cycle.

- *Radio*: a tiny 10^{-10} fraction of the total energy output, but very sensitive to solar events, so much so that during some solar events, the radio noise from the Sun can interfere with communication.

3 The solar wind and its variability

The plasma around the Sun does not remain confined there but streams out into space to form the solar wind. Plasma originating from different regions in the Sun has different properties, and as the Sun evolves so does the solar wind. Fig. 16 reports the outcome of the Ulysses mission. As can be observed, the solar wind has two typical regimes. Above the helmet streamers where the magnetic field lines are closed the solar wind is slow; above coronal holes where the magnetic field is open into space the solar wind is fast. The two types of wind

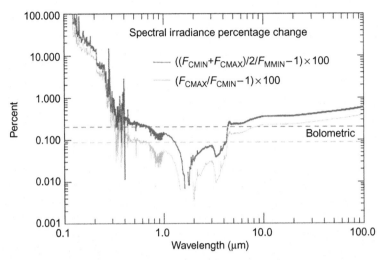

FIG. 15 Fractional change of the total solar irradiance over different energy spectra. The *dashed line* indicates the total irradiance change (*Bolometric*). Two definitions of percent change are reported. *Adapted from Haigh, J.D., 2007. The Sun and the Earth's climate. Living Rev. Sol. Phys. 4 (1), 2.*

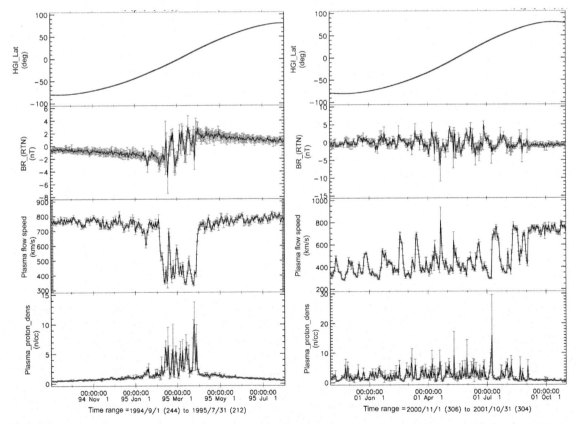

FIG. 16 Solar wind—dependence on solar cycle, data obtained from the Ulysses mission that covered both solar minimum (*left*) and maxiumum (*right*) in the years indicated by the figure. From top to bottom, the heliographic latitude, the radial magnetic field, the proton radial flow speed, and the proton density are reported.

are relatively orderly during solar minimum but at solar maximum the conditions are chaotic. Fig. 16 does not show the longitudinal dependence, but at any given moment the solar wind is very strongly dependent on both latitude and longitude, and at any given point the solar wind depends on time. Monitoring the solar wind and predicting the conditions of the solar wind coming to the Earth is one of the key tasks of space weather.

The main feature of the fast solar wind are:

- it originates from the coronal holes;
- little changes over time from rotation to rotation;
- ranges between 450 and 800 km/s;
- density at 1 AU: 3 ions/cm^3;
- 5% of the ions are helium; and
- proton temperature $T_p = 2 \times 10^5$ K, electrons are colder;

while for the slow solar wind:

- it originates above the helmet streamers;
- it is more unsteady than the fast wind;
- ranges below 450 km/s;
- density at 1 AU: 7–10 ions/cm^3;
- 4% of the ions are helium;
- proton temperature $T_p = 4 \times 10^4$ K.

The solar wind starts at the Sun but it expands in the whole solar system; the speed of the solar wind at first picks up, accelerating away from the Sun. One of the great successes of space weather modeling is the implementation of the first physics-based forecast system based on the ENLIL model operating at the Space Weather Modeling Center (SWMC) of the National Oceanographic and Atmospheric Administration (NOAA). Fig. 17 shows the typical conditions of the solar wind expanding out from the Sun. As can be observed, the density and the velocity vary with the latitude and given the rotation of the Sun, a characteristic spiral is formed. This effect is usually compared with the everyday experience of rotating water fountains used to water lawns. As the

water jet rotates, the water stream assumes a spiraling structure. This spiral has a classic mathematical formula going back to none other than Archimedes, and is thus called Archimedean spiral.

As the solar wind expands out into the solar system, the density, temperature, and magnetic field intensity decrease but the velocity has a more complex behavior. Fig. 18 shows the conditions on a specific day modeled by ENLIL. The figure reports the data up to 10 AU, just outside the orbit of Saturn.

The interplanetary magnetic field also follows the characteristic Archimedean spiral; see Fig. 19. Occasionally regions of the solar wind expanding at different speed come into contact and higher speeds flows push against lower speed flows forming the so-called *coronating interaction regions* (CIR).

Fig. 19 shows the field lines (in black) and the intensity of the magnetic field (in color). But the magnetic field also has a direction or polarity; it is a vector, not a scalar. One of the greatest hurdles to overcome in space weather is the great difficulty of the models in predicting the polarity of the field and especially the vertical direction of the magnetic field in the geomagnetic coordinate system. This value is the most important in determining the effects on the Earth of perturbations in the solar wind, yet it is a parameter we are not very successful in predicting.

After meeting all the planets, the solar wind eventually interacts with the interstellar medium. The solar wind is supersonic so the interaction is very complex. But luckily two missions have already reached far enough into the universe to send back information about that transition: Voyager 1 and Voyager 2. The stuff of myth by now, these missions constitute two memories of more hopeful times when space exploration was the pride of the competing superpowers, the sky was not the limit, and presidents called for tearing down walls, not building new ones. Fig. 20 shows the conditions at the edge of the solar

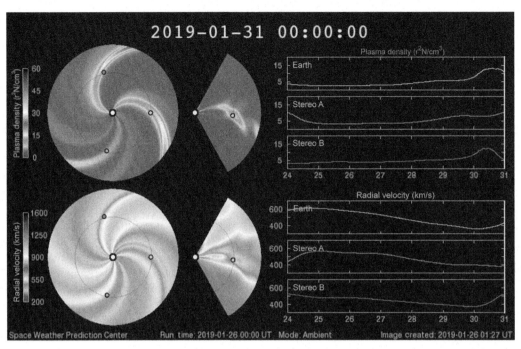

FIG. 17 Solar wind expansion into the solar wind. The conditions at a given time are shown in the *left* (radius and longitude at the ecliptic plane) and *middle* (radial and latitude dependence in the meridional plane where the Earth is) panels. The prediction was made at the time of writing on January 29, 2019. The prediction reports the density and velocity away from the Sun and up to 2 AU. The position of the Earth and of the two spacecraft Stereo A and B are reported along with the predicted solar wind at those locations in time. These predictions can then be validated by measurements done in the subsequent days. *The data are obtained from a prediction made by NOAA at the SWMC for the day February 3, 2019.*

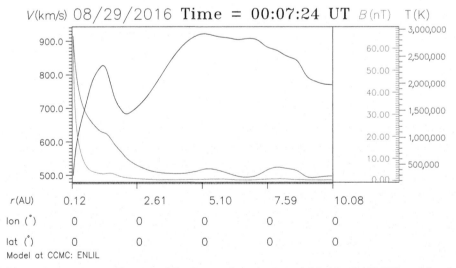

FIG. 18 Solar wind properties as a function of the distance from the Sun modeled by the ENLIL model at the CCMC.

CROT: 2098 **06/16/2010** Time = **00:08:51** UT lat= 0.00°

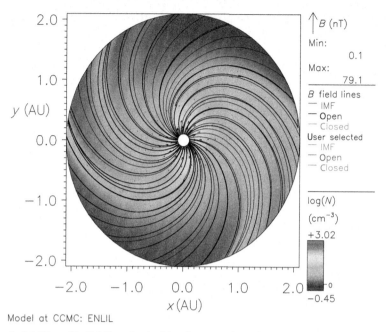

Model at CCMC: ENLIL

FIG. 19   The magnetic field lines (*black*) follow the Archimedean spiral are shown above the density in color log scale. Model by the ENLIL at the CCMC.

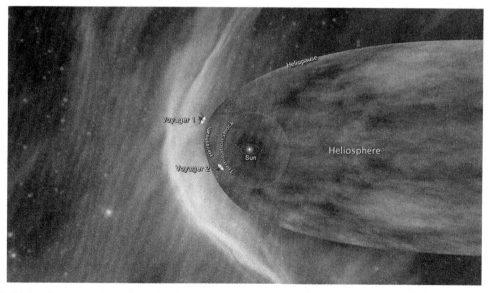

FIG. 20   Heliosphere. Artist's rendition for the Voyager mission at the Jet Propulsion Laboratory.

II. Global complexity

system. The solar wind is supersonic and when it interacts with the interstellar medium, it forms a shock called *termination shock* where the speed drops to subsonic. The heliopause determines the boundary between the plasma arriving from our Sun and the plasma arriving from the nearby stars. If we ever encounter alien neighbors and see a need to establish frontiers, the heliopause would be a reasonable choice. If the interstellar plasma and the Sun move at a relative supersonic speed, another shock bent like a bow and called *bow shock* would be at the nose of the heliopause. The heliopause is not a sphere but rather bean-shaped, because it is compressed at the head in the direction of motion of the Sun with respect to the interstellar medium and elongated in the tail. The tailward region we have never explored.

3.1 Variability of the solar wind at the Earth

When the solar wind arrives at the Earth, it is very variable. Even in the absence of the storms originating on the Sun and described in the next section, the conditions of the wind will depend on the location where it originated from on the Sun. The combined rotation of the Sun and revolution of the Earth around the Sun make this point evolve in time and with it the conditions of the solar wind at its origin. During solar minimum, the helmet streamer region tends to be more equatorial on the Sun and the Earth being on the ecliptic; the solar wind arriving at the Earth would be more likely coming from above the helmet streamers and therefore be slow. But that is no fixed rule and even during solar minimum, the Earth might find itself embedded in the fast solar wind. If the slow solar wind is more variable, the fast wind is also not steady. Predicting the conditions of the solar wind at the Earth is a key task in space weather. Two tools are very useful in this endeavor: models of the solar wind origin coupled with models of the solar wind evolution before it arrives at the Earth and direct

measurements of the solar wind before it reaches the Earth.

On the modeling side, several models are available, but ENLIL has become a reference in the field (Parsons et al., 2011). This uses the information on the photospheric magnetograms supplemented by empirical models to determine the solar wind source at the Sun (the so-called WSA model). ENLIL then evolves the solar wind using the magnetohydrodynamics (MHD) equations in a 3D box around the Sun that expands far out into the solar wind up to the outer radius of 10 AU, just outside the orbit of Saturn. Fig. 19 is an example of this model. ENLIL has been turned into an operational forecast system that operates several times per day making continuously updated predictions of the solar wind in the next days. The prediction is available on the Space Weather Prediction Center of NOAA.

On the observational side, the ACE spacecraft launched in 1997 has been sending us data on the solar wind in real time for decades. The ACE spacecraft is located at the Lagrangian point L_1, an equilibrium point in the Sun-Earth gravitational system that lies between the Sun and the Earth at a distance of some 1.5 million km from the Earth. With solar wind speeds typically in the 400–800 km/s range, this gives 30 min to 1 h advance notice. Given the importance of this, a new spacecraft was launched to replace ACE when it fails: DISCOVR. The ACE real-time data is available on the Space Weather Prediction Center of NOAA. Past ACE data, instead, can be found on the OMNI website of NASA.

An alternative source for solar wind predictions and current conditions is being developed in Europe: the Space Situational Awareness of the European Space Agency. The ESA site provides federated tools from many European centers, including a fully European alternative to ENLIL, EUHFORIA, developed by KULeuven in collaboration with the ROB and the Finnish Meteorological Institute (Poedts, 2018).

4 Space storms

We now come to the *pièce de résistance*. If all there were in space weather were the fast and slow solar wind, we would not be too far from being able to make good predictions, and space would be a safe environment to expand our activities into. But that is not the case. We left the monsters for last. Active regions observed in the previous sections become occasionally frequently during solar maximum, and more infrequently during solar minimum, host to monstrous eruptions of energy and matter with energy budgets that make the combined nuclear arsenals of all world powers look like a joke. There are primarily four agents that make space weather exceedingly dangerous. Solar flares and CMEs are, respectively, bright flashes of light and eruptions of plasma emanating from solar active regions. CMEs have a possibility of arriving at the Earth causing geomagnetic storms. The most famous storm took place from August 28 to September 2, 1859 in a series of succeeding CMEs that reached the Earth on September 1–2, causing serious trouble for telegraph lines. Our technology on the ground (the electrical grid most prominently, but also pipelines and other long conductors) and in space (telecommunication, GPS, the Space Station) is tremendously more advanced and subject to disruption than the horse-driven carriages of that era. Many studies suggest that the grid would be seriously damaged, and most importantly the highest voltage transformers might be damaged beyond repair. Studies suggest that several months would be needed before electricity could be restored, provided the industrial and social infrastructure continues to function. The East Coast of America would be most affected because of the configuration of the Earth's magnetic field. A scenario where millions of people, the New York stock market, hospitals, and food storage facilities remain without electricity for 9 months transcends emergency management and becomes an existential threat for our civilization. This explains why it is important to study space weather. Waiting for another Carrington event would be too late (Eastwood et al., 2017).

In the flares and CMEs, high energy particles are generated, some at such a high energy that their speed is relativistic, traveling to Earth in about 8 min. These are called solar energetic particles (SEP) and pose an existential danger to astronauts and technology in space, but are not dangerous on Earth thanks to the protection provided by the Earth's magnetic field and by the atmosphere itself, which behaves as a shield (Fig. 21).

The last agents of great importance in space weather are the cosmic rays. These are not generated by the Sun but rather come from the interplanetary (interacting coronating regions, shocks in the solar wind), interstellar, and intergalactic cosmos. Cosmic rays are a topic of scientific fascination by themselves, giving us direct access to the most energetic and distant phenomena in the universe, such as accretion disks around black holes, supernova explosions, gamma ray bursts, and extra-galactic jets. But cosmic rays in space are also a cause of serious long-term concern. If SEP can be dangerous in sudden bursts coming from the Sun and posing a danger of death on the scale of minutes, hours, and days, cosmic rays are always present, always chipping away at the materials and at the DNA of people, causing material fatigue and cancer. Space exploration is not a gala dinner.

4.1 Flares and CMEs

Flares are releases of energy, primarily in the form of light, but more generally of electromagnetic radiation, from radio waves to gamma rays in the most energetic cases. Flares are categorized for their intensity according to official scales. Similarly, *CME* are releases of matter from the corona.

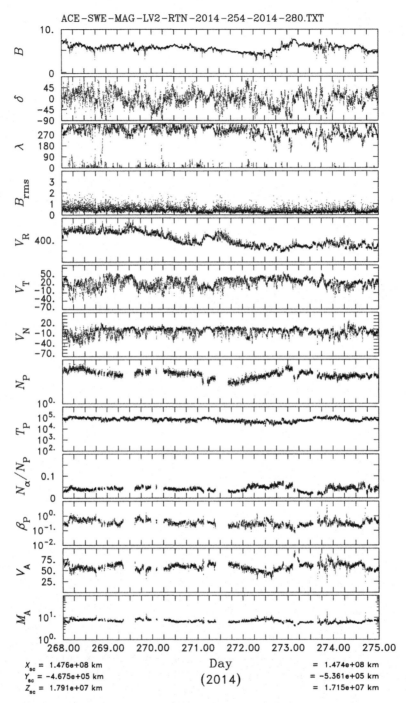

FIG. 21 Solar wind data provided by the ACE spacecraft.

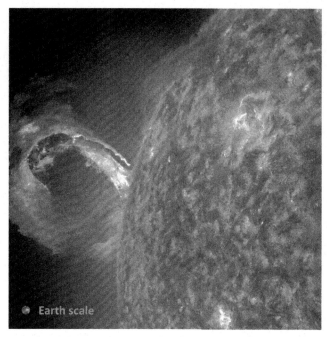

FIG. 22   SDO image of an eruption accompanied by flares. AIA 304 on May 1, 2013 at 02:42:30.

Fig. 22 shows a CME accompanied by a flare from an active region on the limb of the Sun. CMEs on the limb are especially impressive to see against the blackness of the background universe. But by no means are CMEs preferentially happening on the limb. The limb is just relative to the position of the observer.

Fig. 23 shows the expansion of the CME observed in Fig. 22. The time is about 2 h later. The central image from SDO no longer sees the eruption, which lasts only a number of seconds. But the CME then expands out and becomes visible in the coronagraph LASCO on board the spacecraft SOHO. The CME has left the Sun and is now traveling outward into the solar system. When a CME is coming straight toward the viewer, it looks like an expanding blob: these are called *halo CMEs* and their importance derives from the very fact that they are coming straight at the viewer. This CME instead is going off to the side. Tracking CMEs to

determine and predict their effects in the solar system is one of the most important tasks of space weather. If predicting when a CME or a flare happens is currently a daunting task, making predictions of the evolution of a CME, once one is observed to start, is a task attracting attention and meeting significant success. The mentioned models like ENLIL track CMEs as well as the solar wind, and the real-time predictions made on the space weather portals provide the best guess we can make of the effects of a CME.

Predicting when a flare or a CME occurs is beyond our current scientific ability, but they occur often in coincidence with each other and they happen near an active region, being produced by the release of energy stored in the form of magnetic field. The magnetic field changes its structure so much that sometimes matter confined near the solar surface is released, producing CMEs. Sometimes, the same active region can produce several successive flares and CMEs

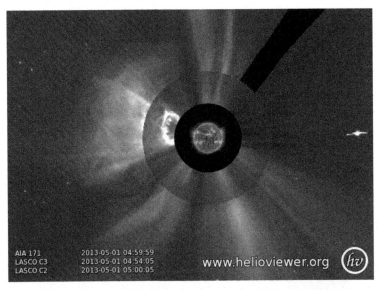

FIG. 23   Combined view of the same CME observed in Fig. 22, but 2 h later. The central image is from the AIA 171 instrument on SDO and shows the inner corona returned to a quiet state, having the CME left the proximity of the Sun. The two views C2 and C3 (C1 became inoperative early in the SOHO mission) are now picking up the CME expanding outward and starting its course within the solar system.

from the same region: this is how the most energetic storms are created. Carrington himself observed several such successive eruptions back in 1859 and we now believe that the so-called superstorms are produced by the piling up of successive CMEs coming toward the Earth and compressing against each other into a surfeit of destructive power. It should be noted that super CMEs will not kill anyone on the ground. Astronauts will have 8 min before the first energetic particles start to do them in, but for those of us on the Earth, a superstorm will just kill off the electrical grid but would not harm people, unlike what movies want you to believe. We just have to be prepared to rebuild quickly what was damaged and have all the needed preventive actions in place (e.g., disconnect the transformers and have a controlled blackout before the storm hits) and repair parts ready. These are prudent precautions to take since the next Carrington event will surely come sooner or later. Superstorms are expected every century,

and the next one is overdue. But study of the past and of nearby stars suggest that millennial super-superstorms much stronger than a Carrington event will also one day come. We have never witnessed one on Earth in a scientific way, but our species is 200,000 years old and our ancestors must have survived many of these millennial storms. So we can survive even super-superstorms and so can our civilization, if we know what to expect: money spent studying space weather is money well spent.

4.2 Solar energetic particles and cosmic rays

SEP are created during the events just described and are a part of the process: the large energy releases, besides producing the intense flashing (flare) and ejections (CME), also accelerate particles, ions, and electrons to high energies. Flares produce impulses of high energy particles while CMEs gradually produce high energy particles via the acceleration caused by the shock forming ahead of the expanding CMEs: these two categories are called therefore *impulsive* and *gradual* SEP. Obviously they were so-called and observed before the understanding of their origin made obvious why they are, respectively, impulsive and gradual.

SEP are sporadic in the sense that they happen only when there is an event on the Sun. Once produced, the SEP travel along the magnetic field lines. This is a typical property of particles immersed in field lines: their motion is one of rotation around field lines, like in a cyclotron device, and for this reason they are called cyclotron motion, and motion along the field lines. A particle cannot wander away from its magnetic field line, but it can freely travel along it. SEP then travel along the magnetic field lines and expand out into the solar system. Magnetic field lines, as we discussed earlier, form an Archimedes' spiral and then SEP also follow the ancient formula of the Syracusan genius:

$$\theta - \theta_0 = -\frac{\Omega_\odot}{V_{SW}}(R - R_\odot) \qquad (1)$$

where the change in angular position from its origin at the Sun's surface (at $R_\odot$) varies with the radius, R, in dependence with the angular rotation of the Sun, $\Omega_\odot$, and speed of the solar wind, V_{SW}. As mentioned earlier, SEP pose an existential danger to astronauts and technology in space, and for this reason they are constantly monitored. The fleet of meteorological instruments of the NOAA includes instruments for this task on board the GOES spacecrafts. Obviously the military also monitors radiation in space, but the military resources are not available for the public. The electron and proton fluxes measured by GOES can be monitored in real time at the NOAA SWPC. A list of the most significant particle radiation events can also be found at the NOAA SWPC. As an example, a significant proton event took place in January 2014 and is reported in Fig. 24.

Cosmic rays complete the story of the variable Sun-Earth space environment. Cosmic rays are not produced by the Sun, but their presence is modulated by the solar magnetic activity. Just like the Earth's magnetic field protects the Earth and its inhabitants from space radiation, so the solar and heliospheric magnetic field protects the solar system from cosmic rays. The intensity and effectiveness of this magnetic field in shielding cosmic particles varies over the solar cycle. Counterintuitively, cosmic rays are more intense during solar minimum. Solar minimum is characterized by an ordered magnetic field with a more dipolar-like structure, and the solar wind magnetic field has more order. This condition is easier for cosmic rays to penetrate. At solar maximum, the magnetic field is more chaotic and in fact more intense on average; this condition is more effective in shielding from cosmic rays. The effect is significant. For example, in a trip of five astronauts to Mars, the effect of cosmic rays would cause cancer in two astronauts during solar minimum, but less than one during solar maximum: there is a chance no astronaut will develop cancer. However, the increased probability of SEP during solar maximum, when the Sun is more active, increases the chances that one SEP storm would hit the mission. If an intense storm happens, all astronauts could die within hours. There is much talk about going to Mars, but as technology stands the chances are the crew will either die of radiation sickness during solar maximum or develop cancer during solar minimum. A responsible leader would not authorize such a suicide mission and would direct the expenditure, instead, in redoubling the study of space weather and to improve space propulsion to cut down on the duration of the trip. For now, Mars is for robots until faster space travel or more effective shielding is developed. A recent study provides more details (Zeitlin et al., 2013).

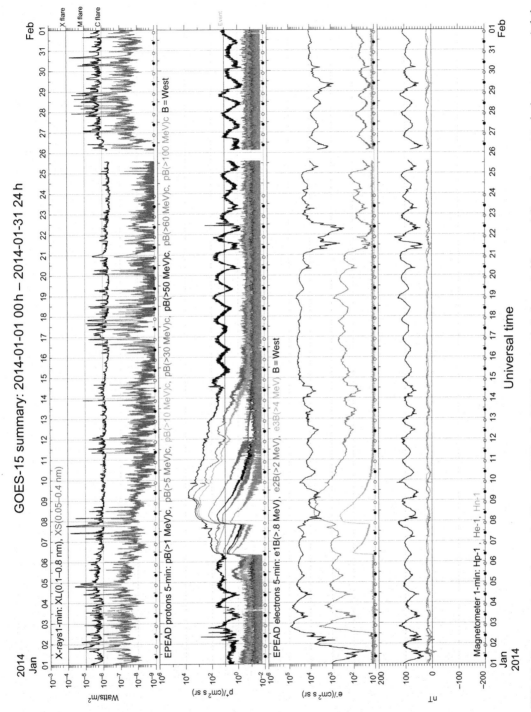

FIG. 24 Summary plot for the month of January 2014 by the GOES15 spacecraft. In the period January 6–9, a proton event took place, accompanied also by X-ray peaks and electron intensification in the highest energy channels.

References

Altschuler, M.D., Newkirk, G., 1969. Magnetic fields and the structure of the solar corona. Sol. Phys. 9 (1), 131–149.

Avrett, E.H., Loeser, R., 2008. Models of the solar chromosphere and transition region from SUMER and HRTS observations: formation of the extreme-ultraviolet spectrum of hydrogen, carbon, and oxygen. Astrophys. J. Suppl. Ser. 175 (1), 229.

Charvátová, I., 2000. Can origin of the 2400-year cycle of solar activity be caused by solar inertial motion? Ann. Geophys. 18 (4), 399–405.

Eastwood, J.P., Biffis, E., Hapgood, M.A., Green, L., Bisi, M.M., Bentley, R.D., Wicks, R., McKinnell, L.-A., Gibbs, M., Burnett, C., 2017. The economic impact of space weather: where do we stand? Risk Anal. 37 (2), 206–218.

Klimchuk, J.A., 2006. On solving the coronal heating problem. Sol. Phys. 234 (1), 41–77.

Lang, K.R., 2001. The Cambridge Encyclopedia of the Sun. Cambridge University Press, Cambridge, ISBN: 0521780934, p. 268 p.

Park, J., Lapenta, G., Gonzalez-Herrero, D., Krall, N., 2019. Discovery of an electron gyroradius scale current layer its relevance to magnetic fusion energy, earth magnetosphere and sunspots. arXiv preprint arXiv:1901.08041.

Parsons, A., Biesecker, D., Odstrcil, D., Millward, G., Hill, S., Pizzo, V., 2011. Wang-Sheeley-Arge-Enlil cone model transitions to operations. Space Weather 9(3). https://doi.org/10.1029/2011SW000663.

Poedts, S., 2018. Forecasting space weather with EUHFORIA in the virtual space weather modeling centre. Plasma Phys. Controll. Fusion 61 (1), 014011.

Zeitlin, C., Hassler, D.M., Cucinotta, F.A., Ehresmann, B., Wimmer-Schweingruber, R.F., Brinza, D.E., Kang, S., Weigle, G., Böttcher, S., Böhm, E., Burmeister, S., Guo, J., Köhler, J., Martin, C., Posner, A., Rafkin, S., Reitz, G., 2013. Measurements of energetic particle radiation in transit to Mars on the Mars science laboratory. Science 340 (6136), 1080–1084. https://doi.org/10.1126/science.1235989.

Further reading

Haigh, J.D., 2007. The Sun and the Earth's climate. Living Rev. Sol. Phys. 4 (1), 2.

McComas, D.J., Ebert, R.W., Elliott, H.A., Goldstein, B.E., Gosling, J.T., Schwadron, N.A., Skoug, R.M., 2008. Weaker solar wind from the polar coronal holes and the whole Sun. Geophys. Res. Lett. 35 (18), L18103.

Storms and substorms—The new whole system approach and future challenges

Naomi Maruyama

CIRES, Univ. of Colorado Boulder and NOAA Space Weather Prediction Center, Boulder, CO, United States

1 Introduction

1.1 A brief description of the evolution of storm time response of the ionosphere prior to the new perspective

Studying the Earth's ionosphere during storms and substorms is critical to understanding how much energy is injected into near-Earth environment from the solar wind and the various processes by which this energy becomes distributed. Since the pioneering work of Anderson (1928), tremendous effort has been devoted to explain the fundamental mechanisms that cause ionospheric behavior during storms and substorms to deviate from that of quiet times. However, it still remains the significant challenge for scientists to establish a consistent understanding of the rapidly expanding ground- and space-based observations in the coupled magnetosphere, ionosphere, and thermosphere (M-I-T) system. Furthermore, the main physical processes occurring in the ionosphere, thermosphere, and magnetosphere during storms and substorms lead to spatial density gradients

and plasma irregularities and instabilities, resulting in scintillations that have significant deleterious impacts on numerous communications and navigation systems (e.g., Groves and Carrano, 2016). This chapter is dedicated to review current understanding and recent advances in the study of Earth's ionosphere during magnetic storms and substorms in the past 14 years. A particular emphasis is placed on a growing awareness of a perspective gained by looking at the ionosphere as a part of the geospace system. Readers should refer to Mendillo (2006), Buonsanto (1999), and Prölss (1995) for further details of the historical development on this topic.

In the early 1990s when the interpretation of observations was limited to a number of points or local measurements from the few ionosonde stations scattered around the world and the still fewer number of incoherent scatter radar (ISR) facilities, storm-time ionospheric changes were characterized as "positive" and "negative" phases. The classical, or "old," positive phase of ionospheric storms was characterized by storm time neutral wind surges pushing ionospheric plasma to a higher altitude where the recombination is slower (e.g., Prölss, 1993). The negative phase of ionospheric storms is attributed to the storm time response of the neutral composition, i.e., the ratio of neutral atomic oxygen density to that of molecular nitrogen, the so-called O/N_2 ratio (which is now widely accepted). The finding is based on the work with the ESRO-4 satellite and ground-based ionosondes (Prölss and von Zahn, 1974; Prölss and Najiita, 1975; Prölss et al., 1975). The physics and chemistry of enhanced loss rates were already well understood at the time. Later the "old positive" phase of ionospheric storms was reproduced by global physics-based model simulations (Fuller-Rowell et al., 1994, 1996). Global model simulations were even more valuable tools when observations were scarcer. Michael Buonsanto led the initiative of the Coupling, Energetics, and Dynamics of Atmospheric Regions (CEDAR) storm study and summarized the results (Buonsanto, 1999). Then game changers appeared: (1) the global mapping of neutral composition change from the Global Ultraviolet Imager (GUVI) instrument on the Thermosphere Ionosphere Mesosphere Energetics and Dynamics (TIMED) satellite (Paxton et al., 1999); (2) "mapping" of the plasma response to storms enabled by the explosion of ground-based observations (e.g., Pi et al., 1997); and (3) dual-frequency, GPS observations of total electron content (TEC) (Foster et al., 2002; Coster et al., 2003).

The Solar Cycle 23 (from year 1996 to 2008) was characterized by many studies with an emphasis on storm time electric fields. It is interesting to note that ionospheric scientists had shown disinterest in the magnetospheric electric field effect because the time scale suggested from old theoretical model calculations was considered too short to significantly impact ionospheric dynamics (Mendillo, 2006). The "biteouts" of the ion density observed at the Defense Meteorological Satellite Program (DMSP) height (~850 km) during the famous

Bastille Day storm on July 15, 2000, clearly demonstrated the importance of the prompt penetration electric field effect on equatorial ionospheric dynamics (Basu et al., 2001). Furthermore, Mannucci et al. (2005) supported this hypothesis by reporting the observed electron content that was lifted above the CHAllenging Minisatellite Payload (CHAMP) satellite altitude (~400 km) during the famous Halloween storm on October 30, 2003. Meanwhile, the long-duration penetration electric field was measured at the Jicamarca ISR at the dip equator and simultaneously at the US ISR chain during the April 2002 storm (Kelley et al., 2003). Fejer et al. (2007) reported the largest daytime prompt penetration electric fields (about 3 mV/m) ever observed over Jicamarca occurred during the November 9 storm main phase. Subauroral Polarization Streams (SAPS) are characterized as the enhancement of the westward (sunward) plasma flow (corresponding to a poleward electric field) in the subauroral ionosphere by Foster and Burke (2002) and Foster and Vo (2002), who pointed out their importance in storm time ionospheric dynamics.

1.2 Discussions on coupling of the ionosphere with the plasmasphere and magnetosphere were initiated

At the same time our understanding of ionospheric storms was developing, discussions on the close coupling of the ionosphere with the plasmasphere and magnetosphere were initiated. These were mainly inspired by the spectacular global views of the plasmasphere from the Extreme Ultraviolet (EUV) Imager instrument (e.g., Sandel et al., 2003; Goldstein et al., 2003) aboard the Imager for Magnetopause-to-Aurora Global Exploration (IMAGE) spacecraft (Burch et al., 2001). The IMAGE Radio Plasma Imager (RPI) instrument (Reinisch et al., 2000, 2001a,b) visualized how field line density profiles of the plasmasphere become eroded initially due to enhanced convection during the main phase of a storm. The depleted flux tubes get refilled from the ionosphere within less than 28 hours during the recovery phase (Reinisch et al., 2004), which is substantially less time than theoretical expectations (the order of three days or longer) (e.g., Rasmussen et al., 1993; Tu et al., 2003). The IMAGE high energetic neutral atom (HENA) measurements (Mitchell et al., 2003) showed global, dynamic views of the ring current (Brandt et al., 2002). The ring current provides earthward Region II field-aligned currents in the dusk magnetic local time (MLT) sector. The field-aligned currents flow through the low conductance dusk ionosphere in the trough, which generates a strong electric field in the subauroral ionosphere. A possible plasma plume connection between the ionosphere and plasmasphere was suggested by the independent IMAGE/EUV and GPS-TEC observations (Foster et al., 2002).

1.3 Questions

The various observations in the previous section challenged scientists with many research questions. The huge plasma biteout at the dip equator observed by DMSP (Basu et al., 2001) raises the question, *"Where does all the plasma go and what is the process?"* The massive plasma redistribution in TEC observed by CHAMP (Mannucci et al., 2005) prompted scientists to ask, *"Where is all the plasma coming from and how?"* Both these questions challenge our current understanding by asking *"Is the electric field solely responsible for redistributing the plasma?"* and *"What is the source of the storm time electric field that could redistribute the observed plasma?"* Furthermore, another question arises, *"Where does the accumulated plasma go?"* Cold plasma in the plasmasphere originates from the ionosphere and cold plasma in both the ionosphere and plasmasphere moves with the same $E \times B$ drift. So *"How are plumes physically related between*

the ionosphere and the plasmasphere?" "How does cold plasma in the ionosphere and plasmasphere impact the magnetosphere and its interaction with the solar wind?" The effort to answer these questions has led scientists to greater understanding on how to change how they approach the analysis of individual observations. The awareness has gradually built up that the ionosphere, thermosphere, and magnetosphere interact with each other so closely that understanding of phenomena observed in one component invariably requires consideration of the others. The awareness has also emerged that multiple simultaneous measurements need to be investigated to understand the whole picture.

1.4 Progress in modeling and observations

Model coupling between different regions of geospace was initiated by the center for integrated space weather modeling (CISM) (e.g., Wang et al., 2004) and the space weather modeling framework (SWMF) (Tóth et al., 2005). New modeling efforts have played a critical role in helping us elucidate the underlying processes within the coupled system. Observational capability has expanded, aiming for higher resolutions both in time and space. For example, one expansion project is the midlatitude Super Dual Auroral Radar Network (SuperDARN) (e.g., Baker et al., 2007), which has been expanding the HF radar network to the midlatitude ionosphere. Another project, Active Magnetosphere and Planetary Electrodynamics Response Experiment (AMPERE) (Anderson et al., 2000; B.J. Anderson et al., 2002; Waters et al., 2001), helps us visualize the global picture of the field aligned current distributions originally illustrated by Iijima and Potemra (1978). Project SuperMAG (Gjerloev, 2009) helps us improve our understanding of the global ionospheric current system during storms and substorms by standardizing data from more than 300 ground-based magnetometers available to users through a worldwide collaboration of organizations and national agencies. New missions have launched: Time History of Events and Macroscale Interactions During Substorms (THEMIS) in 2007 (Angelopoulos, 2008) combined with a ground-based instrument network, THEMIS-All-sky white light imagers (ASI) and fluxgate magnetometers, Van Allen Probes (VAP) in 2012 (e.g., Kletzing et al., 2013; Wygant et al., 2013), Cluster in 2000 (Escoubet et al., 2001), which is the constellation of four spacecraft flying in formation around Earth, and Exploration of energization and Radiation in Geospace (ERG)/Arase (Miyoshi et al., 2018).

The main purpose of this brief review is to report recent new developments that have been brought by this new awareness, expanded observation capability, and progress of modeling capability.

The new awareness and the recent progress need to be summarized, because they have changed the way scientists perceive and conduct research. To gain a better understanding of this awareness, this review compares and contrasts the various old and new studies. The CEDAR strategic plan, "CEDAR The New Dimension" published in 2011 (https://cedarweb.vsp.ucar.edu/wiki/images/1/1e/CEDAR_Plan_June_2011_online.pdf) promoted a new way of viewing the geospace system as a whole. *"The intellectual framework of the system view enables transferable concepts across systems and disciplines to advance and facilitate progress in understanding our whole Sun-Earth system."*

This new perspective inspired scientists, leading to new approaches required when interpreting expanding dataset. Recent progress must be summarized before the coming new era of new missions, Global-scale Observations of the Limb and Disk (GOLD) (Eastes et al., 2013) and Ionospheric Connection Explorer (ICON) (Immel et al., 2018). Understanding the previous research serves for formulating a new mission, such as NASA's forthcoming strategic Ionosphere Thermosphere mission, the Geospace Dynamics

Constellation (GDC, https://science.nasa.gov/heliophysics/resources/stdts/geospace-dynamics-constellation).

The review focuses on recent studies that have been influenced by the system perspective of the coupled M-I-T system. The scope of this overview chapter involves the following three developments. (1) Several excellent reviews on similar subjects have been published (e.g., Mendillo, 2006 on TEC; and Fejer et al., 2016 on poststorm disturbance electric field). This review focuses on the most recent work (2005 and later). A whole system perspective has improved our understanding of the Magnetosphere-Ionosphere-Thermosphere system, with a focus on global-scale phenomena. Our knowledge has improved about storm time response of the coupled system of the magnetosphere-ionosphere-thermosphere. Extended observational capability combined with modeling improvement has brought us to the next level of comprehending the whole geospace system; (2) Our knowledge has broadened and deepened, not only with regards to interplanetary coronal mass ejections (ICME)-driven storms (the most popular type of storm studies), but also corotating interaction regions (CIR)-driven storms and substorms; (3) this review prioritizes the storm time electrodynamics (Section 2.2) and the resultant ionospheric response at midlatitude (Section 2.3). In Section 3, the review is summarized by presenting unsolved questions and future directions.

2 New developments

2.1 Overview

The previous section explained how the stage has been set for the new awareness to emerge. The purpose of this section is to briefly report new developments related to the storm time ionosphere viewed from the coupled M-I-T system over the past approximately 14 years.

Furthermore, this section addresses what has led to the new understanding. During the past 14 years, a great deal of attention has been paid to electrodynamic processes, which connect the ionosphere, thermosphere, and magnetosphere. There has been a renewed interest in the effect of magnetospheric electric fields (Mendillo, 2006).

In many respects, the least understood changes in the ionospheric plasma occur at midlatitudes (Kintner et al., 2008). Here, the dynamic interplay between high- and low-latitude processes and lower atmospheric and magnetospheric forcing is manifest in dramatic changes in the neutral and plasma dynamics driving plasma redistribution. Various local and remote forces influence the motion of ionized and neutral particles, and their interactions. At the same time, steep density gradients and/or equatorward expansion of auroral convection provide sources of plasma irregularities and instability (e.g., Basu et al., 2008). Another reason for the special attention to the midlatitude ionosphere is its importance for space weather application. Steep density gradients evolving rapidly in time and space at midlatitude make the Wide Area Augmentation System (WAAS), which is an extremely accurate navigation system developed for civil aviation used by Federal Aviation Administration (FAA), unusable (e.g., Doherty et al., 2004).

In the following sections, special attention is given to storm time response of electrodynamics (Section 2.2) and the impact of electrodynamics on the ionosphere (Section 2.3).

2.2 Storm time response of ionospheric electrodynamics

2.2.1 Overview

Electric fields are one of the major drivers of ionospheric dynamics. Interestingly, Mendillo (2006) pointed out that the "penetration effect" (a sudden leakage of magnetospheric electric fields that is discussed later in Section 2.2.2) was

once excluded from the ionosphere during storms, that is, that electric-field penetration was considered to be limited to short durations (tens of minutes to an hour or so). Thus, the concept of direct magnetosphere-ionosphere coupling via penetration electric fields has been "reintroduced" to a new generation of researchers during the current era of GPS-based studies (as of 2006). This review paper summarizes the progress on the impact of the storm time electric field on ionospheric storms in the mid- and low-latitude regions since the Mendillo review paper was published. An excellent review on poststorm electric field disturbances was published by Fejer et al. (2016). An effort is made in this review paper to distinguish the high-latitude flywheel effect from the disturbance dynamo effect in discussing electric fields generated by storm time neutral winds. For the storm time response of high-latitude electrodynamics, readers should refer to other reviews (e.g., Lu, 2017).

There are three sources of storm time electric field in the mid- and low-latitude ionosphere: prompt penetration electric field (Section 2.2.2); SAID and SAPS (Section 2.2.3); and neutral wind-generated electric fields (Section 2.2.4). Discerning and quantifying the relative contribution of each component has still been a challenge in interpreting observations and accurately predicting the storm time response of ionospheric electrodynamics, since the electric fields get modified by highly dynamic nonlinear interaction of bulk flows of plasma and neutral gas during storms.

2.2.2 Prompt penetration electric field

What is the prompt penetration electric field?

Global-scale electric fields generated by the interaction between the solar wind and the magnetosphere are not confined to only the high-latitude region. As early as 1968, a clear correlation between the interplanetary magnetic field (IMF) and the magnetic field H component associated with the equatorial electrojet at

Huancayo was revealed (Nishida, 1968). For many years, scientists studied the relationship between rapid changes in IMF B_z and nearly simultaneous short-lived electric fields in the ionosphere (e.g., Kelley et al., 1979), which are called "prompt penetration (PP)" electric fields.

Physical mechanisms of under- vs overshielding

It has long been understood that the plasmasphere and mid-/low-latitude ionosphere are "shielded" from high-latitude electric fields, as first pointed out by Block (1966) and Karlson (1970, 1971) and later discussed by Jaggi and Wolf (1973), Southwood and Wolf (1978), and Wolf (1983), among others. The Alfvén layer is a sharp inner edge of a sheet of ions or electrons in the magnetosphere created by the adiabatic drift in the inner magnetosphere (gradient and curvature drift). Positive charges build up in the dusk MLT sector, while negative charges do so in the dawn sector. The divergence of the ring current causes Birkeland currents to flow from (into) the Alfvén layer into (from) the ionosphere to satisfy the electric current closure in the dusk (dawn) sector. The Region II Birkeland currents from the Alfvén layer that is moving earthward through the magnetosphere from the tail by a large crosstail electric field generate electric fields that oppose the polar-cap and midlatitude electric fields associated with the higher-latitude Region I Birkeland currents. This process is called "shielding." It was first demonstrated by using a computer model of the self-consistently coupled magnetosphere-ionosphere system (Jaggi and Wolf, 1973). If the magnetospheric electric field rapidly increases or decreases, the charges at the Alfvén layer will be temporarily out of balance until the new configuration has established (Kelley et al., 1979). The PP time scale is defined from the enhancement of the polar cap potential (PCP) (or crosstail electric field) until the time when the shielding has established, ranging from 5 to 300 minutes. Conversely, overshielding happens when the IMF B_z turns northward (e.g.,

Kelley et al., 1979). It produces a reversed electric field at the magnetic equator, with a time scale of 10–60 minutes (Spiro et al., 1988; Peymirat et al., 2000; Ebihara et al., 2008, 2014; Kikuchi et al., 2010). The overshielding effect explained the plasmaspheric dynamics and morphology observed by IMAGE EUV (Goldstein et al., 2003).

Shielding efficiency is a key to determining the PP time scale. The shielding time constant depends on plasma sheet conditions and ionospheric conductivities (e.g., Jaggi and Wolf, 1973; Wolf et al., 1982; Senior and Blanc, 1984; Spiro et al., 1988; Peymirat et al., 2000; Garner, 2003). Spiro et al. (1988) suggested the fossil wind mechanism, discussed later, to explain the extremely long overshielding time scale estimated from observations. Furthermore, the magnetospheric reconfiguration effect has also been suggested to impact the PP process, as a result of ineffective shielding (Fejer et al., 1990; Wolf et al., 2007). The magnetospheric reconfiguration process results from the constant stretching of the magnetotail during the early main phase of a storm when the IMF B_z is actually large and negative. The magnitude and lifetime of PP appear to be controlled by the rate and duration of stretching of the magnetic field. Enhanced reconnection at the dayside magnetopause results in constant transfer of magnetic flux to the lobes of the magnetotail, causing stretching of magnetic field lines in the nightside plasma sheet.

Recent studies

Recent observations during super storms have suggested that magnitudes and time scales of the disturbance electric fields can be highly variable (e.g., Fejer et al., 2007). In spite of our knowledge of the basic mechanisms described in the previous section, it is still challenging to fully interpret the individual observations and identify the relative contribution of the sources of the disturbance field, which is essential for prediction. One of the challenges in understanding the response of the storm time electric fields is our insufficient understanding of the storm time electrodynamics, including the magnetosphere-ionosphere coupling, its consequences on the ionosphere-thermosphere, and the feedback between the magnetosphere and ionosphere-thermosphere. Progress in modeling PP has been relatively slow since Spiro et al. (1988) since it has been hampered by numerical limitations, both magnetohydrodynamics (MHD) with sufficiently high resolution and inclusion of nonideal-MHD processes into models. Furthermore, identifying the storm time response of the electric field by analyzing space-based observations has been a challenge because it is difficult to discriminate between event evolution in space and time, which causes the major ambiguity problem with single spacecraft data. The method by D. Anderson et al. (2002) allows us to overcome the problem of limited ISR observations by taking advantage of a ground-based magnetometer pair to continuously monitor equatorial electrojet (EEJ) variabilities. Inferring the daytime $E \times B$ drift has become possible based on the strong correlation between the EEJ and daytime $E \times B$ vertical plasma drift. Several models of the transfer function have been developed to predict the PP effect by using the strikingly linear relationship with the Interplanetary electric field (IEF) (e.g., Nicolls et al., 2007; Anghel et al., 2007). Manoj's model (Manoj and Maus, 2012) has become a real-time prediction model of the equatorial ionospheric electric field at the National Centers for Environmental Information (NECI) by injecting the real-time solar wind parameters from Advanced Composition Explorer (ACE) that are available on the website (http://www.geomag.us/models/PPEFM/RealtimeEF.html). The transfer function was derived from 8 years of IEF data from the ACE satellite, coherent scatter radar data from Jicamarca Unattended Long-term Investigations of the Ionosphere and Atmosphere (JULIA) (e.g., Hysell et al., 1997; Chau and Woodman, 2004), and magnetometer data from the CHAMP satellite.

Long-lasting PP were reported by C.-S. Huang et al. (2005), who reported events that challenge our understanding of the classical PP time scale and shielding time constant as described in the previous section. Maruyama et al. (2007) attributed the discrepancy to the magnetospheric reconfiguration effect (Fejer et al., 1990). Wei et al. (2008a) observationally demonstrated that the time-dependent magnetospheric reconfiguration process is closely related to the development of the equatorial electric field.

Much progress has been made in modeling efforts. Huba et al. (2005) reported on a coupling of the Rice Convection Model (RCM) to a first-principles SAMI3 model of the mid- and low-latitude ionosphere. With a self-consistent coupling, Huba et al. (2005) were able to study the penetration electric field in idealized cases. However, their model lacks a first-principles thermosphere module that is essential to evaluate the disturbance dynamo. The Coupled Magnetosphere-Ionosphere-Thermosphere (CMIT) model (Wang et al., 2004; Hui Wang et al., 2008; Wiltberger et al., 2004) that has been developed by coupling the Lyon-Fedder-Mobary (LFM) global MHD model and the Thermosphere-Ionosphere Nested Grid (TING) model was used to predict the equatorial plasma drift. The superfountain effect, which refers to an extreme case of the equatorial fountain effect (Hanson and Moffett, 1966) associated with enormous uplift caused by severe storm time eastward electric fields (Tsurutani et al., 2004), was reproduced by coupling TIEGCM and SAMI3, using the polar cap potential from Assimilative Mapping of Ionospheric Electrodynamics (AMIE) (Richmond and Kamide, 1988) that generates a penetration effect (Lu et al., 2013). The PP effect in the geomagnetic field perturbations was also reproduced (Marsal et al., 2012) by using the field-aligned currents observed by AMPERE (Anderson et al., 2008), instead of imposing the polar cap potential.

Furthermore, the Prompt Penetration electric fields have also been observed during substorms. The directions of the ionospheric electric field perturbations on the dayside can be either eastward (e.g., Huang et al., 2004) or westward (e.g., Wei et al., 2009). The penetration effect was reported during a Sawtooth event (Huang, 2012). Sawtooth is a type of substorm that is well defined by the gradual decrease and sudden increase in the proton flux measured by the geosynchronous satellites (e.g., Belian et al., 1995; Henderson, 2004). Chakrabarty et al. (2015) identified westward electric field perturbations owing to a pseudo-breakup and a substorm event, resulting in the suppression of equatorial spread F. Furthermore, Hui et al. (2017) reported the penetration field during substorms and argued that the substorm-induced field is found to be strong enough to compete with or almost nullify the effects of storm time IEF_y fields. A typical duration of a substorm is a few hours, while that of a geomagnetic storm generated by CMEs is several hours to days. The mechanism of the substorm PP needs further clarification with regards to what determines the directions of the electric field perturbations. The penetration effect during substorms is caused by the same mechanism as that of magnetic storms: an imbalance between Region I and Region II field-aligned currents; however, exactly how the substorm creates the imbalance of Region I vs Region II field-aligned currents is not clearly understood. One of the major challenges is that substorms can happen during either IMF B_z southward or northward. It is not understood how the imbalance recovers during substorms. Does the ring current become more symmetric during substorms to establish shielding as it does during magnetic storms? Further investigations are needed to quantify the impact of the substorm-driven penetration field on ionospheric dynamics as compared to other forcing such as tidal waves from the lower atmosphere.

Furthermore, PP is also reported during high-speed-stream-driven storms during the declining phase of the solar cycle. Equatorward extensions of solar coronal holes (cooler, less dense regions than the surrounding plasma and are regions of open, unipolar magnetic fields) can have geometries that result in long-lived high-speed wind at Earth (e.g., McAllister et al., 1996). The continuous stream of high-speed solar wind that originated from coronal holes interact with slow-speed solar winds, generating corotating interaction regions (CIRs). These CIRs occur periodically and result in recurrent, long-duration elevated geomagnetic activity with typical durations of a few to several days (e.g., Lei et al., 2008). The penetration effect during CIR events might play a significant role in the positive ionospheric response (e.g., Chen et al., 2015) and in the development of equatorial plasma bubbles (e.g., Tulasi Ram et al., 2015). Furthermore, Silva et al. (2017) reported the penetration effect during a so-called high-intensity long-duration continuous auroral electrojet (AE) activity (HILDCAA) event that was defined by Tsurutani and Gonzalez (1987). While high-speed stream storms tend to last for multiple days, the PP mechanism during high-speed stream storms is expected to be the same as CME-driven magnetic storms. However, it is not understood how the shielding process works during high-speed stream events over multiple days. Will the shielding process establish quicker, since the magnetosphere is more likely to be preconditioned with dense plasma prior to CIR-driven storms than it is prior to CME-driven storms (Borovsky and Denton, 2006b)?

2.2.3 Subauroral ion drift and subauroral polarization streams

SAID vs SAPS

The polarization jet (Galperin et al., 1974) was detected from low-altitude satellite observations.

Spiro et al. (1979) then set a threshold for the polarization jet (a plasma velocity that exceeded $1\,km\,s^{-1}$) and termed such occasions subauroral ion drift (SAID) (Smiddy et al., 1977; Spiro et al., 1979). Foster and Burke (2002) used the term subauroral polarization stream (SAPS) to describe the broad region of fast sunward plasma flow equatorward of the auroral oval, which is frequently observed in the dusk-premidnight subauroral ionosphere by using ISR measurements of plasma velocity. Foster and Vo (2002) performed a statistical analysis of Millstone Hill ISR data to characterize the flows: they become stronger and move to lower latitudes with increasing geomagnetic activity. SAID/SAPS were also observed in the inner magnetosphere (e.g., Anderson et al., 1991; Puhl-Quinn et al., 2007; Nishimura et al., 2008; Califf et al., 2016). SAID is considered to be a more enhanced version of SAPS with a latitudinally narrower spike signature. However, the definitive relationship between SAPS and SAID has still not been clarified (e.g., Mishin et al., 2017). Excellent summaries of SAID and SAPS from the magnetospheric point of views appeared in Kunduri et al. (2018a) and Mishin et al. (2017). This review rather focuses on SAID and SAPS from the ionospheric point of view.

SAID/SAPS generation mechanism: voltage vs current generators

SAID and SAPS are believed to be formed by the necessity for closing the downward Region II field-aligned currents, which are generated by the increased ring current pressure gradient, through the low-conductivity region that lies equatorward of the electron plasma sheet (Anderson et al., 2001). In an effort to explain the fast sunward flow, the paradigm of voltage and current generators has been used. Southwood and Wolf (1978) suggested the voltage is generated due to the gap between the inner boundary of ion and electron trajectories. On the other hand, Anderson et al. (1993, 2001) proposed a model that subauroral field–aligned currents close via Pedersen currents with the outward flowing, Region I currents at higher latitudes. Very recently, however, Mishin et al. (2017)

argued that overall, the observed SAPS and SAPS wave structures (SAPSWS) (Mishin et al., 2003) features disagree with the paradigm of voltage and current generators. Anderson et al. (2001) also suggested a positive feedback of SAPS in the trough: fast sunward flow colliding with the corotating neutral atmosphere generates frictional heating, resulting in substantial increase in the recombination rate between O^+ and N_2. Subsequent reduction in the subauroral F-region conductivities increases the electric field and further frictional heating.

Recent studies

The expansion of the SuperDARN radars to midlatitude contributed significantly to help improve our understanding of the SAPS (e.g., Clausen et al., 2012; Ebihara et al., 2009; Zou and Nishitani, 2014; Kunduri et al., 2017; Kunduri et al., 2018a,b; Maimaiti et al., 2018, 2019; Nishitani et al., 2019); in particular, the HF radar network revealed the highly dynamic variation in latitude and MLT.

Although observations of SAPS have been extensively carried out (e.g., Yeh et al., 1991; Anderson et al., 2001; Foster and Vo, 2002; Oksavik et al., 2006; Hui Wang et al., 2008; Clausen et al., 2012), few modeling efforts have been made to successfully reproduce and understand the dynamical SAPS features in observations (e.g., Garner et al., 2004; Wang et al., 2009; Ebihara et al., 2009; Zheng et al., 2008). Goldstein et al. (2005) presented an equatorial magnetospheric SAPS model based on the average SAPS properties reported by Foster and Vo (2002). Matsui et al. (2008, 2013) developed an empirical model of the inner magnetospheric electric field using Cluster EDI and EFW data. Empirical models of inner magnetospheric electric fields and magnetic fields can provide a statistical, averaged representation of the enhanced flow channels in the inner magnetosphere. On the other hand, ionospheric plasma flow data have been used to develop empirical models based on observations from

SuperDARN (Kunduri et al., 2017, 2018b) and DMSP (Landry and Anderson, 2018). The influence of SAPS on the thermospheric temperature, wind, and composition as well as the ionosphere was studied by using an empirical SAPS model inside TIEGCM (Wenbin Wang et al., 2012). Yu et al. (2015) reproduced the SAPS by using a physics-based ring current model, RAM-SCB, driven by the SWMF (one-way coupling). On the other hand, Raeder et al. (2016) reproduced SAPS by coupling self-consistently between global MHD and another ring current model, RCM, as well as coupling to active ionosphere and thermosphere.

Recent studies examined the impact of SAPS on the ionosphere and thermosphere and possible positive feedback on SAPS. Zheng et al. (2008) demonstrated the positive feedback effect in the M-I coupling of SAPS by using a model simulation. Huba et al. (2017) showed a decrease of conductivity caused by SAPS by using the coupled SAMI3-RCM model. More recent observational study showed that the trough minimum of electron density during nonactive times is located at the poleward edge of SAPS whereas during geomagnetically active times, the minimum density in a trough coincides with the maximum sunward flow of SAPS (Kunduri et al., 2017, 2018a,b). Does this imply that the positive feedback process decreases the trough density by the faster recombination due to frictional heating within SAPS during geomagnetically active times? High ion temperature is required for the recombination to become active (Schunk et al., 1976; Banks and Yasuhara, 1978) because the recombination reaction rate depends on the ion temperature. Future work with physics-based model simulations is needed to quantify the positive feedback effect.

Very recently, SAPS was also observed during nonstorm periods (Lejosne and Mozer, 2016; Kunduri et al., 2017; Maimaiti et al., 2019) even though SAPS has long been considered phenomena associated with elevated geomagnetic activities. It is not clarified, however, whether nonstorm SAPS is really generated in the same manner as storm-time SAPS.

SAPS has been associated with other phenomena in the subauroral region. Gallardo-Lacourt et al. (2017) showed a strong correlation between SAID and auroral streamers, which are the ionospheric representation of the penetration of magnetotail plasma flows into the inner magnetosphere during substorms. They suggested that auroral streamers can intensify SAPS by enhancing the pressure gradients in the ring current, and flow bursts in the plasma sheet can reach the inner magnetosphere and strengthen SAPS flows. A new rare atmospheric phenomenon called Strong Thermal Emission Velocity Enhancement (STEVE) (e.g., MacDonald et al., 2018; Gallardo-Lacourt et al., 2018) appears to be associated with a SAID fast sunward ion flow. However, there has been no definitive conclusion whether or not a STEVE originates in the ionosphere or magnetosphere.

2.2.4 Neutral wind-generated electric fields

Overview

Neutral winds blowing across a magnetic field generate electric currents and fields in the ionosphere where free electrons and ions are present. This process is called the ionospheric wind dynamo (e.g., Richmond, 1995a, b). The dynamo effect at mid and low latitudes during geomagnetically active periods has been attributed to the disturbance dynamo mechanism associated with the gradual action of the Coriolis force on storm-driven equatorward winds at midlatitude (Blanc and Richmond, 1980). On the other hand, the high-latitude dynamo process is called the "flywheel" effect (e.g., Axford and Hines, 1961; Banks, 1972). The inertia of the neutral particles accelerated through coupling with rapidly convecting ions via collisions can help maintain convection in the magnetosphere and ionosphere even when the magnetospheric dynamo source of field-aligned current is suddenly weakened.

Wind-generated electric fields at mid and low latitudes: The disturbance dynamo effect

The disturbance dynamo (DD) mechanism is associated with reversal of the so-called solar quiet Sq current system (see a recent review by Yamazaki and Maute, 2017) generated by storm-time enhancement of mid- and low-latitude neutral winds (Blanc and Richmond, 1980). The disturbance dynamo electric field is developed through the following chain of processes. Electromagnetic energy from the solar wind magnetosphere interaction is deposited into the high-latitude upper atmosphere as Poynting flux. Collisional interaction between neutral particles and rapidly convecting ions in the presence of strong high-latitude electric fields and currents and auroral precipitation heats the ionosphere and thermosphere, which is called Joule heating. The resultant pressure gradient generates an initial equatorward motion of neutral winds. Gradually, westward motion with respect to the earth is built up at midlatitude via the Coriolis force when the equatorward winds transport angular momentum from high latitude. The westward winds flowing above 120-km altitude drive equatorward Pedersen currents, accumulating charge toward the equator. The charge pileup generates a poleward polarization electric field to balance the equatorward wind-driven currents. Additionally, the poleward electric field also produces an eastward Hall current to result in positive charge buildup toward the sunset terminator, which is opposite from that of quiet time. Therefore, the combined effect of these processes tends to result in an "anti-Sq" type of current system.

The time scale of the original DD process takes about 6 hours, the time for the midlatitude zonal wind to be developed. An additional, relatively fast time scale of 2–3 hours was reported by Scherliess and Fejer (1997), and the mechanism is attributed to an equatorward surge of neutral wind (e.g., Fuller-Rowell et al., 2002, 2008). Equatorial electric fields can respond to wind disturbances developed at midlatitude before the equatorial wind surges arrive at the equator. The response observed after 24 hours is attributed to a combined effect of neutral wind and ionospheric conductivity caused by the neutral composition effect (Scherliess and Fejer, 1997). An attempt to separate out the different causes of the disturbance electric fields in observations has usually been made by considering the different time scales. Fejer and Scherliess (1997) developed an empirical model of the disturbance electric fields based on the existing knowledge of the different time scales.

Physics-based model simulations have been a useful approach to combine and separate out the different processes, for the purpose of predicting the disturbance electric fields. The effect of disturbance winds can take several days to die away completely as disturbance winds tend to persist even after the magnetospheric forcing ceases because of considerable inertia of the neutral particles (C.-M. Huang et al., 2005; Huang and Chen, 2008). Maruyama et al. (2005) suggested a difficulty in separating out the causes during the night time and recovery phases of a storm by demonstrating a possible nonlinear interaction and feedback between PP and DD. The comparison between their model results and observations indicated that the interaction becomes stronger for superstorms (Maruyama et al., 2007). On the other hand, it has been challenging to verify the DD mechanism by observations because for a long period of time there have been limited data about electric fields and neutral winds, which prevents us from performing statistical analysis. However, recently, ground-based magnetometers were used to characterize the magnetic field perturbations associated with DD (Yamazaki and Kosch, 2015). Furthermore, Kakad et al. (2011) used post sunset F-layer height changes based on observations from VHF spaced receivers to investigate the statistical properties of the DD effect on the equatorial F-region ionosphere. An excellent summary of the dependences of the DD effect on various geophysical and geomagnetic conditions is found in Fejer et al. (2016).

Wind-generated electric fields at high latitude: The flywheel effect

Neutral wind disturbances last longer than magnetospheric driving forces due to their considerable inertia (e.g., Axford and Hines, 1961; Banks, 1972; Lyons et al., 1985; Deng et al., 1991, 1993; Lu et al., 1995; Odom et al., 1997). This process is called the "flywheel effect." The flywheel effect can sustain the preexisting ion motion even when the externally imposed field aligned currents suddenly diminishes. The flywheel effect is generated through by wind disturbances that have previously been accelerated by storm time convection via ion drag.

The dynamo effect of "fossil winds" was suggested by Spiro et al. (1988) who used the Rice Convection Model (RCM) (Harel et al., 1981) and a simple neutral wind model that did not take into account the Earth's rotation. They were able to get long-lasting equatorial electric field perturbations that agreed with the observations. However, this was possible only when they specified an equatorward displacement of the neutral wind distribution associated with ion convection in the auroral region during a period of southward IMF. These winds may extend equatorward past the shielding region as it contracts during the recovery phase of magnetospheric disturbances, and the electric fields generated by these winds are no longer shielded, but become visible at the equator. Forbes and Harel (1989) also included in the RCM model a neutral wind model that took into account Earth's rotation, but not the extension of the wind driven by ion convection to latitudes equatorward of the ionospheric projection of the Alfvén layer. They discussed how the winds can increase the effective shielding. Richmond et al. (2003) reproduced the fossil wind concept suggested by Spiro et al. (1988), by using a self-consistently coupling model of the

magnetosphere, ionosphere, and thermosphere (Magnetosphere-Thermosphere-Ionosphere-Electrodynamics General Circulation Model, so-called MTIEGCM). The model takes into account more realistic neutral winds driven by Joule heating effects and momentum transport away from the acceleration region of the winds. Richmond et al. (2003) demonstrated that storm time winds tend to enhance steady-state shielding by reducing the net electric field that penetrates to middle and low latitudes.

Recent observations have drawn a renewed interest in the topic. Strong correlation was revealed between SAPS ion drift and neutral wind measurements from DMSP and CHAMP, respectively (Hui Wang et al., 2011, 2012). Different forces have been suggested to explain the acceleration of neutral wind, involving, for example, the ion drag force in the auroral zone (Wang et al., 2018), the ion drag force on a global scale (Hui Wang et al., 2012), the Coriolis force (Lühr et al., 2007; Zhang et al., 2015), and the pressure gradient force. It was generally understood that ion drag in the SAPS region is not expected to play an important role in driving neutral winds because plasma density is too low in the trough region. However, it has been challenging to observationally investigate the definitive causality by quantifying force terms that contribute to accelerating neutral winds in the SAPS region, because of the difficulties in measuring the force terms. Utilizing physics-based models is necessary not only to quantify momentum exchange for driving the subauroral neutral wind, but also to quantify the feedback effect of the ionosphere and magnetosphere due to the flywheel effect. Very recently, Ferdousi et al. (2019) compared observations and physics-based model simulations (RCM-CTIPe) of both plasma and neutral flows during an intermediate storm interval, and demonstrated that the ion drag is the dominant force for driving neutral wind in the SAPS region even though the plasma density is low in the trough. Interestingly, this mechanism of driving winds is not consistent with that of the disturbance dynamo (Blanc and Richmond, 1980), in which the Coriolis force is the main driver of the sunward plasma flow. The flywheel effect was quantified by numerical simulations of the physics-based model. The effect generates active feedback to the plasma flow equatorward of SAPS. Neutral wind feedback amplifies the M-I-T coupling system through an increase in FACs equatorward of SAPS region by 20% during an intermediate storm. This is opposite to the classical flywheel effect in which winds tend to reduce FAC.

Impact of neutral wind-generated electric fields on the magnetosphere

Wind-generated electric fields can impact the magnetosphere. Peymirat et al. (1998, 2002) described, in a self-consistent manner, the first studies that attempted to quantify the effect of active neutral winds on the magnetosphere. Peymirat et al. (1998) coupled an inner magnetospheric plasma convection model (Peymirat and Fontaine, 1994) with the Thermosphere-Ionosphere-Electrodynamics General Circulation Model (TIEGCM), calling it MTIEGCM. Peymirat et al. (2000) demonstrated that active neutral winds can cause pressure changes of ~20% in the magnetosphere. Using the University of Michigan's coupled magnetosphere-ionosphere-thermosphere general circulation model, which includes the neutral wind effect, Ridley et al. (2003) demonstrated that the pressure on the dayside magnetosphere is reduced while the pressure on the nightside is increased by ~10% of the total pressure. Garner et al. (2008) included wind-driven electric fields in the RCM model in a nonself-consistent manner. They showed that wind-generated electric fields alter plasma pressure, Region II FACs, and the resultant electric field. Maruyama et al. (2011) studied the effect of preconditioning of the magnetosphere, ionosphere, and thermosphere on disturbance electric fields by preceding periods

of enhanced convection electric fields. Magnetospheric plasma pressure is increased and distributed more earthward because of an increase in electric fields in the inner magnetosphere ($L < \sim 4$) due to the DD effect as well as the ring current injection that occurred earlier. Subsequently FAC is distributed more earthward, resulting in a stronger and quicker shielding effect. These studies based on model simulation showed the impact of neutral winds on the inner magnetospheric plasma pressure, electric fields, and shielding process. There are very few observational studies that evaluate the impact of neutral wind-generated electric fields on the magnetosphere. Departures from perfect corotation of the plasmasphere have been attributed to the ionospheric DD effect based on estimation of the plasmasphere's rotation rates from IMAGE/EUV observations (e.g., Sandel et al., 2003; Gallagher et al., 2005; Galvan et al., 2010). Recently, Lejosne and Roederer (2016) attributed the observed signature of an azimuthal (longitudinal) dependence in trapped particle distributions below $L \sim 3$ in the radiation belt (zebra stripes) to the effect of F-region zonal plasma drifts generated by neutral winds. The number of stripes indicates how many hours the population spent drifting under quiet conditions. The impact of neutral wind-generated electric fields needs further quantification by using more recent observations in the magnetosphere.

The effect of wind-generated electric fields is not limited to earth. Considerable inertia of the neutral atmospheres may exert a flywheel effect on the magnetosphere of rapidly rotating giant planets (Jupiter and Saturn) in combination with their strong planetary magnetic fields. For example, observed periodicities in the Saturn's magnetospheric structure are thought to be driven by the upper atmosphere (e.g., Jia et al., 2012; Kivelson, 2014; Thomsen et al., 2017). The role of the solar wind in driving dynamics is secondary to that of internal processes at orbital distances of 5 AU and beyond.

2.3 Response of ionosphere to electric fields

2.3.1 Overview

In the previous section, the response of the electrodynamics in the mid- and low-latitude ionosphere to storms and substorms was discussed. Studies have shown that the storm-time electric fields dramatically redistribute plasma through the $\mathbf{E} \times \mathbf{B}$ drift. In addition, it is this $\mathbf{E} \times \mathbf{B}$ drift that connects the cold, dense plasma particles of the ionosphere, plasmasphere, and magnetosphere. In this section, a focus is made on the so-called dusk effect and Storm-Enhanced Density (SED) that have been discussed frequently as a consequence of the interaction of storm-time electric fields with ionospheric plasma. They have played an important role in understanding the storm time ionosphere since they appear to be related to many other phenomena not only in the ionosphere, but also in the plasmasphere and magnetosphere, during geomagnetically active periods. Furthermore, the area of enhanced density is surrounded by steep density gradients. These gradients can have a major societal impact by disrupting radio communication and navigation systems (e.g., Coster et al., 2003; Doherty et al., 2004; Groves and Carrano, 2016). Ionospheric TEC at a given station can change by tens or even hundreds of TEC units during geomagnetic perturbed conditions (e.g., Mendillo et al., 1970; Foster, 1993), which may introduce tens of meters of error in position provided by Global Navigation Satellite System (GNSS) (e.g., Skone and Yousuf, 2007). Furthermore, not only SAPS and trough but also the steep gradients at the edges of the dusk effect and SED are fertile places for plasma irregularities and scintillations (e.g., Basu et al., 2008; Heine et al., 2017). Excellent reviews of the dusk effect and SED have recently been published focusing on the longitudinal variation (Heelis, 2016) and the improvement in observational capabilities (Coster et al., 2016). In this

review, an effort is made to understand the dusk effect and SED from the perspective of the coupled system of the ionosphere, plasmasphere, and magnetosphere.

2.3.2 What are the dusk effect and SED?

The "dusk effect" in ionospheric storms has been defined as enhancements in TEC in the dusk sector (Mendillo et al., 1970). Possible causes of the "dusk effect" have been debated rather vigorously throughout the 1970s. The competing mechanisms of neutral winds and electric fields could each transport plasma to a higher altitude at which reduced recombination results in the positive change in electron density. Anderson (1976) reproduced the sharp density gradient associated with the dusk effect by imposing both an equatorward neutral wind and a westward plasma $\mathbf{E} \times \mathbf{B}$ drift on an ionosphere model including only one flux tube. The idea was inspired by the stagnation of plasma in the trough region reported by Knudsen (1974). An excellent summary of the competing theories and interpretations appeared in the reviews by Prölss (1995), Buonsanto (1999), and Mendillo (2006). Foster (1993) used the term Midlatitude Storm Enhanced Density "SED" to characterize an enhancement of electron density in the premidnight subauroral ionosphere during early stages of magnetic disturbances, based on Millstone Hill ISR observations. These increased-TEC plumes of ionization are seen in the premidnight and afternoon sectors at the equatorward edge of the main ionospheric trough and are observed to stream sunward toward the local noon sector.

2.3.3 Formation mechanisms

Many studies have suggested possible mechanisms to explain the density enhancements associated with the dusk effect and SED. Tsurutani et al. (2004) and Mannucci et al. (2005) showed midlatitude TEC enhancements were accompanied by similar enhancements near the equatorial ionization anomaly (EIA) (e.g., Hanson and Moffett, 1966). They have suggested that both vertical and horizontal transport play an important role in the dusk effect and SED formation. Using numerical simulations, Lin et al. (2005a) have demonstrated the importance of both an $\mathbf{E} \times \mathbf{B}$ drift and an equatorward wind to create midlatitude plasma buildup. They reproduced the massive plasma redistribution resulted from a hugely expanded version of the classical EIA by imposing both a storm time electric field and an equatorward wind. This so-called super fountain effect (e.g., Tsurutani et al., 2004) has been suggested as a possible mechanism for generating the midlatitude plasma pileup associated with the dusk effect and SED. Furthermore, vertically downward plasma flux at the midlatitude shoulders (a sharp drop of ionospheric plasma density at midlatitude) has been reported by Horvath and Lovell (2008), as having possible connections to the EIA including the super fountain effect, the plasma shoulder, and penetration electric fields reaching the equator, as suggested by previous studies (e.g., Lin et al., 2005a; Balan et al., 2009, 2010).

At the same time, the dusk effect and SED have been thought to be connected with high-latitude phenomena. Foster et al. (2005) suggested that a dayside source of a "tongue" of ionization (TOI) is a plume of SED transported from the low-latitude ionosphere in the postnoon sector by subauroral disturbance electric fields. A TOI plume is characterized by a long channel of highly dense F-region plasma, and/ or patches of density extending out of the solar-produced dayside midlatitude ionosphere through the cusp and into the polar cap, all of which are transported by expanded crosspolar cap convection flow (Sato, 1959; Sato and Rourke, 1964; Knudsen, 1974; Sojka et al., 1993; Anderson et al., 1996). The TOI plume is surrounded by the low-density region associated with the midlatitude trough (Rodger, 2008). The dense plasma in the cusp is thought to be a possible source of ion outflow, which is a

process of loading ionospheric plasma, including the heavy ion population into the nightside magnetosphere (e.g., Schunk, 2016). Foster (1993) suggested the following mechanism to explain SED formation. The horizontal transport driven by storm time convection creates a snowplow effect. The convection cell continually encounters fresh corotating ionospheric plasma along its equatorward edge, producing a latitudinally narrow region of denser plasma and increased total electron content that is advected toward higher latitudes in the noon sector. More recent studies (e.g., Foster et al., 2007) suggested that the SED plasma flow is transported by the low-latitude edge of SAPS. Foster et al. (2005) pointed out that the SED appears to be present in the postnoon American sector when Kp is larger than 3. The occurrence statistics of SED is similar to that of SAPS (Foster and Vo, 2002), although SAPS tends to occur where the ionospheric conductance is low in the trough. On the other hand, by using the Time-Dependent Ionospheric Model (TDIM) excluding the coupling with the middle- and low-latitude ionospheres, Heelis et al. (2009) suggested another mechanism of SED formation, in which the prompt penetration electric field caused by equatorward expansion of convection creates vertical drift instead of SAPS because of the tilted magnetic field at midlatitude. In the presence of sunlight, small vertical drifts result in a significant decrease in the chemical losses and dramatic increases in TEC. Indeed, they reproduced a TEC enhancement of a factor of $\sim$2 generated by locally produced plasma. Their hypothesis originates from the previous work of Heelis and Coley (2007), who concluded that midlatitude density enhancements in the topside ionosphere may be unaccompanied by corresponding equatorial density changes. They also suggested that midlatitude features may evolve independent of equatorial features. Several other mechanisms have also been suggested: (1) large upward flows combining both $\mathbf{E} \times \mathbf{B}$ convection and antiparallel flows

(Zou et al., 2013); (2) horizontal advection due to fast flows, such as those associated with SAPS flow (Foster et al., 2007; Liu et al., 2016c); (3) energetic particle precipitation (Yuan et al., 2011); (4) enhanced thermospheric wind in the topside ionosphere (Zou et al., 2013; Sojka et al., 2012); (5) downward ion flux within plumes, possibly caused by altered plasma pressure distribution due to ambipolar diffusion and a wind effect along field lines (Zou et al., 2014). When storm time convection becomes smaller as IMF B_z turns northward, plasma located equatorward of the SED plume (base of the plume) starts to corotate with the Earth. Eventually, plumes become separated from their plasma source, to form what can be termed "fossil plumes" (Coster et al., 2006; Thomas et al., 2013).

2.3.4 Universal time, longitude, season, IMF, and hemispheric dependences

Comparisons of the magnitudes of TEC in SED during different superstorms (e.g., Coster et al., 2005) indicate the possible dependences on universal time, longitude, season, and IMF B_y, which make interpretation of individual SED and TOI events challenging. While the TEC enhancements associated with SEDs and TOIs are conjugate features, the magnitude and structure of the density enhancement do not appear to be conjugate between different hemispheres (Foster and Rideout, 2007; Heelis et al., 2009). For example, Foster and Rideout (2007) showed that the position of the steep gradient region at the poleward edge of the SED was closely aligned in the conjugate hemispheres, whereas the enhancement at the base of the TOI plumes, which provides a plasma source for the erosion events (e.g., Foster et al., 2005), often was significantly nonconjugate. The amount of TEC at the base of the plume is greatest in the American sector, suggesting additional plasma sources resulting from repeatable physical processes in this region, which has been called the Florida effect (e.g., Coster et al., 2007). The density enhancement

should be largest when the longitude region incorporating the magnetic pole is tilted toward the dayside. Therefore, the most effective universal times for TEC enhancement should occur in the northern hemisphere near 1900 UT and in the southern hemisphere near 0700 UT (Heelis et al., 2009). It has been generally agreed from observations of the northern hemisphere that the occurrence of TOIs and patches is clearly increased during local winter while reduced during local summer. On the other hand, a variety of explanations for hemispheric differences in observations of the southern hemisphere has been proposed, none of which, however, has yet been confirmed, mostly due to scarcity of data. Coley and Heelis (1998) and Spicher et al. (2017) found that the occurrence of TOIs and patches was maximum during local winter and minimum during local summer in the southern hemisphere based on the in situ density observations of Dynamics Explorer 2 (DE2) and Swarm, respectively. In contrast, Noja et al. (2013) found a minimum occurrence in local winter and a maximum occurrence around equinoxes and during local summer from CHAMP TEC observations. These measurements indicate that TOIs and patches do not always occur simultaneously or have substantial density asymmetry between the hemispheres, despite the fact that similar TOIs in the two hemispheres do occur. Enhanced westward flow associated with SAPS, which has been thought to play a role in transporting the midlatitude sun-lit plasma into the noon throat region, is observed in magnetically conjugate regions (SAPS channels) (Foster and Rideout, 2007). If the electric fields are the same between the two hemispheres, the neutral atmosphere (temperature, wind, density, and composition) and solar zenith angle are expected to play a role in creating the hemispheric differences. Furthermore, the fact that Flux Transfer Events (FTEs) prefer the winter hemisphere (Raeder et al., 1996; Korotova et al., 2008; Fear et al., 2012) may play a role in generating the hemispheric asymmetry of the polar cap potential distribution that would favor the winter hemisphere for forming TOI plumes. FTEs are bipolar magnetic field changes, which occur as a signature of bursty dayside reconnection in in situ observations (Russell and Elphic, 1978). FTEs lead to poleward motion of flux tubes across the dayside open-closed magnetic field line boundary (OCB). As a result, the OCB shifts equatorward, at least temporarily until magnetic fluxes are supplied from the nightside. In the ionosphere, high density just equatorward of the OCB is carried into the polar cap, forming islands of high plasma density (see Figure 4 of Lockwood and Carlson, 1992), which has been observed in aurora optical and radar measurements. The hemispheric preference of FTEs may depend on IMF B_y as well as season (Jimmy Raeder, private communication), which may also play a role in the polar cap potential distribution. TOIs and patches have been known to depend on IMF B_y. When IMF B_y is negative, the plasma intake into the polar cap is shifted toward the prenoon sector in the northern hemisphere. In contrast, when IMF B_y is positive, the postnoon-shifted cusp inflow region can get easier access to the higher density solar EUV-produced plasma (Jin et al., 2015). This IMF B_y dependence may be opposite in the southern hemisphere. The cusp location that depends on the IMF B_y polarity is often used as a proxy for reconnection activity (e.g., Russell, 2000; Xing et al., 2012). Elucidating how the dusk effect, SEDs, and TOIs depend on various geophysical and geomagnetic conditions will help us identify the definitive causality in the formation processes of their density enhancement.

2.3.5 Ionosphere-plasmasphere-magnetosphere coupling

Recent studies have revealed that plumes of SED and TOI in the ionosphere can be related to those of the plasmasphere. Furthermore, cold, dense plasma in the plasmasphere can influence the magnetosphere.

Outer layers of the plasmasphere become stripped off by magnetospheric convection during geomagnetically active times when solar wind-magnetosphere coupling becomes strong. A plasmaspheric drainage plume is cold, dense (>100 cm^{-3}) plasma that is drained from the outer plasmasphere (e.g., Elphic et al., 1996; Borovsky et al., 1998; Thomsen et al., 1998; Borovsky and Denton, 2008). This process, by which plasmaspheric plasma is drained, is called plasmaspheric erosion (e.g., Chappell et al., 1971; Carpenter et al., 1993; Elphic et al., 1997; Sandel and Denton, 2007). When storm time convection becomes smaller as IMF B_z turns northward, plasmaspheric plumes start to corotate with the Earth as wind-generated electric fields overcome magnetospheric convection during the recovery phase of storms (e.g., Goldstein, 2006). The plumes during recovery phases are considered the plasmaspheric manifestation of the fossil plumes in the ionosphere.

Su et al. (2001) first associated ionospheric TOIs with plasmaspheric plumes observed by Los Alamos National Laboratory (LANL)/Magnetospheric Plasma Analyzer (MPA) observations. Foster et al. (2002) showed that an ionospheric SED plume mapped onto the low-altitude signature of a plasmasphere drainage plume was possibly connected with storm-time erosion of the plasmasphere boundary layer (PBL) (Carpenter and Lemaire, 2004) observed by IMAGE/EUV. Foster et al. (2014) evaluated plume flux from both DMSP in the ionosphere and VAP in the magnetosphere. Several other studies have confirmed that plasmaspheric drainage plumes are nearly magnetically conjugate (e.g., Yizengaw et al., 2008; Huba and Sazykin, 2014). Borovsky et al. (2014) reported long-lived plasmaspheric drainage plumes observed by the MPA instruments onboard the LANL geosynchronous spacecraft ($L=6.6$) for as long as 11 days during high-speed-stream-driven storms. Typical high-speed-stream-driven storms have durations of a few to several days, depending on the duration of the high-speed coronal-hole-origin solar wind that follows the corotating interaction region. It has not been clarified whether or not plasma plumes of the ionosphere and plasmasphere are physically connected during the entire period. Ionospheric anomalies associated with SED and TOI have been observed during CIR-driven storms (Pokhotelov et al., 2009), the magnitude of which is comparable to those of CME-driven storms. However, the typical duration of ionospheric plume observations is several hours (S. Zhang, private communication).

Furthermore, Walsh et al. (2013) reported that plasmaspheric drainage plumes can reach the dayside magnetopause when a TOI in GPS-TEC observations corresponds to the plasmaspheric drainage plumes observed at the dayside magnetopause by the THEMIS spacecraft. Plasmaspheric plumes can influence dayside magnetic reconnection because this reconnection rate is a function of the Alfvén speed, which in turn is inversely proportional to the square root of density (e.g., Borovsky and Denton, 2006a; Borovsky et al., 2008, 2013; Borovsky, 2014; Birn et al., 2001; Cassak and Shay, 2007; Walsh et al., 2013; Ouellette et al., 2016). Dayside reconnection rates control the degree of solar wind-magnetosphere coupling. Thus, this process influences the amount of energy injected into geospace from the solar wind.

Furthermore, cold, dense plasma in the plasmasphere and plumes plays an important role in wave-particle interactions in the inner magnetosphere. Plumes tend to regulate the excitation of electromagnetic ion cyclotron (EMIC) waves (e.g., Denton et al., 2014) since they are excited as a result of overlapping between hot anisotropic H^+ in the ring current and cold dense plasma. Plumes modulate the resonant EMIC wave-particle interactions by energization of cold ions as well as the pitch angle scattering of the ring current hot ions and the radiation belt relativistic electrons (e.g., Spasojević et al., 2004). The locations of the plasmapause and plumes define the region of radiation belt particles

interacting with whistler mode chorus or hiss waves (e.g., Shprits et al., 2006; Li et al., 2006; Khoo et al., 2018). Cold, dense plasma plumes are a storm time phenomenon that connects the ionosphere, plasmasphere, and magnetosphere (e.g., Moldwin et al., 2016). Improving our understanding of plumes will help us better elucidate how the ionosphere is coupled to the magnetosphere during storms and substorms.

3 Summary and future challenges

3.1 Summary

This review reports recent progress in understanding the ionospheric response to storms and substorms with a focus on electrodynamics. The awareness that the ionosphere is part of a coupled system, closely interacting with the thermosphere and magnetosphere, has played a critical role in helping scientists reach new levels in understanding geospace. Observational capabilities have been expanded considerably through increased coverage and resolution in time and space. Various model coupling efforts have improved our ability to describe the coupling processes of the ionosphere, thermosphere, and magnetosphere on a global scale.

3.2 Unsolved questions

In the following section, unsolved questions with regards to storm time response of the coupled magnetosphere-ionosphere-thermosphere system are summarized.

1. Can we attribute all long-lasting penetration events to the magnetospheric reconfiguration effect? Since C.-S. Huang et al. (2005) was published, a number of studies have reported long-lasting penetration electric field events (e.g., Huba et al., 2005; Huang et al., 2007; Maruyama et al., 2007; Wenbin Wang et al., 2008; Wang et al., 2010; Wei et al., 2008b;

Lei et al., 2018; Obana et al., 2019). It appears that the classical shielding time constant has almost no importance. Most of the studies used the magnetospheric reconfiguration effect (Fejer et al., 1990) as an explanation. The magnetospheric reconfiguration effect and the corresponding shielding time constant must be verified by combining the latest observational evidence and physics-based model simulations.

2. How can we quantify the contribution of neutral wind in generating electric fields in both the ionosphere and magnetosphere from recent observations? Most of the previous studies relied on model simulations (e.g., Peymirat et al., 1998, 2000; Ridley et al., 2003). Many magnetosphere models have not yet included the effect of neutral wind in general. It would be interesting to see how many of the recent observational studies, such as those including Zebra stripes (Lejosne and Roederer, 2016) and nonstorm-time SAPS (Maimaiti et al., 2019), can be explained by including the neutral wind effect in magnetospheric models. Furthermore, improved magnetosphere models, which include the feedback effect from the active ionosphere and thermosphere, are expected to improve our predicting capability about the ionosphere.

3. How much do we understand regarding the energetics associated with the dusk effect and SED? Understanding the energetics of the dusk effect and SED will help us improve our understanding of their formation processes. Foster (1993) originally showed a decrease in electron and ion temperatures based on Millstone Hill ISR observations. A recent study also from the Millstone Hill ISR observations showed an increase in electron temperature, resulting in modification of plasma scale height and the subsequent density distribution along a SED field line (Liu et al., 2016a). The location of the trough is at the poleward boundary of the SED density

enhancement, corresponding to the PBL in the magnetosphere (Carpenter and Lemaire, 2004). In the PBL region, plasmaspheric cold dense plasma collocates and interacts with high-energy populations since the plasmasphere, ring current, and radiation belts form a closely coupled system. Subauroral phenomena, such as those including stable auroral red (SAR) arcs (e.g., Barbier, 1960; Kozyra et al., 1997; Mendillo et al., 2016) and STEVE (MacDonald et al., 2018; Gallardo-Lacourt et al., 2018), could be ionospheric manifestations of heat sources associated with crossenergy coupling in the inner magnetosphere.

4. What more is needed to be able to predict the dusk effect and SED? Many studies have been published to explain these phenomena. However, it is still challenging to predict their density enhancement for individual cases. A possible explanation for a considerable amount of their variability could be a role of wind and neutral composition and its UT, longitudinal, and seasonal variation. If SED occurs in association with a negative ionosphere storm, TEC will be smaller than when SED occurs in the absence of a negative storm. An asymmetry of the storm time response of O/N_2 ratio between winter and summer hemispheres (e.g., Lin et al., 2005b; Kil et al., 2011) is explained by the summer to winter circulation transporting a composition bulge of elevated O/N_2 ratio generated at high latitude (Fuller-Rowell et al., 1996). The magnitude of a TEC plume could be bigger if a composition bulge was transported less equatorward for winter. It is not clarified exactly under which circumstances the dusk effect, SED, and TOI lead to plasma irregularities and instabilities although physical mechanisms such as Kelvin-Helmholtz instability (KHI) and gradient drift instability (GDI) have been previously suggested (e.g., Moen et al., 2013; van der Meeren et al., 2014; Wang et al., 2016; Heine et al., 2017).

3.3 Future directions

In this section, three future directions are proposed to serve as concluding remarks; (1) cross-scale system and scale interactions; (2) the whole geospace modeling concept, including lower atmospheric forcing; and (3) multiday forecasting capability.

1. Cross-scale system and scale interactions:

An increasing number of near-earth environmental (geospace) observations provide unprecedented resolution in time and space. For example, joint rocket missions (the Grand Challenge Initiative-Cusp (GCI-Cusp)) have been conducted to examine the role of plasma turbulence processes in the solar-wind-magnetosphere-ionosphere-atmosphere coupling (Moen et al., 2018). In spite of a considerable amount of progress, current modeling capability has not yet reached the level at which turbulence processes are described by global scale M-I-T models. For example, Liu et al. (2016b) and Wiltberger et al. (2017) included in the TIEGCM and CMIT models, respectively, the effect of electrojet turbulence associated with Farley-Buneman instability (FBI; Buneman, 1963; Farley, 1963a,b). However, the manner in which the FBI effect was implemented in their models was based on an analytical formula. It would be better if instability processes were self-consistently calculated within global M-I-T models. For example, in a new modeling work by Tóth et al. (2016), an implicit particle in cell (PiC) code has been embedded into a global MHD model in a self-consistent manner to assess the importance of kinetic effects in controlling the configuration and dynamics of Ganymede's magnetosphere. On the other hand, with regards to lower atmospheric forcing, considerable effort has already been made to evaluate the impact of small-scale gravity waves on tides in the lower and upper atmosphere on a global scale, using both observations (e.g., Forbes et al., 2016) and physics-based models (e.g., Yiğit and Medvedev, 2012).

2. The whole geospace modeling concept, including lower atmospheric forcing:

Varney et al. (2016) and Glocer et al. (2018) successfully demonstrated the impact of the ion outflow effect on the magnetosphere by implementing the ion outflow process in the gap region between the upper boundary of ionospheric models ($\sim$600 km) and the lower boundary of global MHD models ($2\sim3$ R_E). So far, models have been developed for each region of geospace for different energy populations (the ionosphere, plasmasphere, ring current, radiation belt, magnetosphere, and ion outflow, etc.), and efforts have been made to couple these models in order to cover individual regions and populations of plasma in the near-earth environment. However, it may be time to explore a more sophisticated approach to describe plasma in the entire geospace system (coupling is becoming an old technology from the previous decade). This has been done for whole atmosphere models, in which terrestrial weather models have been extended all the way to the exobase in order to include the upper atmosphere (Jin et al., 2012; Liu et al., 2014; Akmaev et al., 2008; Fuller-Rowell et al., 2008).

3. Multiday forecasting capability:

Despite our ability to describe the phenomenology of those features during storms and substorms, which are critical to the understanding of space weather, we are unable at present to accurately predict how these same features are created and dissipated in real time as well as into the future. The definition of predictability was proposed by Thompson (1957), which is *"the extent to which it is possible to predict the atmosphere with a theoretically complete knowledge of the physical laws governing it."* This deficiency arises primarily from the lack of information about forcing, model error, as well as initial condition error. To overcome the problems, a number of pioneering efforts have been made to implement data assimilation into not only global ionosphere thermosphere models (e.g., Schunk et al., 2004; Scherliess et al., 2017; Codrescu et al., 2018; Hsu et al., 2018; Sutton, 2018) but also whole atmosphere models (e.g., Houjun Wang et al., 2011; Pedatella et al., 2018). Exploring an innovative method to effectively integrate a growing number of data into physics-based models, data assimilation, or the frontier technology of machine learning (e.g., Camporeale et al., 2018), will be the key to our future progress. Understanding and quantifying the relative roles between internal coupling processes and external influences in determining and specifying the behavior of the coupled M-I-T system during storms and substorms presents one of our outstanding challenges.

Acknowledgments

The author would like to thank the editors who gave the opportunity to write this paper. The author had wanted to show this paper to Michael C. Kelley, who inspired me through his text book and passed away in 2018. The author had also wanted to show this paper to Rashid Akmaev, a dear colleague, who passed away in 2018. The author would like to acknowledge the following colleagues who gave insightful discussions: Art Richmond, Dave Anderson, Bela Fejer, Dick Wolf, the organizers of CEDAR Grand Challenge workshop: Mike Ruohoniemi, Tony Mannucci, Phil Erikson, Stan Sazykin, Simon Shepherd, and Jo Baker. This work was supported by NASA grants (NNX16AB83G, NNX15AI91G, 80NSSC17K0720, 80NSSC17K0718, 80NSSC19K0084, 80NSSC19K0277), NSF grants (AGS-1452298, AGS-1552248), and AFOSR FA9550-18-1-0483.

References

Akmaev, R.A., Fuller-Rowell, T.J., Wu, F., Forbes, J.M., Zhang, X., Anghel, A.F., Iredell, M.D., Moorthi, S., Juang, H.-M., 2008. Tidal variability in the lower thermosphere: comparison of Whole Atmosphere Model (WAM) simulations with observations from TIMED. Geophys. Res. Lett. 35, L03810. https://doi.org/10.1029/2007GL032584.

Anderson, B.J., Takahashi, K., Toth, B.A., 2000. Sensing global Birkeland currents with Iridium® engineering magnetometer data. Geophys. Res. Lett. 27, 4045–4048. https://doi.org/10.1029/2000GL000094.

Anderson, B.J., Takahashi, K., Kamei, T., Waters, C.L., Toth, B.A., 2002. Birkeland current system key parameters derived from Iridium observations: method and initial validation results. J. Geophys. Res. 107 (A6), SMP 11-1–11-13, https://doi.org/10.1029/2001JA000080.

Anderson, B.J., Korth, H., Waters, C.L., Green, D.L., Stauning, P., 2008. Statistical Birkeland current distributions from magnetic field observations by the Iridium constellation. Ann. Geophys. 26, 671–687. https://doi.org/10.5194/angeo-26-671-2008.

Anderson, C.N., 1928. Correlation of long wave transatlantic radio transmission with other factors affected by solar activity. Proc. Inst. Radio Eng. 16, 297–347.

Anderson, D.N., 1976. Modeling the midlatitude F-region ionosphere storm using east-west drift and a meridional wind. Planet. Space Sci. 24, 69–77. https://doi.org/10.1016/0032-0633(76)90063-5.

Anderson, D.N., Decker, D.T., Valladares, C.E., 1996. Modeling boundary blobs using time varying convection. Geophys. Res. Lett. 23 (5), 579–582. https://doi.org/10.1029/96GL00371.

Anderson, D., Anghel, A., Yumoto, K., Ishitsuka, M., Kudeki, E., 2002. Estimating daytime vertical $\mathbf{E} \times \mathbf{B}$ drift velocities in the equatorial F-region using ground-based magnetometer observations. Geophys. Res. Lett. 29, 1596. https://doi.org/10.1029/2001GL014562.

Anderson, P.C., Heelis, R.A., Hanson, W.B., 1991. The ionospheric signatures of rapid subauroral ion drifts. J. Geophys. Res. 968, 5785–5792. https://doi.org/10.1029/90JA02651.

Anderson, P.C., Hanson, W.B., Heelis, R.A., Craven, J.D., Baker, D.N., Frank, L.A., 1993. A proposed production model of rapid subauroral ion drifts and their relationship to substorm evolution. J. Geophys. Res. 98 (A4), 6069–6078. https://doi.org/10.1029/92JA01975.

Anderson, P.C., Carpenter, D.L., Tsuruda, K., Mukai, T., Rich, F.J., 2001. Multisatellite observations of rapid subauroral ion drifts (SAID). J. Geophys. Res. 106 (A12), 29585–29599. https://doi.org/10.1029/2001JA000128.

Angelopoulos, V., 2008. The THEMIS mission. Space Sci. Rev. 141, 5–34. https://doi.org/10.1007/s11214-008-9336-1.

Anghel, A., Anderson, D., Maruyama, N., Chau, J., Yumoto, K., Bhattacharyya, A., Alex, S., 2007. Interplanetary electric fields and their relationship to low-latitude electric fields under disturbed conditions. J. Atmos. Sol. Terr. Phys. 69, 1147–1159. https://doi.org/10.1016/j.jastp.2006.08.018.

Axford, W.I., Hines, C.O., 1961. A unifying theory of high-latitude geophysical phenomena and geomagnetic storms. Can. J. Phys. 39, 1433.

Baker, J.B.H., Greenwald, R.A., Ruohoniemi, J.M., Oksavik, K., Gjerloev, J.W., Paxton, L.J., Hairston, M.R., 2007. Observations of ionospheric convection from the Wallops Super-DARN radar at middle latitudes. J. Geophys. Res. 112, A01303. https://doi.org/10.1029/2006JA011982.

Balan, N., Shiokawa, K., Otsuka, Y., Watanabe, S., Bailey, G.J., 2009. Super plasma fountain and equatorial ionization anomaly during penetration electric field. J. Geophys. Res. 114, A03310. https://doi.org/10. 1029/2008JA013768.

Balan, N., Shiokawa, K., Otsuka, Y., Kikuchi, T., Vijaya Lekshmi, D., Kawamura, S., Yamamoto, M., Bailey, G.J., 2010. A physical mechanism of positive ionospheric storms at low latitudes and midlatitudes. J. Geophys. Res. 115, A02304. https://doi.org/10.1029/2009JA014515.

Banks, P.M., 1972. Magnetospheric processes and the behavior of the neutral atmosphere. In: Space Research XII. vol. 2, pp. 1051–1067.

Banks, P.M., Yasuhara, F., 1978. Electric fields and conductivity in the nighttime E-region: a new magnetosphere-ionosphere-atmosphere coupling effect. Geophys. Res. Lett. 5 (12), 1047–1050. https://doi.org/10.1029/ GL005i012p01047.

Barbier, D., 1960. L'arc auroral stable. Ann. Geophys. 16, 544–549.

Basu, S., Basu, S., Groves, K.M., Yeh, H.-C., Su, S.-Y., Rich, F.J., Sultan, P.J., Keskinen, M.J., 2001. Response of the equatorial ionosphere in the South Atlantic Region to the great magnetic storm of July 15, 2000. Geophys. Res. Lett. 28, 3577–3580. https://doi.org/10.1029/2001GL013259.

Basu, S., et al., 2008. Large magnetic storm-induced nighttime ionospheric flows at midlatitudes and their impacts on GPS-based navigation systems. J. Geophys. Res. 113, A00A06. https://doi.org/10.1029/2008JA013076.

Belian, R.D., Cayton, T.E., Reeves, G.D., 1995. Quasi-periodic, substorm associated, global flux variations observed at geosynchronous orbit. In: Ashour-Abdalla, M., Chang, T., Dusenbery, P. (Eds.), Space Plasmas: Coupling Between Small and Medium Scale Processes. In: Geophys. Monogr. Ser., vol. 86. AGU, Washington, DC, p. 143.

Birn, J., et al., 2001. Geospace Environment Modeling (GEM) magnetic reconnection challenge. J. Geophys. Res. 106 (A3), 3715–3719. https://doi.org/10. 1029/1999JA900449.

Blanc, M., Richmond, A.D., 1980. The ionospheric disturbance dynamo. J. Geophys. Res. 85, 1669–1686. https:// doi.org/10.1029/JA085iA04p01669.

Block, L.P., 1966. On the distribution of electric fields in the magnetosphere. J. Geophys. Res. 71, 855–864. https:// doi.org/10.1029/JZ071i003p00855.

Borovsky, J.E., 2014. Feedback of the magnetosphere. Science 343 (6175), 1086–1087. https://doi.org/10.1126/science. 1250590.

Borovsky, J.E., Denton, M.H., 2006a. The effect of plasmaspheric drainage plumes on solar-wind/magnetosphere coupling. Geophys. Res. Lett. 33, L20101. https://doi. org/10.1029/2006GL026519.

Borovsky, J.E., Denton, M.H., 2006b. Differences between CME-driven storms and CIR-driven storms.

J. Geophys. Res. 111, A07S08. https://doi.org/10. 1029/2005JA011447.

Borovsky, J.E., Denton, M.H., 2008. A statistical look at plasmaspheric drainage plumes. J. Geophys. Res. 113, A09221. https://doi.org/10.1029/2007JA012994.

Borovsky, J.E., Thomsen, M.F., McComas, D.J., Cayton, T.E., Knipp, D.J., 1998. Magnetospheric dynamics and mass flow during the November 1993 storm. J. Geophys. Res. 103 (A11), 26373–26394. https://doi.org/10.1029/97JA03051.

Borovsky, J.E., Hesse, M., Birn, J., Kuznetsova, M.M., 2008. What determines the reconnection rate at the dayside magnetosphere? J. Geophys. Res. 113, A07210. https:// doi.org/10.1029/2007JA012645.

Borovsky, J.E., Denton, M.H., Denton, R.E., Jordanova, V.K., Krall, J., 2013. Estimating the effects of ionospheric plasma on solar wind/magnetosphere coupling via mass loading of dayside reconnection: ion-plasma-sheet oxygen, plasmaspheric drainage plumes, and the plasma cloak. J. Geophys. Res. Space Phys. 118, 5695–5719. https://doi.org/10.1002/jgra.50527.

Borovsky, J.E., Welling, D.T., Thomsen, M.F., Denton, M.H., 2014. Long-lived plasmaspheric drainage plumes: where does the plasma come from? J. Geophys. Res. Space Phys. 119, 6496–6520. https://doi.org/10. 1002/2014JA020228.

Brandt, P.C., Ohtani, P.S., Mitchell, D.G., Fok, M.-C., Roelof, E.C., DeMajistre, R., 2002. Geophys. Res. Lett. 29 (20), 1954.

Buneman, O., 1963. Excitation of field aligned sound waves by electron streams. Phys. Rev. Lett. 10, 285. https://doi. org/10.1103/PhysRevLett.10.285.

Buonsanto, M.J., 1999. Ionospheric storm—a review. Space Sci. Rev. 88 (3), 563–601. https://doi.org/10.1023/ A:1005107532631.

Burch, J.L., Mende, S.B., Mitchell, D.G., et al., 2001. Views of Earth's magnetosphere with the IMAGE satellite. Science 291, 619–624. https://doi.org/10.1126/science.291.5504. 619.

Califf, S., Li, X., Wolf, R.A., Zhao, H., Jaynes, A.N., Wilder, F.D., Malaspina, D.M., Redmon, R., 2016. Large-amplitude electric fields in the inner magnetosphere: Van Allen Probes observations of subauroral polarization streams. J. Geophys. Res. Space Phys 121. https://doi.org/10.1002/2015JA022252.

Camporeale, E., Wing, S., Johnson, J.R. (Eds.), 2018. Machine Learning Techniques for Space Weather. Elsevier. https://doi.org/10.1016/C2016-0-01976-9.

Carpenter, D.L., Lemaire, J., 2004. The plasmasphere boundary layer. Ann. Geophys. 22, 4291–4298. https://doi.org/ 10.5194/angeo-22-4291-2004.

Carpenter, D.L., Giles, B.L., Chappell, C.R., Decreau, P.M.E., Anderson, R.R., Persoon, A.M., Smith, A.J., Corcuff, Y., Canu, P., 1993. Plasmasphere dynamics in the duskside bulge region: a new look at an old topic. J. Geophys.

Res. 98 (A11), 19243–19271. https://doi.org/10.1029/93JA00922.

Cassak, P.A., Shay, M.A., 2007. Scaling of asymmetric magnetic reconnection in collisional plasmas. Phys. Plasmas 14, 102144. https://doi.org/10.1063/1.2795630.

Chakrabarty, D., Rout, D., Sekar, R., Narayanan, R., Reeves, G.D., Pant, T.K., Veenadhari, B., Shiokawa, K., 2015. Three different types of electric field disturbances affecting equatorial ionosphere during a long-duration prompt penetration event. J. Geophys. Res. Space Phys. 120 (6), 4993–5008.

Chappell, C.R., Harris, K.K., Sharp, G.W., 1971. The dayside plasmasphere. J. Geophys. Res. 76 (31), 7632–7647. https://doi.org/10.1029/JA076i031p07632.

Chau, J.L., Woodman, R.F., 2004. Daytime vertical and zonal velocities from 150-km echoes: their relevance to F-region dynamics. Geophys. Res. Lett. 31. https://doi.org/10.1029/2004GL020800.

Chen, Y., Wang, W., Burns, A.G., Liu, S., Gong, J., Yue, X., Jiang, G., Coster, A., 2015. Ionospheric response to CIR-induced recurrent geomagnetic activity during the declining phase of solar cycle 23. J. Geophys. Res. Space Phys. 120 (2), 1394–1418. https://doi.org/10.1002/2014JA020657.

Clausen, L.B.N., et al., 2012. Large-scale observations of a subauroral polarization stream by midlatitude SuperDARN radars: instantaneous longitudinal velocity variations. J. Geophys. Res. 117, A05306. https://doi.org/10.1029/2011JA017232.

Codrescu, S.M., Codrescu, M.V., Fedrizzi, M., 2018. An Ensemble Kalman Filter for the thermosphere-ionosphere. Space Weather 16, 57–68. https://doi.org/10.1002/2017SW001752.

Coley, W.R., Heelis, R.A., 1998. Seasonal and universal time distribution of patches in the northern and southern polar caps. J. Geophys. Res. 103 (A12), 29229–29237. https://doi.org/10.1029/1998JA900005.

Coster, A.J., Foster, J.C., Erickson, P.J., 2003. Monitoring the ionosphere with GPS. GPS World 14, 40.

Coster, A., Colerico, M., Foster, J., Rideout, B., Rich, F., Yeh, H.-C., 2005. Statistical Study of SED Events During Magnetic Storms. Presented at Workshop on Penetration Electric Fields and Their Effects in the Inner Magnetosphere and Ionosphere, MIT Haystack Observatory, Nov 7–9, 2005, http://www.haystack.edu/atm/science/magneto/pef/workshop/.

Coster, A.J., Colerico, M., Foster, J.C., Ruohoniemi, J.M., 2006. Observations of the Tongue of Ionization With GPS TEC and SuperDARN. Haystack Observatory, Westford, MA.

Coster, A.J., Colerico, M.J., Foster, J.C., Rideout, W., Rich, F., 2007. Longitude sector comparisons of storm enhanced density. Geophys. Res. Lett. 34, L18105. https://doi.org/10.1029/2007GL030682.

Coster, A.J., Erickson, P.J., Foster, J.C., Thomas, E.G., Ruohoniemi, J.M., Baker, J., 2016. Solar cycle 24 observations of storm-enhanced density and the tongue of ionization. In: Fuller-Rowell, T., Yizengaw, E., Doherty, P.H., Basu, S. (Eds.), In: Ionospheric Space Weather: Longitude Dependence and Lower Atmosphere Forcing, Geophysical Monograph Series, vol. 220. Am. Geophys. Union, Washington, DC, pp. 71–84.

Deng, W., Killeen, T.L., Burns, A.G., Roble, R.G., 1991. The flywheel effect: ionospheric currents after a geomagnetic storm. Geophys. Res. Lett. 18, 1845–1848. https://doi.org/10.1029/91GL02081.

Deng, W., Killeen, T.L., Burns, A.G., Roble, R.G., Slavin, J.A., Wharton, L.E., 1993. The effects of neutral inertia on ionospheric currents in the high-latitude thermosphere following a geomagnetic storm. J. Geophys. Res. 98 (A5), 7775–7790. https://doi.org/10.1029/92JA02268.

Denton, R.E., Jordanova, V.K., Fraser, B.J., 2014. Effect of spatial density variation and O+ concentration on the growth and evolution of electromagnetic ion cyclotron waves. J. Geophys. Res. Space Phys. 119, 8372–8395. https://doi.org/10.1002/2014JA020384.

Doherty, P., Coster, A., Murtagh, M., 2004. Space weather effects of October–November 2003. GPS Solutions 8 (4), 267–271. https://doi.org/10.1007/s10291-004-0109-3.

Eastes, R.W., Mcclintock, W.E., Codrescu, M.V., Aksnes, A., Anderson, D.N., Andersson, L., Baker, D.N., et al., 2013. Global-Scale Observations of the Limb and Disk (Gold): new observing capabilities for the ionosphere-thermosphere, midlatitude ionospheric dynamics and disturbances. In: Kintner Jr., P.M., et al., (Ed.), Midlatitude Ionospheric Dynamics and Disturbances. Geophys. Monogr. Ser., vol. 181. AGU, Washington, DC, pp. 319–326.

Ebihara, Y., Nishitani, N., Kikuchi, T., Ogawa, T., Hosokawa, K., Fok, M.-C., 2008. Two-dimensional observations of overshielding during a magnetic storm by the Super Dual Auroral Radar Network (SuperDARN) Hokkaido radar. J. Geophys. Res. 113, A01213. https://doi.org/10.1029/2007JA012641.

Ebihara, Y., Nishitani, N., Kikuchi, T., Ogawa, T., Hosokawa, K., Fok, M.-C., Thomsen, M.F., 2009. Dynamical property of storm time subauroral rapid flows as a manifestation of complex structures of the plasma pressure in the inner magnetosphere. J. Geophys. Res. Space Phys. 114, A01306. https://doi.org/10.1029/2008JA013614.

Ebihara, Y., Tanaka, T., Kikuchi, T., 2014. Counter equatorial electrojet and overshielding after substorm onset: global MHD simulation study. J. Geophys. Res. Space Phys. 119, 7281–7296. https://doi.org/10.1002/2014JA020065.

Elphic, R.C., Weiss, L.A., Thomsen, M.F., McComas, D.J., Moldwin, M.B., 1996. Evolution of plasmaspheric ions at geosynchronous orbit during times of high geomagnetic activity. Geophys. Res. Lett. 23, 2189–2192. https://doi.org/10.1029/96GL02085.

Elphic, R.C., Thomsen, M.F., Borovsky, J.E., 1997. The fate of the outer plasmasphere. Geophys. Res. Lett. 24 (4), 365–368. https://doi.org/10.1029/97GL00141.

Escoubet, C.P., Fehringer, M., Goldstein, M., 2001. The Cluster mission. Ann. Geophys. 19, 1197–1200. https://doi.org/10.5194/angeo-19-1197-2001.

Farley, D.T., 1963a. Two-stream plasma instability as a source of irregularities in the ionosphere. Phys. Rev. Lett. 10 (7), 279–282. https://doi.org/10.1103/PhysRevLett.10.279.

Farley, D.T., 1963b. A plasma instability resulting in field-aligned irregularities in the ionosphere. J. Geophys. Res. 68 (22), 6083–6097. https://doi.org/10.1029/JZ068i022p06083.

Fear, R.C., Palmroth, M., Milan, S.E., 2012. Seasonal and clock angle control of the location of flux transfer event signatures at the magnetopause. J. Geophys. Res. 117, A04202. https://doi.org/10.1029/2011JA017235.

Fejer, B.G., Scherliess, L., 1997. Empirical models of storm time equatorial electric fields. J. Geophys. Res. 102 (A11), 24047–24056. https://doi.org/10.1029/97JA02164.

Fejer, B.G., Spiro, R.W., Wolf, R.A., Foster, J.C., 1990. Latitudinal variation of perturbation electric fields during magnetically disturbed periods: 1986 SUNDIAL observations and model results. Ann. Geophys. 8 (6), 441–454.

Fejer, B.G., Jensen, J.W., Kikuchi, T., Abdu, M.A., Chau, J.L., 2007. Equatorial ionospheric electric fields during the November 2004 magnetic storm. J. Geophys. Res. 112, A10304. https://doi.org/10.1029/2007JA012376.

Fejer, B.G., Blanc, M., Richmond, A.D., 2016. Post-storm middle and low-latitude ionospheric electric fields effects. Space Sci. Rev. 206, 407–429. https://doi.org/10.1007/s11214-016-0320-x.

Ferdousi, B., Nishimura, Y., Maruyama, N., Lyons, L.R., 2019. Subauroral neutral wind driving and its feedback to SAPS during the March 17, 2013 geomagnetic storm. J. Geophys. Res. Space Phys. https://doi.org/10.1029/2018JA026193.

Forbes, J.M., Harel, M., 1989. Magnetosphere-thermosphere coupling: an experiment in interactive modeling. J. Geophys. Res. 94 (A3), 2631–2644. https://doi.org/10.1029/JA094iA03p02631.

Forbes, J.M., Bruinsma, S.L., Doornbos, E., Zhang, X., 2016. Gravity wave-induced variability of the middle thermosphere. J. Geophys. Res. Space Phys. 121, 6914–6923. https://doi.org/10.1002/2016JA022923.

Foster, J.C., 1993. Storm-time plasma transport at middle and high latitudes. J. Geophys. Res. 98, 1675–1689. https://doi.org/10.1029/92JA02032.

Foster, J.C., Burke, W.J., 2002. SAPS: a new categorization for sub-auroral electric fields. EOS Trans. AGU 83 (36), 393. https://doi.org/10.1029/2002EO000289.

Foster, J.C., Rideout, W., 2007. Storm enhanced density: magnetic conjugacy effects. Ann. Geophys. 25 (8), 1791–1799. https://doi.org/10.5194/angeo-25-1791-2007.

Foster, J.C., Vo, H.B., 2002. Average characteristics and activity dependence of the subauroral polarization stream. J. Geophys. Res. 107 (A12), 1475. https://doi.org/10.1029/2002JA009409.

Foster, J.C., Coster, A.J., Erickson, P.J., Goldstein, J., Rich, F.J., 2002. Ionospheric signatures of plasmaspheric tails. Geophys. Res. Lett. 29 (13), 1623. https://doi.org/10.1029/2002GL015067.

Foster, J.C., Coster, A.J., Erickson, P.J., Holt, J.M., Lind, F.D., Rideout, W., McCready, M., et al., 2005. Multiradar observations of the polar tongue of ionization. J. Geophys. Res. 110, A09S31. https://doi.org/10.1029/2004JA010928.

Foster, J.C., Rideout, W., Sandel, B., Forrester, W.T., Rich, F.J., 2007. On the relationship of SAPS to storm enhanced density. J. Atmos. Space Terr. Phys. 69, 303–313.

Foster, J.C., Erickson, P.J., Coster, A.J., Thaller, S., Tao, J., Wygant, J.R., Bonnell, J., 2014. Stormtime observations of plasmasphere erosion flux in the magnetosphere and ionosphere. Geophys. Res. Lett. 41, 762–768. https://doi.org/10.1002/2013GL059124.

Fuller-Rowell, T.J., Codrescu, M.V., Moffett, R.J., Quegan, S., 1994. Response of the thermosphere and ionosphere to geomagnetic storms. J. Geophys. Res. 99 (A3), 3893–3914. https://doi.org/10.1029/93JA02015.

Fuller-Rowell, T.J., Codrescu, M.V., Rishbeth, H., Moffett, R.J., Quegan, S., 1996. On the seasonal response of the thermosphere and ionosphere to geomagnetic storms. J. Geophys. Res. 101 (A2), 2343–2353. https://doi.org/10.1029/95JA01614.

Fuller-Rowell, T.J., Millward, G.H., Richmond, A.D., Codrescu, M.V., 2002. Storm-time changes in the upper atmosphere at low latitudes. J. Atmos. Sol. Terr. Phys. 64, 1383–1391. https://doi.org/10.1016/S1364-6826(02)00101-3.

Fuller-Rowell, T.J., Richmond, A., Maruyama, N., 2008. Global modeling of storm time thermospheric dynamics and electrodynamics. In: Kintner Jr., P.M., Coster, A.J., Fuller-Rowell, T.J., Mannucci, A.J., Mendillo, M., Heelis, R. (Eds.), Midlatitude Ionospheric Dynamics and Disturbances. In: Geophysical Monograph Series, vol. 181. American Geophysical Union, Washington, DC, pp. 187–200.

Gallagher, D.L., Adrian, M.L., Liemohn, M.W., 2005. Origin and evolution of deep plasmaspheric notches. J. Geophys. Res. 110, A09201. https://doi.org/10.1029/2004JA010906.

Gallardo-Lacourt, B., Nishimura, Y., Lyons, L.R., Mishin, E.V., Ruohoniemi, J.M., Donovan, E.F., et al., 2017. Influence of auroral streamers on rapid evolution of ionospheric SAPS flows. J. Geophys. Res. Space Phys. 122, 12,406–12,420. https://doi.org/10.1002/2017JA024198.

Gallardo-Lacourt, B., Liang, J., Nishimura, Y., Donovan, E., 2018. On the origin of STEVE: particle precipitation or ionospheric skyglow? Geophys. Res. Lett. 45, 7968–7973. https://doi.org/10.1029/2018GL078509.

Galperin, Y.I., Ponomarev, V.N., Zosimova, A.G., 1974. Plasma convection in the polar ionosphere. Ann. Geophys. 30 (1), 1–7.

Galvan, D.A., Moldwin, M.B., Sandel, B.R., Crowley, G., 2010. On the causes of plasmaspheric rotation variability: IMAGE EUV observations. J. Geophys. Res. 115, A01214. https://doi.org/10.1029/2009JA014321.

Garner, T.W., 2003. Numerical experiments on the inner magnetospheric electric field. J. Geophys. Res. 108 (A10), 1373. https://doi.org/10.1029/2003JA010039.

Garner, T.W., Wolf, R.A., Spiro, R.W., Burke, W.J., Fejer, B.G., Sazykin, S., Roeder, J.L., Hairston, M.R., 2004. Magnetospheric electric fields and plasma sheet injection to low L-shells during the 4–5 June 1991 magnetic storm: comparison between the rice convection model and observations. J. Geophys. Res. 109, A02214. https://doi.org/10.1029/2003JA010208.

Garner, T.W., Crowley, G., Wolf, R.A., 2008. Impact of the neutral wind dynamo on the development of the region 2 dynamo. In: Kintner, P.M., Coster, A.J., Fuller-Rowell, T.J., Mannucci, A.J., Mendillo, M., Heelis, R. (Eds.), Midlatitude Ionospheric Dynamics and Disturbances Geophysical Monograph Series, 181, pp. 179–186.

Gjerloev, J.W., 2009. A global ground-based magnetometer initiative. Eos Trans. AGU 90 (27), 230–231.

Glocer, A., Tóth, G., Fok, M.-C., 2018. Including kinetic ion effects in the coupled global ionospheric outflow solution. J. Geophys. Res. Space Phys. 123, 2851–2871. https://doi.org/10.1002/2018JA025241.

Goldstein, J., 2006. Plasmasphere response: tutorial and review of recent imaging results. Space Sci. Rev. 124, 203–216. https://doi.org/10.1007/s11214-006-9105-y.

Goldstein, J., Spiro, R.W., Sandel, B.R., Wolf, R.A., Su, S.-Y., Reiff, P.H., 2003. Overshielding event of 28–29 July, 2000. Geophys. Res. Lett. 30 (8), 1421. https://doi.org/10.1029/2002GL016644.

Goldstein, J., Burch, J.L., Sandel, B.R., 2005. Magnetospheric model of subauroral polarization stream. J. Geophys. Res. 110, A09222. https://doi.org/10.1029/2005JA011135.

Groves, K.M., Carrano, C.S., 2016. Space weather effects on communication and navigation. In: Khazanov, G. (Ed.), Space Weather Fundamentals. CRC Press, pp. 353–387.

Hanson, W.B., Moffett, R.J., 1966. Ionization transport effects in the equatorial F region. J. Geophys. Res. 71, 5559–5572. https://doi.org/10.1029/JZ071i023p05559.

Harel, M., Wolf, R.A., Reiff, P.H., Spiro, R.W., Burke, W.J., Rich, F.J., Smiddy, M., 1981. Quantitative simulation of a magnetospheric substorm, 1, model logic and overview. J. Geophys. Res. 86 (A4), 2217–2241. https://doi.org/10.1029/JA086iA04p02217.

Heelis, R.A., 2016. Longitude and hemispheric dependencies in storm-enhanced density. In: Fuller-Rowell, T., Yizengaw, E., Doherty, P.H., Basu, S. (Eds.), Ionospheric Space Weather: Longitude Dependence and Lower Atmosphere Forcing. In: Geophysical Monograph Series, vol. 220. American Geophysical Union, Washington, DC, pp. 61–70.

Heelis, R.A., Coley, W.R., 2007. Variations in the low- and middle-latitude topside ion concentration observed by DMSP during superstorm events. J. Geophys. Res. 112, A08310. https://doi.org/10.1029/2007JA012326.

Heelis, R.A., Sojka, J.J., David, M., Schunk, R.W., 2009. Storm time density enhancements in the middle-latitude dayside ionosphere. J. Geophys. Res. 114, A03315. https://doi.org/10.1029/2008JA013690.

Heine, T.R.P., Moldwin, M.B., Zou, S., 2017. Small-scale structure of the midlatitude storm enhanced density plume during the 17 March 2015 St. Patrick's Day storm. J. Geophys. Res. Space Phys. 122, 3665–3677. https://doi.org/10.1002/2016JA022965.

Henderson, M.G., 2004. The May 2–3 1986 CDAW–9C interval: a sawtooth event. Geophys. Res. Lett. 31 (1) L11804. https://doi.org/10.1029/2004GL019941.

Horvath, I., Lovell, B.C., 2008. Formation and evolution of the ionospheric plasma density shoulder and its relationship to the superfountain effects investigated during the 6 November 2001 great storm. J. Geophys. Res. 113, A12315. https://doi.org/10.1029/2008JA013153.

Hsu, C.-.T., Matsuo, T., Yue, X., Fang, T.-.W., Fuller-Rowell, T., Ide, K., Liu, J.-.Y., 2018. Assessment of the impact of FORMOSAT-7/COSMIC-2 GNSS RO observations on midlatitude and low-latitude ionosphere specification: observing system simulation experiments using Ensemble Square Root Filter. J. Geophys. Res. Space Phys. 123, 2296–2314. https://doi.org/10.1002/2017JA025109.

Huang, C.-.M., Chen, M.-.Q., 2008. Formation of maximum electric potential at the geomagnetic equator by the disturbance dynamo. J. Geophys. Res. Space Physics. 113, A03301. https://doi.org/10.1029/2007JA012843.

Huang, C.-.M., Richmond, A.D., Chen, M.-.Q., 2005. Theoretical effects of geomagnetic activity on low-latitude ionospheric electric fields. J. Geophys. Res. 110, A05312. https://doi.org/10.1029/2004JA010994.

Huang, C.-.S., 2012. Statistical analysis of dayside equatorial ionospheric electric fields and electrojet currents produced by magnetospheric substorms during sawtooth events. J. Geophys. Res. 117, A02316. https://doi.org/10.1029/2011JA017398.

Huang, C.-.S., Foster, J.C., Goncharenko, L.P., Reeves, G.D., Chau, J.L., Yumoto, K., Kitamura, K., 2004. Variations of low-latitude geomagnetic fields and Dst index caused by magnetospheric substorms. J. Geophys. Res. 109, A05219. https://doi.org/10.1029/2003JA010334.

Huang, C.-.S., Foster, J.C., Kelley, M.C., 2005. Long-duration penetration of the interplanetary electric field to the low-latitude ionosphere during the main phase of magnetic storms. J. Geophys. Res. 110, A11309. https://doi.org/10.1029/2005JA011202.

Huang, C.-S., Sazykin, S., Chau, J.L., Maruyama, N., Kelley, M.C., 2007. Penetration electric fields: efficiency and characteristic time scale. J. Atmos. SolarTerr. Phys. 30 (4), 1158. https://doi.org/10.1016/j.jastp.2006.08.016.

Huba, J.D., Sazykin, S., 2014. Storm time ionosphere and plasmasphere structuring: SAMI3-RCM simulation of the 31 March 2001 geomagnetic storm. Geophys. Res. Lett. 41, 8208–8214. https://doi.org/10.1002/2014GL062110.

Huba, J.D., Joyce, G., Sazykin, S., Wolf, R., Spiro, R., 2005. Simulation study of penetration electric field effects on the low- to mid-latitude ionosphere. Geophys. Res. Lett. 32, L23101. https://doi.org/10.1029/2005GL024162.

Huba, J.D., Sazykin, S., Coster, A., 2017. SAMI3-RCM simulation of the 17 March 2015 geomagnetic storm. J. Geophys. Res. Space Phys. 122, 1246–1257. https://doi.org/10.1002/2016JA023341.

Hui, D., Chakrabarty, D., Sekar, R., Reeves, G.D., Yoshikawa, A., Shiokawa, K., 2017. Contribution of storm time substorms to the prompt electric field disturbances in the equatorial ionosphere. J. Geophys. Res. Space Phys. 122, 5568–5578. https://doi.org/10.1002/2016JA023754.

Hysell, D.L., Larsen, M.F., Woodman, R.F., 1997. JULIA radar studies of electric fields in the equatorial electrojet. Geophys. Res. Lett. 24, 1687–1690.

Iijima, T., Potemra, T.A., 1978. Large-scale characteristics of field-aligned currents associated with substorms. J. Geophys. Res. Space Phys. 83, 599–615. https://doi.org/10.1029/JA083iA02p00599.

Immel, T.J., England, S.L., Mende, S.B., et al., 2018. The ionospheric connection explorer mission: mission goals and design. Space Sci. Rev. 214 (1), 13. https://doi.org/10.1007/s11214-017-0449-2S.

Jaggi, R.K., Wolf, R.A., 1973. Self-consistent calculation of the motion of a sheet of ions in the magnetosphere. J. Geophys. Res. 78, 2852–2866. https://doi.org/10.1029/JA078i016p02852.

Jia, X., Kivelson, M.G., Gombosi, T.I., 2012. Driving Saturn's magnetospheric periodicities from the upper atmosphere/ionosphere. J. Geophys. Res. 117, A04215. https://doi.org/10.1029/2011JA017367.

Jin, H., Miyoshi, Y., Pancheva, D., Mukhtarov, P., Fujiwara, H., Shinagawa, H., 2012. Response of migrating tides to the stratospheric sudden warming in 2009 and their effects on the ionosphere studied by a whole atmosphere-ionosphere model GAIA with COSMIC and TIMED/SABER observations. J. Geophys. Res. 117, A10323. https://doi.org/10.1029/2012JA017650.

Jin, Y., Moen, J.I., Miloch, W.J., 2015. On the collocation of the cusp aurora and the GPS phase scintillation: a statistical study. J. Geophys. Res. Space Phys. 120, 9176–9191. https://doi.org/10.1002/2015JA021449.

Kakad, B., Tiwari, D., Pant, T.K., 2011. Study of disturbance dynamo effects at nighttime equatorial F region in Indian longitude. J. Geophys. Res. 116, A12318. https://doi.org/10.1929/2011JA016626.

Karlson, E.T., 1970. On the equilibrium of the magnetopause. J. Geophys. Res. 75, 2438–2448. https://doi.org/10.1029/JA075i013p02438.

Karlson, E.T., 1971. Plasma flow in the magnetosphere, I. A two-dimensional model of stationary flow. Cosmic Electrodyn. 1, 474–495.

Kelley, M.C., Fejer, B.G., Gonzales, C.A., 1979. An explanation for anomalous ionospheric electric fields associated with a northward turning of the interplanetary magnetic field. Geophys. Res. Lett. 6 (4), 301–304. https://doi.org/10.1029/GL006i004p00301.

Kelley, M.C., Makela, J.J., Chau, J.L., Nicolls, M.J., 2003. Penetration of the solar wind electric field into the magnetosphere/ionosphere system. Geophys. Res. Lett. 30 (4), 1158. https://doi.org/10.1029/2002GL016321.

Khoo, L.Y., Li, X., Zhao, H., Sarris, T.E., Xiang, Z., Zhang, K., Kellerman, A.C., Blake, J.B., 2018. On the initial enhancement of energetic electrons and the innermost plasmapause locations: coronal mass ejection-driven storm periods. J. Geophys. Res. Space Physics 123, 9252–9264. https://doi.org/10.1029/2018JA026074.

Kikuchi, T., Ebihara, Y., Hashimoto, K.K., Kataoka, R., Hori, T., Watari, S., Nishitani, N., 2010. Penetration of the convection and overshielding electric fields to the equatorial ionosphere during a quasiperiodic DP 2 geomagnetic fluctuation event. J. Geophys. Res. 115, A05209. https://doi.org/10.1029/2008JA013948.

Kil, H., Kwak, Y.-S., Paxton, L.J., Meier, R.R., Zhang, Y., 2011. O and N^2 disturbances in the F region during the 20 November 2003 storm seen from TIMED/GUVI. J. Geophys. Res. 116, A02314. https://doi.org/10.1029/2010JA016227.

Kintner, P.M., Coster, A.J., Fuller-Rowell, T., Mannucci, A.J., Mendillo, M., Heelis, R., 2008. Midlatitude ionospheric dynamics and disturbances: introduction. In: Kintner Jr., P.M., Coster, A.J., Fuller-Rowell, T., Mannucci, A.J., Mendillo, M., Heelis, R. (Eds.), Midlatitude Ionospheric Dynamics and Disturbances. In: Geophysical Monograph Series, vol. 181. American Geophysical Union, Washington, DC, pp. 1–7. https://doi.org/10.1029/181GM02.

Kivelson, M.G., 2014. Planetary magnetodiscs: some unanswered questions. In: Szego, K., et al. (Eds.), The Magnetodiscs and Aurorae of Giant Planets. Space Sciences Series of ISSI, vol. 50. Springer, New York, NY. https://doi.org/10.1007/978-1-4939-3395-2_2 Reprinted from Space Science Reviews, 187(1-4), 5–21, https://doi.org/10.1007/s11214-014-0046-6.

Kletzing, C.A., Kurth, W.S., Acuna, M., MacDowall, R.J., Torbert, R.B., Averkamp, T., et al., 2013. The electric and magnetic field instrument suite and integrated science (EMFISIS) on RBSP. Space Sci. Rev. 179 (1–4), 127–181. https://doi.org/10.1007/s11214-013-9993-6.

Knudsen, W.C., 1974. Magnetospheric convection and the high-latitude F_2 ionosphere. J. Geophys. Res. 79 (7), 1046–1055. https://doi.org/10.1029/JA079i007p01046.

Korotova, G.I., Sibeck, D.G., Rosenberg, T., 2008. Seasonal dependence of Interball flux transfer events. Geophys. Res. Lett. 35, L05106. https://doi.org/10.1029/2008GL033254.

Kozyra, J.U., Nagy, A.F., Slater, D.W., 1997. High-altitude energy source(s) for stable auroral red arcs. Rev. Geophys. 35, 155–190. https://doi.org/10.1029/96RG03194.

Kunduri, B.S.R., Baker, J.B.H., Ruohoniemi, J.M., Thomas, E.G., Shepherd, S.G., Sterne, K.T., 2017. Statistical characterization of the large-scale structure of the subauroral polarization stream. J. Geophys. Res. Space Phys. 122, 6035–6048. https://doi.org/10.1002/2017JA024131.

Kunduri, B.S.R., Baker, J.B.H., Ruohoniemi, J.M., Sazykin, S., Oksavik, K., Maimaiti, M., et al., 2018a. Recent developments in our knowledge of inner magnetosphere-ionosphere convection. J. Geophys. Res. Space Phys. 123, 7276–7282. https://doi.org/10.1029/2018JA025914.

Kunduri, B.S.R., Baker, J.B.H., Ruohoniemi, J.M., Nishitani, N., Oksavik, K., Erickson, P.J., et al., 2018b. A new empirical model of the subauroral polarization stream. J. Geophys. Res. Space Physics 123, 7342–7357. https://doi.org/10.1029/2018JA025690.

Landry, R.G., Anderson, P.C., 2018. An auroral boundary-oriented model of subauroral polarization streams (SAPS). J. Geophys. Res. Space Phys. https://doi.org/10.1002/2017JA024921.

Lei, J., Thayer, J.P., Forbes, J.M., Wu, Q., She, C., Wan, W., Wang, W., 2008. Ionosphere response to solar wind high-speed streams. Geophys. Res. Lett. 35, L19105. https://doi.org/10.1029/2008GL035208.

Lei, J., Huang, F., Chen, X., Zhong, J., Ren, D., Wang, W., Yue, X., et al., 2018. Was magnetic storm the only driver of the long-duration enhancements of daytime total electron content in the Asian-Australian sector between 7 and 12 September 2017? J. Geophys. Res. 123 (A4), 3217–3232. https://doi.org/10.1029/2017JA025166.

Lejosne, S., Mozer, F.S., 2016. Typical values of the electric drift $E \times B/B^2$ in the inner radiation belt and slot region as determined from Van Allen Probe measurements. J. Geophys. Res. Space Phys. 121, 12,014–12,024. https://doi.org/10.1002/2016JA023613.

Lejosne, S., Roederer, J.G., 2016. The "zebra stripes": an effect of F region zonal plasma drifts on the longitudinal distribution of radiation belt particles. J. Geophys. Res. Space Phys. 121, 507–518. https://doi.org/10.1002/2015JA021925.

Li, X., Baker, D.N., O'Brien, T.P., Xie, L., Zong, Q.G., 2006. Correlation between the inner edge of outer radiation belt electrons and the innermost plasmapause location. Geophys. Res. Lett. 33, L14107. https://doi.org/10.1029/2006GL026294.

Lin, C.H., Richmond, A.D., Heelis, R.A., Bailey, G.J., Lu, G., Liu, J.Y., Yeh, H.C., Su, S.-Y., 2005a. Theoretical study of the low- and midlatitude ionospheric electron density enhancement during the October 2003 superstorm: relative importance of the neutral wind and the electric field. J. Geophys. Res. 110, A12312. https://doi.org/10.1029/2005JA011304.

Lin, C.H., Richmond, A.D., Liu, J.Y., Yeh, H.C., Paxton, L.J., Lu, G., Tsai, H.F., Su, S.-Y., 2005b. Large-scale variations of the low-latitude ionosphere during the October–November 2003 superstorm: observational results. J. Geophys. Res. 110, A09S28. https://doi.org/10.1029/2004JA010900.

Liu, H.-L., McInerney, J.M., Santos, S., Lauritzen, P.H., Taylor, M.A., Pedatella, N.M., 2014. Gravity waves simulated by high-resolution Whole Atmosphere Community Climate Model. Geophys. Res. Lett. 41, 9106–9112. https://doi.org/10.1002/2014GL062468.

Liu, J., Wang, W., Burns, A., Yue, X., Zhang, S., Zhang, Y., Huang, C., 2016a. Profiles of ionospheric storm-enhanced density during the 17 March 2015 great storm. J. Geophys. Res. Space Phys. 121, 727–744. https://doi.org/10.1002/2015JA021832.

Liu, J., Wang, W., Oppenheim, M., Dimant, Y., Wiltberger, M., Merkin, S., 2016b. Anomalous electron heating effects on the E region ionosphere in TIEGCM. Geophys. Res. Lett. 43 (6), 2351–2358.

Liu, J., Wang, W., Burns, A., Solomon, S.C., Zhang, S., Zhang, Y., Huang, C., 2016c. Relative importance of horizontal and vertical transports to the formation of ionospheric storm-enhanced density and polar tongue of ionization. J. Geophys. Res. Space Phys. 121, 8121–8133. https://doi.org/10.1002/2016JA022882.

Lockwood, M., Carlson Jr., H., 1992. Production of Polar electron density patches by transient magnetopause reconnection. Geophys. Res. Lett. 19, 1731–1734. https://doi.org/10.1029/92GL01993.

Lu, G., 2017. Space Sci. Rev. 206, 431. https://doi.org/10.1007/s11214-016-0269-9.

Lu, G., Richmond, A.D., Emery, B.A., Roble, R.G., 1995. Magnetosphere-ionosphere-thermosphere coupling: effect of neutral winds on energy transfer and field-aligned current. J. Geophys. Res. 100 (A10), 19643–19659. https://doi.org/10.1029/95JA00766.

Lu, G., Huba, J.D., Valladares, C., 2013. Modeling ionospheric super-fountain effect based on the coupled TIMEGCM-SAMI3. J. Geophys. Res. Space Phys. 118, 2527–2535. https://doi.org/10.1002/jgra.50256.

Lühr, H., Rentz, S., Ritter, P., Liu, H., Häusler, K., 2007. Average thermospheric wind patterns over the polar regions, as observed by CHAMP. Ann. Geophys. 25, 1093–1101. https://doi.org/10.5194/angeo-25-1093-2007.

Lyons, L.R., Killeen, T.L., Walterscheid, R.L., 1985. The neutral wind "fly wheel" as a source of quiet time polar cap currents. Geophys. Res. Lett. 12 (2), 101–104. https://doi.org/10.1029/GL012i002p00101.

MacDonald, E.A., Donovan, E., Nishimura, Y., Case, N., Gillies, D.M., Gallardo-Lacourt, B., et al., 2018. New science in plain sight: citizen scientists lead to the discovery of optical structure in the upper atmosphere. Sci. Adv. 4 (3), eaaq0030. https://doi.org/10.1126/sciadv.aaq0030.

Maimaiti, M., Ruohoniemi, J.M., Baker, J.B.H., Ribeiro, A.J., 2018. Statistical study of nightside quiet time midlatitude ionospheric convection. J. Geophys. Res. Space Phys. 123, 2228–2240. https://doi.org/10.1002/2017JA024903.

Maimaiti, M., Baker, J.B.H., Ruohoniemi, J.M., Kunduri, B., 2019. Morphology of nightside subauroral ionospheric convection: monthly, seasonal, Kp, and IMF dependencies. J. Geophys. Res. https://doi.org/10.1029/2018JA026268.

Mannucci, A.J., Tsurutani, B.T., Iijima, B.A., Komjathy, A., Saito, A., Gonzalez, W.D., Guarnieri, F.L., Kozyra, J.U., Skoug, R., 2005. Dayside global ionospheric response to the major interplanetary events of October 29–30, 2003 "Halloween Storms". Geophys. Res. Lett. 32, L12S02. https://doi.org/10.1029/2004GL021467.

Manoj, C., Maus, S., 2012. A real-time forecast service for the ionospheric equatorial zonal electric field. Space Weather 10, S09002. https://doi.org/10.1029/2012SW000825.

Marsal, S., Richmond, A.D., Maute, A., Anderson, B.J., 2012. Forcing the TIEGCM model with Birkeland currents from the active magnetosphere and planetary electrodynamics response experiment. J. Geophys. Res. 117, A06308. https://doi.org/10.1029/2011JA017416.

Maruyama, N., Richmond, A.D., Fuller-Rowell, T.J., Codrescu, M.V., Sazykin, S., Toffoletto, F., Spiro, R.W., Millward, G.H., 2005. Interaction between direct penetration and disturbance dynamo electric fields in the storm-time equatorial ionosphere. Geophys. Res. Lett. https://doi.org/10.1029/2005GL023763.

Maruyama, N., Sazykin, S., Spiro, R.W., Anderson, D., Anghel, A., Wolf, R.A., Toffoletto, F., Fuller-Rowell, T.J., Codrescu, M.V., Richmond, A.D., Millward, G.H., 2007. Modeling storm-time electrodynamics of the low latitude ionosphere-thermosphere system: can long lasting disturbance electric fields be accounted for? J. Atmos. Sol. Terr. Phys. 69 (10), 1182–1199. https://doi.org/10.1016/j.jastp.2006.08.020.

Maruyama, N., Fuller-Rowell, T.J., Codrescu, M.V., Anderson, D., Richmond, A.D., Maute, A., et al., 2011. Modeling the storm time electrodynamics. In: Abdu, M.A., Pancheva, D. (Eds.), Aeronomy of the Earth's Atmosphere and Ionosphere. In: IAGA, Special Sopron Book Series, vol. 2. Springer, Dordrecht, pp. 455–464. https://doi.org/10.1007/978-94-007-0326-1.

Matsui, H., Puhl-Quinn, P.A., Jordanova, V.K., Khotyaintsev, Y., Lindqvist, P.-.A., Torbert, R.B., 2008. Derivation of inner magnetospheric electric field (UNH-IMEF) model using Cluster data set. Ann. Geophys. 26, 2887–2898.

Matsui, H., Torbert, R.B., Spence, H.E., Khotyaintsev, Y.V., Lindqvist, P.-.A., 2013. Revision of empirical electric field modeling in the inner magnetosphere using Cluster data. J. Geophys. Res. Space Phys. 118, 4119–4134. https://doi.org/10.1002/jgra.50373.

McAllister, A.H., Dryer, M., McIntosh, P., Singer, H., Weiss, L., 1996. A large polar crown coronal mass ejection and a "problem" geomagnetic storms: April 14–23, 1994. J. Geophys. Res. 101, 13,497–13,515. https://doi.org/10.1029/96JA00510.

Mendillo, M., 2006. Storms in the ionosphere: patterns and processes for total electron content. Rev. Geophys. 44, RG4001. https://doi.org/10.1029/2005RG000193.

Mendillo, M., Papagiannis, M.D., Klobuchar, J.A., 1970. Ionospheric storms at midlatitudes. Radio Sci. 5, 895–898. https://doi.org/10.1029/RS005i006p00895.

Mendillo, M., Finan, R., Baumgardner, J., Wroten, J., Martinis, C., Casillas, M., 2016. A stable auroral red (SAR) arc with multiple emission features. J. Geophys. Res. Space Phys. 121. https://doi.org/10.1002/2016JA023258.

Mishin, E., Burke, W., Huang, C., Rich, F., 2003. Electromagnetic wave structures within subauroral polarization streams. J. Geophys. Res. 108 (A8), 1309. https://doi.org/10.1029/2002JA009793.

Mishin, E., Nishimura, Y., Foster, J., 2017. SAPS/SAID revisited: a causal relation to the substorm current wedge. J. Geophys. Res. Space Phys. 122, 8516–8535. https://doi.org/10.1002/2017JA024263.

Mitchell, D.G., Brandt, P.C., Roelof, E.C., Hamilton, D.C., Retterer, K.C., Mende, S., 2003. Global imaging of O⁺ from IMAGE/HENA. In: Burch, J.L. (Ed.), Magnetospheric Imaging—The Image Prime Mission. Springer, Dordrecht. https://doi.org/10.1007/978-94-010-0027-7_4.

Miyoshi, Y., Shinohara, I., Takashima, T., Asamura, K., Higashio, N., Mitani, T., et al., 2018. Geospace exploration project ERG. Earth Planets Space 70, 101. https://doi.org/10.1186/s40623-018-0862-0.

Moen, J., Oksavik, K., Alfonsi, L., Daabakk, Y., Romano, V., Spogli, L., 2013. Space weather challenges of the polar cap ionosphere. J. Space Weather Space Clim. 3(A02). https://doi.org/10.1051/swsc/2013025.

Moen, J., Spicher, A., Rowland, D.E., Kletzing, C., LaBelle, J., 2018. Grand challenge initiative—Cusp for SIOS, State of Environmental Science in Svalbard (SESS) 2018 report.

Moldwin, M.B., Zou, S., Heine, T., 2016. The story of plumes: the development of a new conceptual framework for understanding magnetosphere and ionosphere coupling. Ann. Geophys. 34, 1243–1253. https://doi.org/10.5194/angeo-34-1243-2016.

Nicolls, M.J., Kelley, M.C., Chau, J.L., Veliz, O., Anderson, D., Anghel, A., 2007. The spectral properties of low latitude daytime electric fields inferred from magnetometer observations. J. Atmos. Sol. Terr. Phys. 69, 1160–1173. https://doi.org/10.1016/j.jastp.2006.08.015.

Nishida, A., 1968. Coherence of geomagnetic *DP 2* fluctuations with interplanetary magnetic variations. J. Geophys. Res. 73 (17), 5549–5559. https://doi.org/10.1029/JA073i017p05549.

Nishimura, Y., Wygant, J., Ono, T., Iizima, M., Kumamoto, A., Brautigam, D., Friedel, R., 2008. SAPS measurements around the magnetic equator by CRRES. Geophys. Res. Lett. 35, https://doi.org/10.1029/2008GL033970 L10104.

Nishitani, et al., 2019. Review of the accomplishments of midlatitude Super Dual Auroral Radar Network (SuperDARN) HF radars. Prog. Earth Planet. Sci. https://doi.org/10.1186/s40645-019-0270-5.

Noja, M., Stolle, C., Park, J., Lühr, H., 2013. Long-term analysis of ionospheric polar patches based on CHAMP TEC data. Radio Sci. 48, 289–301. https://doi.org/10.1002/rds.20033.

Obana, Y., Maruyama, N., Shinbori, A., Hashimoto, K.K., Fedrizzi, M., Nosé, M., Otsuka, Y., et al., 2019. Response of the ionosphere-plasmasphere coupling to the September 2017 storm: what erodes the plasmasphere so severely? Space Weather. https://doi.org/10.1029/2019SW002168.

Odom, C.D., Larsen, M.F., Christensen, A.B., Anderson, P.C., Hecht, J.H., Brinkman, D.G., Walterscheid, R.L., Lyons, L.R., Pfaff, R., Emery, B.A., 1997. ARIA II neutral flywheel-driven field-aligned currents in the postmidnight sector of the auroral oval: a case study. J. Geophys. Res. 102, 9749–9759. https://doi.org/10.1029/97JA00098.

Oksavik, K., Greenwald, R.A., Ruohoniemi, J.M., Hairston, M.R., Paxton, L.J., Baker, J.B.H., Gjerloev, J.W., Barnes, R.J., 2006. First observations of the temporal/spatial variation of the sub-auroral polarization stream from the SUPERDARN wallops HF radar. Geophys. Res. Lett. 33, L12104. https://doi.org/10.1029/2006GL026256.

Ouellette, J.E., Lyon, J.G., Brambles, O.J., Zhang, B., Lotko, W., 2016. The effects of plasmaspheric plumes on dayside reconnection. J. Geophys. Res. Space Phys. 121 (5), 4111–4118. https://doi.org/10.1002/2016JA022597.

Paxton, L.J., Christensen, A.B., Humm, D.C., et al., 1999. Global Ultraviolet Imager (GUVI): measuring composition and energy inputs for the NASA TIMED mission. Proc. SPIE 3756, 265–276. https://doi.org/10.1117/12.366380.

Pedatella, N.M., Liu, H.-L., Marsh, D.R., Raeder, K., Anderson, J.L., Chau, J.L., et al., 2018. Analysis and hindcast experiments of the 2009 sudden stratospheric warming in WACCMX+DART. J. Geophys. Res. Space Phys. 123, 3131–3153. https://doi.org/10.1002/2017JA025107.

Peymirat, C., Fontaine, D., 1994. Numerical simulation of magnetospheric convection including the effect of field-aligned currents and electron precipitation. J. Geophys. Res. 99 (A6), 11155–11176. https://doi.org/10.1029/93JA02546.

Peymirat, C., Richmond, A.D., Emery, B.A., Roble, R.G., 1998. A magnetosphere-thermosphere-ionosphere electrodynamics general-circulation model. J. Geophys. Res. 103 (A8), 17467–17477. https://doi.org/10.1029/98JA01235.

Peymirat, C., Richmond, A.D., Kobea, A.T., 2000. Electrodynamic coupling of high and low latitudes: simulations of shielding/overshielding effects. J. Geophys. Res. 105 (A10), 22991–23003. https://doi.org/10.1029/2000JA000057.

Peymirat, C., Richmond, A.D., Roble, R.G., 2002. Neutral wind influence on the electrodynamic coupling between the ionosphere and the magnetosphere. J. Geophys. Res. 107 (A1), 1006. https://doi.org/10.1029/2001JA900106.

Pi, X., Mannucci, A.J., Lindqwister, U.J., Ho, C.M., 1997. Monitoring of global ionospheric irregularities using the Worldwide GPS Network. Geophys. Res. Lett. 24, 2283–2286. https://doi.org/10.1029/97GL02273.

Pokhotelov, D., Mitchell, C.N., Jayachandran, P.T., MacDougall, J.W., Denton, M.H., 2009. Ionospheric response to the corotating interaction region-driven geomagnetic storm of October 2002. J. Geophys. Res. 114, A12311. https://doi.org/10.1029/2009JA014216.

Prölss, G.W., 1993. Common origin of positive ionospheric storms at middle latitudes and the geomagnetic activity effect at low latitudes. J. Geophys. Res. 98 (A4), 5981–5991. https://doi.org/10.1029/92JA02777.

Prölss, G.W., 1995. Ionospheric *F*-region storms. In: Volland, H. (Ed.), Handbook of Atmospheric Electrodynamics. In: vol. 2. CRC Press, Boca Raton, FL, pp. 195–248(Chapter 8).

Prölss, G.W., Najiita, K., 1975. Magnetic storm associated changes in the electron content at low latitudes. J. Atmos. Terr. Phys. 37, 635–643. https://doi.org/10.1016/0021-9169(75)90058-6.

Prölss, G.W., von Zahn, U., 1974. Esro 4 gas analyzer results: 2. Direct measurements of changes in the neutral composition during an ionospheric storm. J. Geophys. Res. 79 (16), 2535–2539. https://doi.org/10.1029/JA079i016p02535.

Prölss, G.W., von Zahn, U., Raitt, W.J., 1975. Neutral atmospheric composition, plasma density, and electron temperature at *F* region heights. J. Geophys. Res. 80, 3715–3718.

Puhl-Quinn, P.A., Matsui, H., Mishin, E., Mouikis, C., Kistler, L., Khotyaintsev, Y., Décréau, P.M.E., Lucek, E., 2007. Cluster and DMSP observations of SAID electric fields. J. Geophys. Res. 112, A05219. https://doi.org/10.1029/2006JA012065.

Raeder, J., Berchem, J., Ashour-Abdalla, M., 1996. The importance of small scale processes in global MHD simulations: some numerical experiments. In: Chang, T., Jasperse, J.R. (Eds.), The Physics of Space Plasmas. In: MIT Cent. For Theoret. Geo/Cosmo Plasma Phys., vol. 14. Cambridge, MA, p. 403.

Raeder, J., Cramer, W.D., Jensen, J., Fuller-Rowell, T., Maruyama, N., Toffoletto, F., Vo, H., 2016. Sub-auroral polarization streams: a complex interaction between the magnetosphere, ionosphere, and thermosphere. J. Phys. Conf. Ser. 767 (1), 12021. https://doi.org/10.1088/1742-6596/767/1/012021.

Rasmussen, C.E., Guiter, S.M., Thomas, S.G., 1993. A two-dimensional model of the plasmasphere: refilling time constants. Planet. Space Sci. 41 (1), 35–43. https://doi.org/10.1016/0032-0633(93)90015-T.

Reinisch, B.W., Haines, D., Bibl, K., et al., 2000. The radio plasma imager investigation on the IMAGE spacecraft. Space Sci. Rev. 91, 319–359. https://doi.org/10.1023/A:1005252602159.

Reinisch, B.W., Huang, X., Haines, D.M., et al., 2001a. First results from the radio plasma imager in IMAGE. Geophys. Res. Lett. 28, 1167–1170. https://doi.org/10.1029/2000GL012398.

Reinisch, B.W., Huang, X., Song, P., Sales, G., Fung, S.F., Green, J.L., Gallagher, D.L., Vasyliunas, V.M., 2001b. Plasma density distribution along the magnetospheric field: RPI observations from IMAGE. Geophys. Res. Lett. 28, 4521–4524. https://doi.org/10.1029/2001GL013684.

Reinisch, B.W., Huang, X., Song, P., Green, J.L., Fung, S.F., Vasyliunas, V.M., Gallagher, D.L., Sandel, B.R., 2004. Plasmaspheric mass loss and refilling as a result of a magnetic storm. J. Geophys. Res. 109, A01202. https://doi.org/10.1029/2003JA009948.

Richmond, A.D., 1995a. Ionospheric electrodynamics. In: Volland, H. (Ed.), Handbook of Atmospheric Electrodynamics. In: vol. 2. CRC Press, Boca Raton, FL, pp. 249–290.

Richmond, A.D., 1995b. The ionospheric wind dynamo: effects of its coupling with different atmospheric regions. In: Johnson, R.M., Killeen, T.L. (Eds.), The Upper Mesosphere and Lower Thermosphere: A Review of Experiment and Theory. American Geophysical Union, Washington, DC, pp. 49–65.

Richmond, A.D., Kamide, Y., 1988. Mapping electrodynamics features of the high-latitude ionosphere from localized observations: technique. J. Geophys. Res. 93 (A6), 5741–5759. https://doi.org/10.1029/JA093iA06p05741.

Richmond, A.D., Peymirat, C., Roble, R.G., 2003. Long-lasting disturbances in the equatorial ionospheric electric field simulated with a coupled magnetosphere-ionosphere-thermosphere model. J. Geophys. Res. 108 (A3), 1118. https://doi.org/10.1029/2002JA009758.

Ridley, A.J., Richmond, A.D., Gombosi, T.I., De Zeeuw, D.L., Clauer, C.R., 2003. Ionospheric control of the magnetospheric configuration: thermospheric neutral winds. J. Geophys. Res. 108 (A8), 1328. https://doi.org/10.1029/2002JA009464.

Rodger, A., 2008. The mid-latitude trough—revisited. In: Kintner Jr., P.M, et al. (Eds.), Midlatitude Ionospheric Dynamics and Disturbances. Geophysical Monograph Series, vol. 181. AGU, Washington, DC, p. 25.

Russell, C.T., 2000. The polar cusp. Adv. Space Res. 25 (7–8), 1413–1424. https://doi.org/10.1016/S0273-1177(99)00653-5.

Russell, C.T., Elphic, R.C., 1978. Initial ISEE magnetometer results: magnetopause observations. Space Sci. Rev. 22, 681. https://doi.org/10.1007/BF00212619.

Sandel, B.R., Denton, M.H., 2007. Global view of refilling of the plasmasphere. Geophys. Res. Lett. 34, L17102. https://doi.org/10.1029/2007GL030669.

Sandel, B.R., Goldstein, J., Gallagher, D.L., Spasojević, M., 2003. Extreme ultraviolet imager observations of the structure and dynamics of the plasmasphere. Space Sci. Rev. 109, 25–46. https://doi.org/10.1023/B:SPAC.0000007511.47727.5b.

Sato, T., 1959. Morphology of ionospheric F_2 disturbances in the polar regions. Rept. Ionos. Space Res. Jpn. 13, 91–104.

Sato, T., Rourke, G.F., 1964. F-Region enhancements in the Antarctic. J. Geophys. Res. 69 (21), 4591–4607. https://doi.org/10.1029/JZ069i021p04591.

Scherliess, L., Fejer, B.G., 1997. Storm time dependence of equatorial disturbance dynamo zonal electric fields. J. Geophys. Res. 102 (A11), 24037–24046. https://doi.org/10.1029/97JA02165.

Scherliess, L., Schunk, R.W., Gardner, L.C., Eccles, J.V., Zhu, L., Sojka, J.J., 2017. The USU-GAIM-FP data assimilation model for ionospheric specifications and forecasts. In: 2017 XXXIInd General Assembly and Scientific Symposium of the International Union of Radio Science (URSI GASS), Montreal, QC, 2017, pp. 1–4. https://doi.org/10.23919/URSIGASS.2017.8104978.

Schunk, R.W., 2016. Modeling magnetosphere-ionosphere coupling via ion outflow: past, present, and future. In: Chappell, C.R., Schunk, R.W., Banks, P.M., Burch, J.L., Thorne, R.M. (Eds.), Magnetosphere-Ionosphere Coupling in the Solar System. Geophysical Monograph Series, vol. 222, pp. 167–177.

Schunk, R.W., Banks, P.M., Raitt, W.J., 1976. Effects of electric fields and other processes upon the nighttime high-latitude F layer. J. Geophys. Res. 81 (19), 3271–3282. https://doi.org/10.1029/JA081i019p03271.

Schunk, R.W., Scherliess, L., Sojka, J.J., Thompson, D.C., Anderson, D.N., Codrescu, M., Minter, C., Fuller-Rowell, T.J., et al., 2004. Global assimilation of ionospheric measurements (GAIM). Radio Sci. 39, RS1S02. https://doi.org/10.1029/2002RS002794.

Senior, C., Blanc, M., 1984. On the control of magnetospheric convection by the spatial distribution of ionospheric conductivities. J. Geophys. Res. 89 (A1), 261–284. https://doi.org/10.1029/JA089iA01p00261.

Shprits, Y.Y., Li, W., Thorne, R.M., 2006. Controlling effect of the pitch angle scattering rates near the edge of the loss cone on electron lifetimes. J. Geophys. Res. 111, A12206. https://doi.org/10.1029/2006JA011758.

Silva, R.P., Sobral, J.H.A., Koga, D., Souza, J.R., 2017. Evidence of prompt penetration electric fields during HILDCAA events. Ann. Geophys. 35, 1165–1176. https://doi.org/10.5194/angeo-35-1165-2017.

Skone, S., Yousuf, R., 2007. Performance of satellite-based navigation for marine users during ionospheric disturbances. Space Weather 5, S01006. https://doi.org/10.1029/2006SW000246.

Smiddy, M., Kelley, M.C., Burke, W., Rich, F., Sagalyn, R., Shuman, B., Hays, R., Lai, S., 1977. Intense poleward-directed electric fields near the ionospheric projection of the plasmapause. Geophys. Res. Lett. 4, 543–546. https://doi.org/10.1029/GL004i011p00543.

Sojka, J.J., Bowline, M.D., Schunk, R.W., Decker, D.T., Valladares, C.E., Sheehan, R., Anderson, D.N., Heelis, R.A., 1993. Modeling polar cap F-region patches using time varying convection. Geophys. Res. Lett. 20 (17), 1783–1786. https://doi.org/10.1029/93GL01347.

Sojka, J.J., David, M., Schunk, R.W., Heelis, R.A., 2012. A modeling study of the longitudinal dependence of storm time midlatitude dayside total electron content enhancements. J. Geophys. Res. 117, A02315. https://doi.org/10.1029/2011JA017000.

Southwood, D.J., Wolf, R.A., 1978. An assessment of the role of precipitation in magnetospheric convection. J. Geophys. Res. 83, 5227–5232. https://doi.org/10.1029/JA083iA11p05227.

Spasojević, M., Frey, H.U., Thomsen, M.F., Fuselier, S.A., Gary, S.P., Sandel, B.R., Inan, U.S., 2004. The link between a detached subauroral proton arc and a plasmaspheric plume. Geophys. Res. Lett. 31, L04803. https://doi.org/10.1029/2003GL018389.

Spicher, A., Clausen, L.B.N., Miloch, W.J., Lofstad, V., Jin, Y., Moen, J.I., 2017. Interhemispheric study of polar cap patch occurrence based on Swarm in situ data. J. Geophys. Res. Space Phys. 122, 3837–3851. https://doi.org/10.1002/2016JA023750.

Spiro, R.W., Heelis, R.A., Hanson, W.B., 1979. Rapid subauroral ion drifts observed by Atmospheric Explorer C. Geophys. Res. Lett. 6, 657–660. https://doi.org/10.1029/GL006i008p00657.

Spiro, R.W., Wolf, R.A., Fejer, B.G., 1988. Penetration of high-latitude electric-field effects to low latitudes during SUNDIAL 1984. Ann Geophys. 6 (1), 39–49.

Su, Y.-J., Thomsen, M.F., Borovsky, J.E., Foster, J.C., 2001. A linkage between polar patches and plasmaspheric drainage plumes. Geophys. Res. Lett. 28, 111. https://doi.org/10.1029/2000GL012042.

Sutton, E.K., 2018. A new method of physics-based data assimilation for the quiet and disturbed thermosphere. Space Weather 16, 736–753. https://doi.org/10.1002/2017SW001785.

Thomas, E.G., Baker, J.B.H., Ruohoniemi, J.M., Clausen, L.B.N., Coster, A.J., Foster, J.C., Erickson, P.J., 2013. Direct observations of the role of convection electric field in the formation of a polar tongue of ionization from storm enhanced density. J. Geophys. Res. Space Phys. 118, 1180–1189. https://doi.org/10.1002/jgra.50116.

Thompson, P., 1957. Uncertainty in the initial state as a factor in the predictability of large scale atmospheric flow patterns. Tellus 9, 275–295. https://doi.org/10.1111/j.2153-3490.1957.tb01885.x.

Thomsen, M.F., Borovsky, J.E., McComas, D.J., Elphic, R.C., Maurice, S., 1998. Magnetospheric response to the CME passage of January 10–11, 1997, as seen at geosynchronous orbit. Geophys. Res. Lett. 25, 2545–2548. https://doi.org/10.1029/98GL00514.

Thomsen, M.F., Jackman, C.M., Cowley, S.W.H., Jia, X., Kivelson, M.G., Provan, G., 2017. Evidence for periodic variations in the thickness of Saturn's nightside plasma sheet. J. Geophys. Res. Space Phys. 122 (1), 280–292. https://doi.org/10.1002/2016JA023368.

Tóth, G., Sokolov, I.V., Gombosi, T.I., et al., 2005. Space weather modeling framework: a new tool for the space science community. J. Geophys. Res. 110, A12226. https://doi.org/10.1029/2005JA011126.

Tóth, G., et al., 2016. Extended magnetohydrodynamics with embedded particle-in-cell simulation of Ganymede's magnetosphere. J. Geophys. Res. Space Phys. 121, 1273–1293. https://doi.org/10.1002/2015JA021997.

Tsurutani, B.T., Gonzalez, W.D., 1987. The cause of high intensity long-duration continuous AE activity (HILDCAAs): interplanetary Alfven wave trains. Planet. Space Sci. 35, 405–412. https://doi.org/10.1016/0032-0633(87)90097-3.

Tsurutani, B.T., et al., 2004. Global dayside ionospheric uplift and enhancement associated with interplanetary electric fields. J. Geophys. Res. 109 A08302. https://doi.org/10.1029/2003JA010342.

Tu, J.-N., Horwitz, J.L., Song, P., Huang, X.-Q., Reinisch, B.W., Richards, P.G., 2003. Simulating plasmaspheric field-aligned density profiles measured with IMAGE/RPI: effects of plasmasphere refilling and ion heating. J. Geophys. Res. 108 (A1), 1017. https://doi.org/10.1029/2002JA009468.

Tulasi Ram, S., Sandeep, K., Su, S.-Y., Veenadhari, B., Ravindran, S., 2015. The influence of Corotating Interaction Region (CIR) driven geomagnetic storms on the development of equatorial plasma bubbles (EPBs) over wide range of longitudes. Adv. Space Res. 55, 535–544.

van der Meeren, C., Oksavik, K., Lorentzen, D., Moen, J.I., Romano, V., 2014. GPS scintillation and irregularities at the front of an ionization tongue in the nightside polar ionosphere. J. Geophys. Res. Space Phys. 119, 8624–8636. https://doi.org/10.1002/2014JA020114.

Varney, R.H., Wiltberger, M., Zhang, B., Lotko, W., Lyon, J., 2016. Influence of ion outflow in coupled geospace simulations: 1. Physics-based ion outflow model development and sensitivity study. J. Geophys. Res. Space Phys. 121, 9671–9687. https://doi.org/10.1002/2016JA022777.

Walsh, B.M., Sibeck, D.G., Nishimura, Y., Angelopoulos, V., 2013. Statistical analysis of the plasmaspheric plume at the magnetopause. J. Geophys. Res. Space Phys. 118, 4844–4851. https://doi.org/10.1002/jgra.50458.

Wang, Houjun, Fuller-Rowell, T.J., Akmaev, R.A., Hu, M., Kleist, D.T., Iredell, M.D., 2011. First simulations with a whole atmosphere data assimilation and forecast system: the January 2009 major sudden stratospheric warming. J. Geophys. Res. 116, A12321. https://doi.org/10.1029/2011JA017081.

Wang, Hui, Ridley, A.J., Lühr, H., Liemohn, M.W., Ma, S.Y., 2008. Statistical study of the subauroral polarization stream: its dependence on the cross-polar cap potential and subauroral conductance. J. Geophys. Res. 113, A12311. https://doi.org/10.1029/2008JA013529.

Wang, Hui, Ma, S.-.Y., Ridley, A.J., 2009. Comparative study of subauroral polarization streams with DMSP observation and ram simulation. Chin. J. Geophys. 52 (3), 531–540. https://doi.org/10.1002/cjg2.1374.

Wang, Hui, Lühr, H., Häusler, K., Ritter, P., 2011. Effect of subauroral polarization streams on the thermosphere: a statistical study. J. Geophys. Res. 116, A03312. https://doi.org/10.1029/2010JA016236.

Wang, Hui, Lühr, H., Ritter, P., Kervalishvili, G., 2012. Temporal and spatial effects of subauroral polarization streams on the thermospheric dynamics. J. Geophys. Res. 117, A11307. https://doi.org/10.1029/2012JA018067.

Wang, Hui, Zhang, K., Zheng, Z., Ridley, A., 2018. Subauroral polarization streams effect on thermospheric disturbance winds at middle latitudes: universal time effect. Ann. Geophys. 36, 509–525. https://doi.org/10.5194/angeo-36-509-2018.

Wang, Wenbin, Lei, J., Burns, A.G., Wiltberger, M., Richmond, A.D., Solomon, S.C., Killeen, T.L., Talaat, E.R., Anderson, D.N., 2008. Ionospheric electric field variations during a geomagnetic storm simulated by a coupled magnetosphere ionosphere thermosphere (CMIT) model. Geophys. Res. Lett. 35(18).

Wang, Wenbin, Lei, J., Burns, A.G., Solomon, S.C., Wiltberger, M., Xu, J., Zhang, Y., Paxton, L., Coster, A., 2010. Ionospheric response to the initial phase of geomagnetic storms: common features. J. Geophys. Res. Space Phys. 115(A7).

Wang, Wenbin, Talaat, E.R., Burns, A.G., Emery, B., Hsieh, S., Lei, J., Xu, J., 2012. Thermosphere and ionosphere response to subauroral polarization streams (SAPS) model simulations. J. Geophys. Res. 117, A07301. https://doi.org/10.1029/2012JA017656.

Wang, Y.L., Raeder, J., Russell, C.T., 2004. Plasma depletion layer: magnetosheath flow structure and forces. Ann. Geophys. 22, 1001. https://doi.org/10.5194/angeo-22-1001-2004.

Wang, Y., Zhang, Q.-H., Jayachandran, P.T., Lockwood, M., Zhang, S.-R., Moen, J., Xing, Z.-Y., Ma, Y.-Z., Lester, M., 2016. A comparison between large-scale irregularities and scintillations in the polar ionosphere. Geophys. Res. Lett. 43, 4790–4798. https://doi.org/10.1002/2016GL069230.

Waters, C.L., Anderson, B.J., Liou, K., 2001. Estimation of global field aligned currents using Iridium magnetometer data. Geophys. Res. Lett. 28, 2165–2168. https://doi.org/10.1029/2000GL012725.

Wei, Y., Hong, M., Wan, W., Du, A., Pu, Z., Thomsen, M.F., Ren, Z., Reeves, G.D., 2008a. Coordinated observations of magnetospheric reconfiguration during an overshielding event. Geophys. Res. Lett. 35, L15109. https://doi.org/10.1029/2008GL033972.

Wei, Y., Hong, M., Wan, W., Du, A., Lei, J., Zhao, B., Wang, W., Ren, Z., Yue, X., 2008b. Unusually long-lasting multiple penetration of interplanetary electric field to equatorial ionosphere under oscillating IMF Bz. Geophys. Res. Lett. 35, L02102. https://doi.org/10.1029/2007GL032305.

Wei, Y., Pu, Z., Hong, M., Zong, Q., Ren, Z., Fu, S., Xie, L., Alex, S., Cao, X., Wang, J., Chu, X., 2009. Westward ionospheric electric field perturbations on the dayside associated with substorm processes. J. Geophys. Res. Space Phys. 114(A12). https://doi.org/10.1029/2009JA014445.

Wiltberger, M., Wang, W., Burns, A., Solomon, S., Lyon, J.G., Goodrich, C.C., 2004. Initial results from the coupled magnetosphere ionosphere thermosphere model: magnetospheric and ionospheric responses. J. Atmos. Sol. Terr. Phys. 66, 1411–1423. https://doi.org/10.1016/j.jastp.2004.03.026.

Wiltberger, M., Merkin, V., Zhang, B., Toffoletto, F., Oppenheim, M., Wang, W., Lyon, J.G., Liu, J., Dimant, Y., Sitnov, M.I., Stephens, G.K., 2017. Effects of electrojet turbulence on a magnetosphere-ionosphere simulation of a geomagnetic storm. J. Geophys. Res. Space Phys. 122 (5), 5008–5027. https://doi.org/10.1002/2016JA023700.

Wolf, R.A., 1983. The quasi-static (slow-flow) region of the magnetosphere. In: Carovillano, R.L., Forbes, J.M. (Eds.), Solar-Terrestrial Physics, Principles and Theoretical Foundations. D. Reidel, Dordrecht, pp. 303–329.

Wolf, R.A., Harel, M., Spiro, R.W., Voigt, G.-H., Reiff, P.H., Chen, C.K., 1982. Computer simulation of inner magnetospheric dynamics for the magnetic storm of July 29, 1977. J. Geophys. Res. 87 (A8), 5949–5962. https://doi.org/10.1029/JA087iA08p05949.

Wolf, R.A., Spiro, R.W., Sazykin, S., Toffoletto, F.R., 2007. How the Earth's inner magnetosphere works: an evolving picture. J. Atmos. Sol. Terr. Phys. 69 (3), 288–302. https://doi.org/10.1016/j.jastp.2006.07.026.

Wygant, J.R., Bonnell, J.W., Goetz, K., et al., 2013. The electric field and waves instruments on the Radiation Belt Storm

Probes Mission. Space Sci. Rev. 179 (1), 183–220. https://doi.org/10.1007/s11214-013-0013-7.

Xing, Z.Y., Yang, H.G., Han, D.S., Wu, Z.S., Hu, Z.J., Zhang, Q.H., Kamide, Y., et al., 2012. Poleward moving auroral forms (PMAFs) observed at the Yellow River Station: a statistical study of its dependence on the solar wind conditions. J. Atmos. Sol. Terr. Phys. 86, 25–33. https://doi.org/10.1016/j.jastp.2012.06.004.

Yamazaki, Y., Kosch, M.J., 2015. The equatorial electrojet during geomagnetic storms and substorms. J. Geophys. Res. Space Phys. 120, 2276–2287. https://doi.org/10.1002/2014JA020773.

Yamazaki, Y., Maute, A., 2017. Sq and EEJ—a review on the daily variation of the geomagnetic field caused by ionospheric dynamo currents. Space Sci. Rev. 206, 299–405. https://doi.org/10.1007/s11214-016-0282-z.

Yeh, H.-.C., Foster, J.C., Rich, F.J., Swider, W., 1991. Storm time electric field penetration observed at mid-latitude. J. Geophys. Res. 96 (A4), 5707–5721. https://doi.org/10.1029/90JA02751.

Yiğit, E., Medvedev, A.S., 2012. Gravity waves in the thermosphere during a sudden stratospheric warming. Geophys. Res. Lett. 39, L21101. https://doi.org/10.1029/2012GL053812.

Yizengaw, E., Dewar, J., MacNeil, J., Moldwin, M.B., Galvan, D., Sanny, J., Berube, D., Sandel, B., 2008. The occurrence of ionospheric signatures of plasmaspheric plumes over different longitudinal sectors. J. Geophys. Res. 113, A08318. https://doi.org/10.1029/2007JA012925.

Yu, Y., Jordanova, V., Zou, S., Heelis, R., Ruohoniemi, M., Wygant, J., 2015. Modeling subauroral polarization streams during the 17 March 2013 storm. J. Geophys. Res. Space Phys. 120, 1738–1750. https://doi.org/10.1002/2014JA020371.

Yuan, Z., Zhao, L., Xiong, Y., Deng, X., Wang, J., 2011. Energetic particle precipitation and the influence on the sub-ionosphere in the SED plume during a super geomagnetic storm. J. Geophys. Res. 116, A09317. https://doi.org/10.1029/2011JA016821.

Zhang, S.-.R., et al., 2015. Thermospheric poleward wind surge at midlatitudes during great storm intervals. Geophys. Res. Lett. 42, 5132–5140. https://doi.org/10.1002/2015GL064836.

Zheng, Y., Brandt, P.C., Lui, A.T.Y., Fok, M.-.C., 2008. On ionospheric trough conductance and subauroral polarization streams: simulation results. J. Geophys. Res. 113, A04209. https://doi.org/10.1029/2007JA012532.

Zou, S., Ridley, A.J., Moldwin, M.B., Nicolls, M.J., Coster, A.J., Thomas, E.G., Ruohoniemi, J.M., 2013. Multi-instrument observations of SED during 24–25 October 2011 storm: implications for SED formation processes. J. Geophys. Res. Space Phys. 118, 7798–7809. https://doi.org/10.1002/2013JA018860.

Zou, S., Moldwin, M.B., Ridley, A.J., Nicolls, M.J., Coster, A.J., Thomas, E.G., Ruohoniemi, J.M., 2014. On the generation/decay of the storm-enhanced density plumes: role of the convection flow and field-aligned ion flow. J. Geophys. Res. Space Phys. https://doi.org/10.1002/2014JA020408.

Zou, Y., Nishitani, N., 2014. Study of mid-latitude ionospheric convection during quiet and disturbed periods using the SuperDARN Hokkaido radar. Adv. Space Res. 54 (3), 473–480. https://doi.org/10.1016/j.asr.2014.01.011.

Geomagnetically induced currents

Mirko Piersanti[a], Brett Carter[b]

[a]National Institute of Nuclear Physics, University of Rome "Tor Vergata", Rome, Italy
[b]SPACE Research Center, RMIT University, Melbourne, VIC, Australia

1 Introduction

Coronal mass ejections (CMEs) represent the most geoeffective disturbances of solar origin (Gosling, 1997; Shen et al., 2017). When the CMEs impact the Earth's magnetosphere, a storm sudden commencement or sudden impulse (SSC or SI), a sudden increase in the magnetic field strength detected at the surface, almost instantaneously occurs. The SI is the result of an increase in the magnetopause current, which is caused by the magnetopause compression (e.g., Villante and Piersanti, 2008, 2009, 2011; Piersanti et al., 2012; Piersanti and Villante, 2016, and reference therein) induced by a shock wave driven by the CME sheath region (Araki and Shinbori, 2016; Lugaz et al., 2015; Oliveira and Raeder, 2015). Generally, CMEs bring magnetic clouds, which are structures characterized by intense magnetic fields (i.e., interplanetary magnetic field—IMF) whose north-south component ($B_{z,\ IMF}$) rotates at high rates during the interplanetary (IP) travel (Gosling, 1997; Shen et al., 2017). When interacting with the northward geomagnetic field, if $B_{z,\ IMF}$ is directed southward it can give rise to a reconnection process (Gonzalez et al., 1994; Souza et al., 2017; Piersanti et al., 2017b). The energy transmitted

during this process depends upon how long $B_{z,\ IMF}$ remains southward (Gonzalez et al., 1994, 1999). In this case, a geomagnetic storm occurs (Gonzalez et al., 1994, 1999; Piersanti et al., 2017b; Vellante et al., 2014a, b). During severe solar disturbances, modern technology (e.g., telecommunications, power grids, oil pipelines) can be damaged. At the ground the most dangerous disturbance developing during a geomagnetic storm is represented by geomagnetically induced currents (GICs). They are the direct consequence of the induction of an electric field on the Earth's surface by the highly temporal variations of both the magnetosphere and ionosphere currents (Pirjola, 1982, 2000, 2002; Viljanen and Pirjola, 1994; Pulkkinen et al., 2012). In general, their effects are evident at higher latitudes (Ngwira et al., 2015; Pulkkinen et al., 2012; Viljanen and Pirjola, 1994; Viljanen et al., 2004), but recent research has shown that also low-latitude and equatorial regions can be significantly affected by GICs (e.g., Ngwira et al., 2013, 2015; Carter et al., 2015, 2016; Zhang et al., 2015; Adebesin et al., 2016; Doumbia et al., 2017; MacManus et al., 2017; Marshall et al., 2017; Rodger et al., 2017; Kasran et al., 2018; Liu et al., 2018, and references therein).

The Dynamical Ionosphere
https://doi.org/10.1016/B978-0-12-814782-5.00010-8

This chapter provides the reader with the basic theory and observations associated with GICs. Section 2 gives the reader the basic concepts and nomenclature about geomagnetic storms. Section 3 introduces the main theoretical concepts for the evaluation of GICs. Section 4 shows the example of GIC calculation for the June 22, 2015 event. Section 5 illustrates a direct comparison between the major storm of the last solar maximum and the June event in terms of GIC effects. Finally, Section 6 summarizes and concludes the chapter.

2 Geomagnetic storm

In general, a CME which drives a geomagnetic storm is characterized by the interaction of two distinct regions with the Earth's magnetosphere: the IP shock causing the increase of the horizontal component of the geomagnetic field (H); the magnetic cloud that, due to the reconnection process, produces a strong depression of H at the ground. The first is the initial phase of the storm, called SI (or SSC). The second is called the *main phase* and can span from one to a few hours, or even a few days, until it recovers to its initial condition (*recovery phase*) (Lakhina and Tsurutani, 2016). The decrease of the geomagnetic field measured at ground is caused by the intensification of the ring current flowing around the Earth in the westward direction (Daglis et al., 1997). Geomagnetic storm intensities are measured by the SYM-H index which is evaluated from ground magnetometers observations located at middle-to-low latitudes (Iyemori, 1990). As a consequence, geomagnetic storms are classified according to the minimum value of the SYM-H index at the end of their main phase (Gonzalez et al., 1994). The most intense storm ever recorded is the Carrington event which occurred between September 1 and 2, 1859. It was characterized by an SI amplitude of ~120 nT and a decrease in the H component of the geomagnetic field of ~1600 nT,

lasting for ~1.5 h (Tsurutani et al., 2003). The occurrence of extreme storms, like the Carrington one, could be catastrophic to modern technology (Lakhina and Tsurutani, 2016; Ngwira et al., 2014).

3 Geomagnetically induced currents evaluation

Essentially, the origin of phenomena that give rise to GICs begin at the Sun. In fact, the dynamic coupling between the disturbed solar wind (SW) and the geomagnetic field produces strong variations of both the magnetospheric and ionospheric current systems, leading to high rate changes in the ground magnetic field. More simply, a time-varying magnetic field external to the Earth induces telluric currents-electric currents in the conducting ground. These currents create a secondary (internal) magnetic field (Fig. 1). As a consequence of Faraday's law of induction, an electric field, related to time variations of the magnetic field, is induced at ground (Pulkkinen et al., 2009). Such electric field, acting as a voltage source across the power network, generated electrical currents, known as GICs, which flow in any conducting structure (e.g., a power or pipeline grid grounded in the Earth).

The physics behind the GICs flow is related to the Faraday's law of induction:

$$\vec{\nabla} \times \vec{E} = -\frac{\partial \vec{B}}{\partial t} \tag{1}$$

This law directly relates the temporal variation of the geomagnetic field to the formation of the geoelectric field ($\vec{E}$) that drives GICs at the ground according to Ohm's law $\vec{J} = \sigma \vec{E}$. It is important to highlight that $\vec{E}$ depends only on the magnetospheric-ionospheric current system and on the Earth's geology (Pirjola, 1982). Since $\vec{J}$ is difficult to evaluate since σ is a tensor, typically the current (GIC) flowing through a particular network is evaluated. Following the

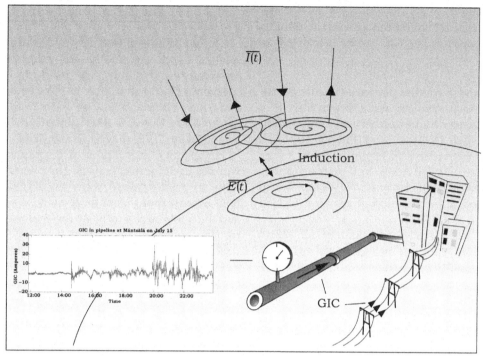

FIG. 1   GIC generation: the time variability of the ionospheric currents ($I(t)$) generates an electric field ($E(t)$) driving GIC. Box on the *left* shows a real GIC observation from a Finnish pipeline. *Adapted from Wikimedia Foundation Inc., 2018. Geomagnetically induced current. October 28, 2018, 19:08 UTC, Wikipedia: The Free Encyclopedia. Wikimedia Foundation Inc. Encyclopedia online. Available from: https://en.wikipedia.org/wiki/Geomagnetically_induced_current (Retrieved 10 April 2019).*

approach of Pirjola (2000, 2002), the GIC calculation necessitates two steps:

(1) estimation of the geoelectric field through the evaluation of the magnetospheric and ionospheric currents, and the knowledge of the conductivity at ground; and

(2) calculation of the flowing GIC through the determined geoelectric field and the knowledge of the particular power network.

Since the geoelectric field is the primary driver of GICs, it is, therefore, the principal quantity that determines their magnitude. To determine $\vec{E}$, the simplest model assumption is a plane wave propagating vertically downwards through a uniform (or a layered) medium, the

conductivity of which is σ (Cagniard, 1953). Assuming ω as the frequency of the planar wave and X and Y as the horizontal directions, $\vec{E}$ is given by

$$E_{x,y} = \sqrt{\frac{\omega}{\mu_0 \sigma}} e^{\frac{i\pi}{4}} B_{y,x} \tag{2}$$

where $B_{y,\,x}$ are the horizontal geomagnetic field components and μ_0 is the magnetic permeability of free space. Eq. (2) is referred to as the basic equation of magnetotellurics. This model is well established and is a commonly used method in GIC applications (e.g., Ngwira et al., 2008; Pirjola, 2002; Pulkkinen et al., 2012; Viljanen et al., 2006). Once $\vec{E}$ is known, it is relatively easy to take the second step for the evaluation of the

GIC flowing through a power network (Pirjola, 2000). In fact, assuming the geoelectric field as spatially constant, the GIC can be calculated as

$$GIC = aE_x(t) + bE_y(t) \qquad (3)$$

where a and b are the network-specific coefficients at each network node depending only on the resistance and geometrical composition of the system (Viljanen and Pirjola, 1994) an GIC is the current in [A]. Typically, a and b vary in the range of 0–200 A km/V. If no transformer data are known, then $a = b = 50$ A km/V can give a good approximation of the sum of GIC (Pulkkinen et al., 2012).

In the last decade, Marshall et al. (2010) defined a GIC index, which gives a measure of the GIC conditions, similar to dB/dt observations. Those indices can be easily obtained from the horizontal magnetic field measurements through a frequency domain filter normalized in amplitude and characterized by 45 degrees phase.

The following section presents the analysis we made to evaluate GIC during the 2015 June event using Eqs. (1), (3).

4 The 2015 June event

The analysis proposed here makes use of both satellite and ground-based magnetometer observatories. Namely, we used the WIND spacecraft (L1 position) for the SW observations and the International Real-Time Magnetic Observatory Network (INTERMAGNET) (Love and Chulliat, 2013) magnetometer data.

Fig. 2 shows the locations of the ground observatories used in this analysis. The blue triangles (gray in print version) show the INTERMAGNET observatories used for the GIC evaluation, and the white circles indicate the chosen stations for the ionospheric current evaluation. The dashed lines indicate the locations of the 0, ±30, ±60, and ±75 degrees magnetic latitudes estimated using Baker and Wing (1989) model.

Fig. 3 shows the CME observations by the WIND spacecraft. Four IP shocks were observed at 16:05 UT on June 21 (IP1), 05:02 UT (IP2) and 18:07 UT (IP3) on June 22, and 13:12 UT (IP4) on June 24, respectively. The first shock was caused by the June 18 CME, while the second shock was driven by a CME from June 19 (Piersanti et al.,

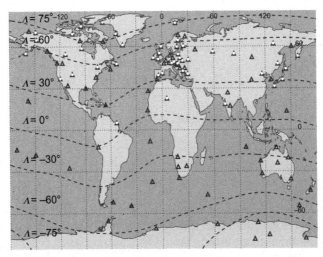

FIG. 2 The locations of the INTERMAGNET magnetic observatories used in this analysis. The magnetic latitudes 0, ±30, ±60, and ±75 degrees are shown for reference. The *white circles* indicate the observatories used to evaluate the ionospheric current system associated the SI.

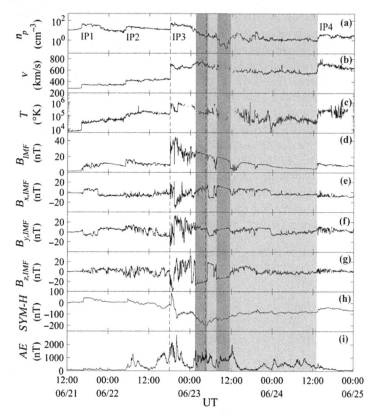

FIG. 3 Solar wind parameters observed by WIND: (a) SW density; (b) SW velocity; (c) SW temperature; (d) IMF intensity; (e–g) IMF x, y, z components in GSE coordinate system. Panels h and i show SYM-H and AE indices, respectively, between June 21 and June 24, 2015. The two *dashed lines* indicate the IP of June 22 at 17:59 UT (IP3) and the minimum values reached by SYM-H during the storm main phase on June 23 at 04:27 UT. The *white area* behind the IP3 shock is the sheath, while the *red shaded region* (*dark gray* in print version) corresponds to the overall ejecta interval. The *green shaded regions* (*gray* in print version) show two small magnetic clouds and/or fluxropes identified within the CME. *Adapted from Piersanti, M., Alberti, T., Bemporad, A., Berrilli, F., Bruno, R., Capparelli, V., Carbone, V., Cesaroni, C., Consolini, G., Cristaldi, A., Del Corpo, A., Del Moro, D., Di Matteo, S., Ermolli, I., Fineschi, S., Giannattasio, F., Giorgi, F., Giovannelli, L., Guglielmino, S.L., Laurenza, M., Lepreti, F., Marcucci, M.F., Martucci, M., Mergè, M., Pezzopane, M., Pietropaolo, E., Romano, P., Sparvoli, R., Spogli, L., Stangalini, M., Vecchio, A., Vellante, M., Villante, U., Zuccarello, F., Heilig, B., Reda, J., Lichtenberger, J., 2017b. Comprehensive analysis of the geoeffective solar event of 21 June 2015: effects on the magnetosphere, plasmasphere, and ionosphere systems. Sol. Phys. 292 (11), 169. https://doi.org/10.1007/s11207-017-1186-0.*

2017b). In addition, the CME (and its preceding shock—IP3) was produced by the June 21 CME and the fourth shock (IP4) was associated with the June 22 CME (Piersanti et al., 2017b). The CME boundaries are determined by the decrease in the temperature coupled with a smooth rotation of the magnetic field.

At the ground, on June 22 at 18:34 UT, SYM-H shows a large SI, directly related to the increase of the SW speed and density (Fig. 3h). After the first rapid decrease of SYM-H (at 20:17 UT), a large negative peak (SYM-H = −208 nT) is observed on June 23 at 04:27 UT. This structure resembles the $B_{z,\ IMF}$ component behavior, characterized by two periods of nearly stable negative values. As a consequence, the SW plasma can flow inside the Earth's magnetosphere because of the occurrence of the magnetic reconnection process between the IMF and Earth's magnetic field (Piersanti et al., 2017b). At high latitude, the geomagnetic activity is characterized by large bursts of activity (AE-index, panel i). This is the evidence of the activity in the geomagnetic tail, due to the so-called loading-unloading process (Consolini and De Michelis, 2005). Moreover, the high-latitude geomagnetic activity continues also during the first part of the geomagnetic storm recovery phase,

probably due to the successive negative turnings of the $B_{z,\ IMF}$ (green-shaded regions in Fig. 3), occurring on June 23 after the 10:00 UT (Piersanti et al., 2017b).

To estimate GIC, we used the time derivative of the horizontal geomagnetic field observed at ground stations in Fig. 2 (Viljanen et al., 2001). Fig. 4 shows the largest temporal variation in the magnetic field (Max dB/dt) as a function of magnetic latitude spanning June 22 and June 23, 2015. In Fig. 4A, the points are colored according to the storm time (i.e., the number of hours following the shock arrival). In Fig. 4B, the points are colored according to the

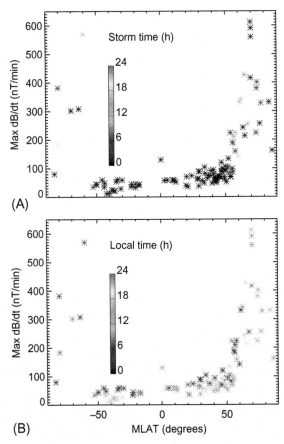

(A)

(B)

FIG. 4 The maximum dB/dt observed during June 22–23, 2015 versus magnetic latitude, colored according to (A) storm time (i.e., the number of hours from IP3) and (B) local time.

corresponding local time of the station. As expected (Ngwira et al., 2013; Love et al., 2016; Pulkkinen et al., 2012; Carter et al., 2016), the Max dB/dt latitudinal behavior shows larger values at latitudes higher than ∼55 degrees. It is worth noticing that Max dB/dt (Fig. 4A) coincides to four moments in terms of the storm time: (1) black points corresponding to the SI (IP3), (2) purple points corresponding to ∼4 h into the storm, (3) light-blue (gray in print version) points corresponding to ∼10 h into storm, and (4) green points (gray in print version) corresponding to ∼16 h into the storm.

The high-latitude stations typically compose groups (2) and (3), while group (1) tends to span across all latitudes (particularly low- and mid-latitudes). In terms of local time, group (1) comprises all local times, and groups (2) and (3) tend to correspond to either mid-morning or mid-evening. One particular feature worth highlighting is the enhanced dB/dt measured at the equator (HUA station) at the shock arrival: ∼130 nT/min compared to ∼60 nT/min observed at mid-latitude stations to the north and south. The corresponding local time of this enhancement is close to noon.

Fig. 4 provides indications about which phase of June 22 storm was the most favorable for GIC generation. It can be easily seen that the SI represents the principal driver of GIC at both lower ($MLAT \sim 0$ degrees) and higher (60 degrees $<$ $|MLAT| < 70$ degrees) latitudes. On the other hand, during the main phase the perturbation leading GIC formation affected medium latitudes too (45 degrees $< |MLAT| < 70$ degrees).

4.1 June 22 sudden impulse: IP3

As stated before, an SI generally precedes the main phase of a geomagnetic storm and indicates the arrival of the IP fast shocks or discontinuities of the incoming SW impinging and compressing the magnetosphere (Araki, 1994; Piersanti and Villante, 2016). At the ground the SI presents a complex behavior, depending upon both LT and geomagnetic latitude. The total disturbance field (D_{SI}) associated to an SI can be decomposed into different subfields, namely, $D_{SI} = DL + DP$ (Araki, 1994). They consist of a step-like structure of magnetospheric origin dominant at low latitudes (DL field, where L stands for low latitude) and a double-pulse structure of ionospheric origin (DP field, where P stands for polar latitude), dominant at high latitudes; the first and the second pulse are called preliminary impulse (PI_{IC}) and main impulse (MI_{IC}), respectively. The DP fields generate in ionosphere a double cell vortices (one in the morning sector and one in the afternoon sector) for both the PI_{IC} and MI_{IC} (Araki, 1994; Piersanti and Villante, 2016; Piersanti et al., 2017a, and reference therein).

In order to evaluate the ionospheric current pattern associated to the SI of June 22, 2015 at 18:34 UT (IP3), we applied the Piersanti and Villante (2016) model to 49 ground magnetic observatories (Fig. 2, white circles) in the northern hemisphere. Fig. 5 shows the results obtained for both the PI_{IC} and MI_{IC}. Panels A and B show the direction of the ionospheric current for the PI_{IC} and the MI_{IC}, respectively, expressed in nT. To obtain the relative current values, the Biot-Savart law has to be inverted (Piersanti and Villante, 2016) at the ionospheric E-layer altitude ($h \sim 90$ km). Panels C and D give a representation of the latitudinal behavior of the amplitude of both the PI_{IC} and MI_{IC}.

The ionospheric current system is consistent with a morning counterclockwise (CCW) and an afternoon clockwise (CW) vortices for the PI_{IC} and a morning CW and an afternoon CCW vortexes for the MI_{IC}. These results are in agreement with Araki (1994) and Piersanti and Villante (2016). Both PI_{IC} and MI_{IC} field amplitudes show an exponential decrease with latitudes (black dashed lines).

The behavior of the GIC as a function of latitude during the June 22, 2018 SI is presented in Fig. 6, which is similar to Fig. 4, but only for the time interval within 5 min of the IP3 arrival.

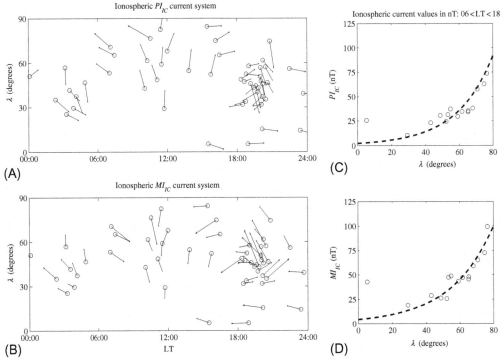

FIG. 5 The direction of the ionospheric currents for the PI_{IC} (A) and for the MI_{IC} (B), as a function of latitude and local time after a 90 degrees rotation of the disturbance magnetic field. The characteristics of the PI_{IC} (C) and MI_{IC} (D) amplitude fields as a function of latitude in the dayside sector (06 <LT < 18); *dashed lines* represent the exponential fits and *black circles* represent the morning PI_{IC} and MI_{IC}. Adapted from Piersanti, M., Alberti, T., Bemporad, A., Berrilli, F., Bruno, R., Capparelli, V., Carbone, V., Cesaroni, C., Consolini, G., Cristaldi, A., Del Corpo, A., Del Moro, D., Di Matteo, S., Ermolli, I., Fineschi, S., Giannattasio, F., Giorgi, F., Giovannelli, L., Guglielmino, S.L., Laurenza, M., Lepreti, F., Marcucci, M.F., Martucci, M., Mergè, M., Pezzopane, M., Pietropaolo, E., Romano, P., Sparvoli, R., Spogli, L., Stangalini, M., Vecchio, A., Vellante, M., Villante, U., Zuccarello, F., Heilig, B., Reda, J., Lichtenberger, J., 2017b. Comprehensive analysis of the geoeffective solar event of 21 June 2015: effects on the magnetosphere, plasmasphere, and ionosphere systems. Sol. Phys. 292 (11), 169. https://doi.org/10.1007/s11207-017-1186-0.

Interestingly, there appears to be a sharp increase in the dB/dt observed beyond 60 degrees MLAT, particularly for stations in the early morning and late evening sectors. Clearly, both Figs. 4 and 6 show an equatorial dB/dt spike of 130 nT/min, although it is interesting to note in Fig. 6 that, during the IP arrival, this equatorial dB/dt spike magnitude is only matched/exceeded by stations at higher latitudes than 50 degrees; in the case of several high-latitude stations, the equatorial dB/dt spike is actually larger. Last but not least, the

close correspondence between the behavior of dB/dt in Fig. 6 and both PI_{IC} and MI_{IC} in Fig. 5C and D shows the direct "cause-effect" correlation between ionospheric current generation and GIC production during an SI. Interestingly, the local enhancement of dB/dt at the magnetic equator on the dayside, confirmed by the corresponding local increase of both PI_{IC} and MI_{IC} amplitudes, shows evidence of the action of the equatorial electrojet in gaining GIC amplitude at very low latitudes (Carter et al., 2015, 2016).

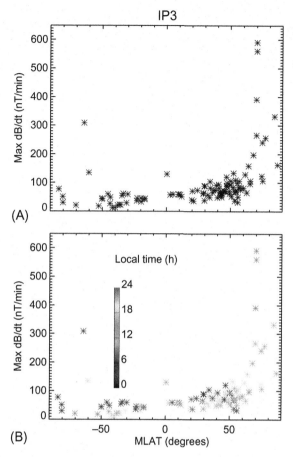

FIG. 6 Same as Fig. 4, but for the 5 min surrounding the arrival of the IP3; 18:37 UT ±2.5 min, June 22, 2015.

5 A comparison with 2015 St. Patrick's Day storm

Carter et al. (2016) gave a brief evaluation of the magnetic field variations observed during the 2015 St. Patrick's Day storm (March 17). Their analysis showed that dB/dt was enhanced at magnetic latitudes higher than 50 degrees, in good agreement with others' reports for other geomagnetic storms (Pulkkinen et al., 2012). The June 2015 event investigated in this study also shows similar trends, although some notable differences can be observed in Fig. 6. First, the maximum dB/dt observed across all stations

was more than 600 nT/min at 70 degrees MLAT (FCC). Alaskan stations JCO and DED registered slightly lower dB/dt values; 590 and 560 nT/min, respectively. The maximum value reported by Carter et al. (2016) was only ~400 nT/min. Further, the maximum dB/dt values observed by the Alaskan stations, JCO and DED, for the June storm were measured at the moment of the shock arrival, and not during the subsequent substorm activity; although significant dB/dt levels were observed during substorm activity later into the event (i.e., groups (2) and (3)). This result shows that significant GIC risk is present during SI arrivals, in addition to substorms.

Carter et al. (2016) reported dB/dt values of the order of 20–50 nT/min across mid-latitude stations that were associated with the St. Patrick's Day IP arrival. In the case of the June 2015 event examined here, the maximum dB/dt values reached significantly larger, approaching 100 nT/min at MLATs 30–50 degrees. This difference in dB/dt between the St. Patrick's Day storm and the June storm can be understood in terms of the SW dynamic pressure change caused by the IP. The dynamic pressure increased by ~20 nPa for the St. Patrick's Day storm, whereas the dynamic pressure increased by ~45 nPa for the June event examined here. Interestingly, these large mid-latitude dB/dt values are not restricted to the daytime hours, with stations in both the early morning and late evening sectors detecting large dB/dt values.

This result obtained for the June 22, 2015 storm is similar to the examples of the March 18, 2002 event, shown by Carter et al. (2015) (see their Fig. 1) and of St Patrick's Day storm event shown by Carter et al. (2016). In that work, the authors highlighted that the both EEJ in Southeast Asia and the evening sector auroral electrojet (over North America) responded to the shock arrival with a sudden decrease in the eastward current (to almost zero for the EEJ) followed by an abrupt increase to a larger eastward current than preshock conditions. It was this sudden increase in the eastward EEJ from its initial depression that caused the enhanced dB/dt at the equator. In that same study, it was shown that the westward current in the auroral electrojet in the morning sector (over Europe) underwent a decrease in strength and was followed by a recovery (see their Fig. 6). Together, these observations showed a clear connection between the electrojet currents in both the evening and morning sectors in the auroral region and the EEJ at the equator on the dayside.

Unfortunately, for the June 2015 event analyzed here, nearby magnetometer data from outside of the EEJ region near HUA was not available. Although, the very high dB/dt observed in the morning sector in Alaska (JCO and DED) suggests that this coupling between high and low latitudes at the moment of the IP arrival exists for the June event as well, but with more significant GIC consequences than during the 2015 St. Patrick's Day IP arrival.

6 Summary and Conclusion

In the ever more technological society in which we are living, the consideration of space weather in the protection of critical infrastructure both in space and on Earth is becoming more important. The space weather predictions, in terms of IMF evolution at Earth during high solar activity, affect, also, the evaluation of electric currents flowing in the magnetosphere-ionosphere system and at ground (Pulkkinen, 2015).

In general, geomagnetic field variations in the near-Earth space environment generate a surface geoelectric field via the electromagnetic induction process (Viljanen and Pirjola, 1994; Pirjola, 2000). The geoelectric field in turn drives GICs, which represents a significant challenge for society, given our strong dependence on stable electricity supply (Knipp, 2015). In fact, a recent analysis using a global economics model has shown that a 10% reduction in electricity supply to Earth's most populated and highly industrialized regions due to a severe geomagnetic storm can impact the global economy on the same scale as wars and global financial crises (Schulte in den Bäumen et al., 2014).

Despite the fact that the bare IMF specification allows for geomagnetic storm strength predictions with a 13-day lead time, in the GIC context it is not sufficient to capture only the overall average features of the IMF; in fact, Huttunen et al. (2008) show that the turbulent IMF fluctuations are one of the major drivers of GICs. In addition, from the space weather point of view, the reconstructed equivalent ionospheric current systems (ECSs) during active magnetic conditions are of interest too, because they can bring new understanding to GICs dynamics. In fact, recent discoveries show the role played by small spatial-scale ($\sim$100 km) ionospheric features in generating extreme geoelectric fields (Pulkkinen, 2015; Ngwira et al., 2015). The determination of ECSs from measurements of ground magnetic field disturbances is of primary interest in ionosphere-magnetosphere research (Untiedt and Baumjohann, 1993). Ionospheric ECSs are a convenient tool to study the characteristics of the ionosphere. Recently, Piersanti and Villante (2016) developed a model to evaluate the ionospheric current pattern from ground observations of the geomagnetic field during an SI for both the PI_{IC} and MI_{IC}.

Research attention has been focused on quantifying and modeling the effects of GICs in the high-latitude region, which is appropriate given that GICs are known to have the highest intensity in the auroral regions, beneath the auroral electrojets (Viljanen and Pirjola, 1994; Pulkkinen et al., 2012). However, at equatorial latitudes, the EEJ has been suspected to play a significant role in the generation of GICs during geomagnetic storms, much like the auroral electrojets at high-latitude regions (Pulkkinen et al., 2012; Moldwin and Tsu, 2016). On this context, Carter et al. (2015, 2016) confirmed that the EEJ at the equator caused enhanced GIC activity during SI events compared to nearby low-latitude regions. Importantly, their analysis showed that equatorial GIC activity was not limited to geomagnetic storms, but was also evident for IP shock arrivals that did not necessarily cause a geomagnetic storm.

The analysis presented here represents an example of GIC evaluation during a geomagnetic storm. The event selected was the June 22, 2015 storm. Our analysis shows that the largest GIC amplitudes (Fig. 4) are at latitudes higher than $\sim$55 degrees. The large SI during this event represented the principal driver of GIC at both lower ($MLAT \sim 0$ degrees) and higher (60 degrees $< |MLAT| < 70$ degrees) latitudes. This aspect verifies that high GIC values are not restrained to geomagnetic storm alone, but also to IP impinging the magnetopause, especially at low latitudes, where the EEJ plays a key role in amplifying GIC values. Moreover, the close correspondence between the behavior of GIC amplitude (Fig. 6) and SI currents amplitude (Fig. 5C and D) confirms the direct "cause-effect" relation between ionospheric current generation and GIC production.

Lastly, a direct comparison between GIC amplitude during 2015 June and the St. Patrick, shows that the former observed at low latitudes (MLATs 30–50 degrees) larger dB/dt values ($\sim$100 nT/min). This difference can be explained in terms of the SW dynamic pressure enhancement during the IP passage. In fact, while the dynamic pressure increased by $\sim$20 nPa for the St. Patrick's Day storm, it presented an increase of $\sim$45 nPa for the June event. Interestingly, these large mid-latitude dB/dt values are not restricted to the daytime hours, with stations in both the early morning and late evening sectors detecting large dB/dt values.

Acknowledgments

The geomagnetic activity and solar wind data were obtained from NASA's CDAWEB online facility (https://cdaweb.sci.gsfc.nasa.gov). The results presented in this chapter rely on data collected at magnetic observatories. We thank the national institutes that support them and INTERMAGNET for promoting high standards of magnetic observatory practice (www.intermagnet.org). M. Piersanti thanks the Italian Space Agency (ASI) for the financial support under the contract ASI "LIMADOU scienza" No. 2016-16-H0. This research was also supported by the Australian Research Council Linkage grant (project LP160100561) awarded to B.A. Carter.

References

Adebesin, B.O., Pulkkinen, A., Ngwira, C.M., 2016. The interplanetary and magnetospheric causes of extreme dB/dt at equatorial locations. Geophys. Res. Lett. 43 (22), 11501–11509. https://doi.org/10.1002/2016GL071526.

Araki, T., 1994. A physical model of the geomagnetic sudden commencement. In: Engebreston, M.J., Takahashi, K., Scholer, M. (Eds.), Geophysical Monograph Series,. In: vol. 81. AGU, Washington, DC, pp. 183–200.

Araki, T., Shinbori, A., 2016. Relationship between solar wind dynamic pressure and amplitude of geomagnetic sudden commencement (SC). Earth Planets Space 68 (9), 1–7. https://doi.org/10.1186/s40623-016-0444-y.

Baker, K.B., Wing, S., 1989. A new magnetic coordinate system for conjugate studies at high latitudes. J. Geophys. Res. 94 (A7), 9139–9143. https://doi.org/10.1029/JA094iA07p09139.

Cagniard, L., 1953. Basic theory of the magneto-telluric method of geophysical prospecting. Geophysics 18 (3), 605–635. https://doi.org/10.1190/1.1437915.

Carter, B.A., Yizengaw, E., Pradipta, R., Halford, A.J., Norman, R., Zhang, K., 2015. Interplanetary shocks and the resulting geomagnetically induced currents at the equator. Geophys. Res. Lett. 42 (16), 6554–6559. https://doi.org/10.1002/2015GL065060.

Carter, B.A., Yizengaw, E., Pradipta, R., Weygand, J.M., Piersanti, M., Pulkkinen, A., Moldwin, M.B., Norman, R., Zhang, K., 2016. Geomagnetically induced currents around the world during the 17 march 2015 storm. J. Geophys. Res-Space 121 (10), 10496–10507. https://doi.org/10.1002/2016JA023344.

Consolini, G., De Michelis, P., 2005. Local intermittency measure analysis of AE index: the directly driven and unloading component. Geophys. Res. Lett. 32. https://doi.org/10.1029/2004GL022063.

Daglis, I.A., Axford, W.I., Sarris, E.T., et al., 1997. Particle acceleration in geospace and its association with solar events. Sol. Phys. 172, 287. https://doi.org/10.1023/A:1004911013182.

Doumbia, V., Boka, K., Kouassi, N., Grodji, O.D.F., Amory-Mazaudier, C., Menvielle, M., 2017. Induction effects of geomagnetic disturbances in the geo-electric field variations at low latitudes. Ann. Geophys. 35 (1), 39–51. https://doi.org/10.5194/angeo-35-39-2017.

Gonzalez, W.D., Joselyn, J.A., Kamide, Y., Kroehl, H.W., Rostoker, G., Tsurutani, B.T., Vasyliunas, V.M., 1994. What is a geomagnetic storm? J. Geophys. Res. 99 (A4), 5771–5792. https://doi.org/10.1029/93JA02867.

Gonzalez, W.D., Tsurutani, B.T., Clúa de Gonzalez, A.L., 1999. Interplanetary origin of geomagnetic storms. Space Sci. Rev. 88 (3–4), 529–562. https://doi.org/10.1023/A:1005160129098.

Gosling, J.T., 1997. Coronal mass ejections: an overview. In: - Crooker, N., Jocelyn, J.A., Feynman, J. (Eds.), Coronal Mass Ejections. In: Geophysical Monograph Series, vol. 99. American Geophysical Union, Washington, DC, pp. 9–16.

Huttunen, K.E.J., Kilpua, S.P., Pulkkinen, A., Viljanen, A., Tanskanen, E., 2008. Solar wind drivers of large geomagnetically induced currents during the solar cycle 23.

Space Weather 6, S10002. https://doi.org/10.1029/2007SW000374.

Iyemori, T., 1990. Storm-time magnetospheric currents inferred from mid-latitude geomagnetic field variations. J. Geomagn. Geoelectr. 42 (11), 1249–1265. https://doi.org/10.5636/jgg.42.1249.

Kasran, F.A.M., Jusoh, M.H., Rahim, S.A.E.A., Abdullah, N., 2018. Geomagnetically induced currents (GICs) in equatorial region. In: 2018 IEEE Eight International Conference on System Engineering and Technology (ICSET), pp. 112–117.

Knipp, D.J., 2015. Synthesis of geomagnetically induced currents: commentary and research. Space Weather 13, 727–729. https://doi.org/10.1002/2015SW001317.

Lakhina, G.S., Tsurutani, B.T., 2016. Geomagnetic storms: historical perspective to modern view. Geosci. Lett. 3 (5), 1–11. https://doi.org/10.1186/s40562-016-0037-4.

Liu, C., Ganebo, Y.S., Wang, H., Li, X., 2018. Geomagnetically induced currents in Ethiopia power grid: calculation and analysis. IEEE Access 6, 64649–64658. https://doi.org/10.1109/ACCESS.2018.2877618.

Love, J.J., Chulliat, A., 2013. An international network of magnetic observatories. EOS Trans. AGU 94 (42), 373–374. https://doi.org/10.1002/2013EO420001.

Love, J.J., Coïsson, P., Pulkkinen, A., 2016. Global statistical maps of extreme-event magnetic observatory 1 min first differences in horizontal intensity. Geophys. Res. Lett. 43, 4126–4135. https://doi.org/10.1002/2016GL068664.

Lugaz, N., Farrugia, C.J., Smith, C.W., Paulson, K., 2015. Shocks inside CMEs: a survey of properties from 1997 to 2006. J. Geophys. Res. Space Phys. 120 (4), 2409–2427. https://doi.org/10.1002/2014JA020848.

MacManus, D.H., Rodger, C.J., Dalzell, M., Thomson, A.W.P., Clilverd, M.A., Petersen, T., Wolf, M.M., Thomson, N.R., Divett, T., 2017. Long-term geomagnetically induced current observations in New Zealand: earth return corrections and geomagnetic field driver. Space Weather 15 (8), 1020–1038. https://doi.org/10.1002/2017SW001635.

Marshall, R.A., Waters, C.L., Sciffer, M.D., 2010. Spectral analysis of pipe-to-soil potentials with variations of the Earth's magnetic field in the Australian region. Space Weather 8(5). https://doi.org/10.1029/2009SW000553.

Marshall, R.A., Kelly, A., Walt, T.V.D., Honecker, A., Ong, C., Mikkelsen, D., Spierings, A., Ivanovich, G., Yoshikawa, A., 2017. Modeling geomagnetic induced currents in Australian power networks. Space Weather 15 (7), 895–916. https://doi.org/10.1002/2017SW001613.

Moldwin, M.B., Tsu, J.S., 2016. Stormtime Equatorial Electrojet Ground Induced Currents: Increasing Power Grid Space Weather Impacts at Equatorial Latitudes. In Ionospheric Space Weather, Geophysical Monograph Series. AGU, Washington, DC (Chapter 3).

Ngwira, C.M., Pulkkinen, A., McKinnell, L.-A., Cilliers, P.J., 2008. Improved modeling of geomagnetically induced currents in the South African power network. Space Weather 6(11). https://doi.org/10.1029/2008SW000408.

Ngwira, C.M., Pulkkinen, A., Wilder, F.D., Crowley, G., 2013. Extended study of extreme geoelectric field event scenarios for geomagnetically induced current applications. Space Weather 11 (3), 121–131. https://doi.org/10.1002/swe.20021.

Ngwira, C.M., Pulkkinen, A., Kuznetsova, M.M., Glocer, A., 2014. Modeling extreme Carrington-type space weather events using three-dimensional global MHD simulations. J. Geophys. Res. Space Phys. 119, 4456–4474. https://doi.org/10.1002/2013JA019661.

Ngwira, C.M., Pulkkinen, A.A., Bernabeu, E., Eichner, J., Viljanen, A., Crowley, G., 2015. Characteristics of extreme geoelectric fields and their possible causes: localized peak enhancements. Geophys. Res. Lett. 42 (17), 6916–6921. https://doi.org/10.1002/2015GL065061.

Oliveira, D.M., Raeder, J., 2015. Impact angle control of interplanetary shock geoeffectiveness: a statistical study. J. Geophys. Res. Space Phys. 120 (6), 4313–4323. https://doi.org/10.1002/2015JA021147.

Piersanti, M., Villante, U., 2016. On the discrimination between magnetospheric and ionospheric contributions on the ground manifestation of sudden impulses. J. Geophys. Res. Space Phys. 121, 6674–6691. https://doi.org/10.1002/2015JA021666.

Piersanti, M., Villante, U., Waters, C., Coco, I., 2012. The 8 June 2000 ULF wave activity: a case study. J. Geophys. Res. 117, A02204. https://doi.org/10.1029/2011JA016857.

Piersanti, M., Cesaroni, C., Spogli, L., Alberti, T., 2017a. Does TEC react to a sudden impulse as a whole? The 2015 Saint Patrick's Day storm event. Adv. Space Res. 60 (8), 1807–1816. https://doi.org/10.1016/j.asr.2017.01.021.

Piersanti, M., Alberti, T., Bemporad, A., Berrilli, F., Bruno, R., Capparelli, V., Carbone, V., Cesaroni, C., Consolini, G., Cristaldi, A., Del Corpo, A., Del Moro, D., Di Matteo, S., Ermolli, I., Fineschi, S., Giannattasio, F., Giorgi, F., Giovannelli, L., Guglielmino, S.L., Laurenza, M., Lepreti, F., Marcucci, M.F., Martucci, M., Mergè, M., Pezzopane, M., Pietropaolo, E., Romano, P., Sparvoli, R., Spogli, L., Stangalini, M., Vecchio, A., Vellante, M., Villante, U., Zuccarello, F., Heilig, B., Reda, J., Lichtenberger, J., 2017b. Comprehensive analysis of the geoeffective solar event of 21 June 2015: effects on the magnetosphere, plasmasphere, and ionosphere systems. Sol. Phys. 292 (11), 169. https://doi.org/10.1007/s11207-017-1186-0.

Pirjola, R., 1982. Electromagnetic induction in the earth by a plane wave or by fields of line currents harmonic in time and space. Geophysica 18 (1–2), 1–161.

Pirjola, R., 2000. Geomagnetically induced currents during magnetic storms. IEEE Trans. Plasma Sci. 28 (6), 1867–1873. https://doi.org/10.1109/27.902215.

Pirjola, R., 2002. Review on the calculation of surface electric and magnetic fields and of geomagnetically induced currents in ground-based technological systems. Surv. Geophys. 23 (1), 71–90. https://doi.org/10.1023/A:1014816009303.

Pulkkinen, A., 2015. Geomagnetically induced currents modeling and forecasting. Space Weather 13, 734–736. https://doi.org/10.1002/2015SW001316.

Pulkkinen, A., Hesse, M., Habib, S., VanderZel, L., Damsky, B., Policelli, F., Fugate, D., Jacobs, W., 2009. Solar shield: forecasting and mitigating space weather effects on high-voltage power transmission systems. Nat. Hazards 53 (2), 333–345. https://doi.org/10.1007/s11069-009-9432-x.

Pulkkinen, A., Bernabeu, E., Eichner, J., Beggan, C., Thomson, A.W.P., 2012. Generation of 100-year geomagnetically induced current scenarios. Space Weather 10(4). https://doi.org/10.1029/2011SW000750.

Rodger, C.J., MacManus, D.H., Dalzell, M., Thomson, A.W.P., Clarke, E., Petersen, T., Clilverd, M.A., Divett, T., 2017. Long-term geomagnetically induced current observations from New Zealand: peak current estimates for extreme geomagnetic storms. Space Weather 15 (11), 1447–1460. https://doi.org/10.1002/2017SW001691.

Schulte in den Bäumen, H., Moran, D., Lenzen, M., Cairns, I., Steenge, A., 2014. How severe space weather can disrupt global supply chains. Nat. Hazards Earth Syst. Sci. 14, 2749–2759. https://doi.org/10.5194/nhessd-2-4463.

Shen, C., Chi, Y., Wang, Y., Xu, M., Wang, S., 2017. Statistical comparison of the ICMEs geoeffectiveness of different types and different solar phases from 1995 to 2014. J. Geophys. Res. Space Phys. 122. https://doi.org/10.1002/2016JA023768.

Souza, V.M., Koga, D., Gonzalez, W.D., Cardoso, F.R., 2017. Observational aspects of magnetic reconnection at the Earth's magnetosphere. Braz. J. Phys. 47 (4), 447–459. https://doi.org/10.1007/s13538-017-0514-z.

Tsurutani, B.T., Gonzalez, W.D., Lakhina, G.S., Alex, S., 2003. The extreme magnetic storm of 12 September 1859. J. Geophys. Res. 108, 1268. https://doi.org/10.1029/2002JA009504,A7.

Untiedt, J., Baumjohann, W., 1993. Studies of polar current systems using the IMS Scandinavian magnetometer array. Space Sci. Rev. 63, 245–390.

Vellante, M., Piersanti, M., Heilig, B., Reda, J., Corpo, A.D., 2014a. Magnetospheric plasma density inferred from field line resonances: effects of using different magnetic field models. In: 2014 XXXIth URSI General Assembly and Scientific Symposium (URSI GASS), pp. 1–4.

Vellante, M., Piersanti, M., Pietropaolo, E., 2014b. Comparison of equatorial plasma mass densities deduced from field line resonances observed at ground for dipole and IGRF models. J. Geophys. Res. Space 119 (4), 2623–2633. https://doi.org/10.1002/2013JA019568.

Viljanen, A., Pirjola, R., 1994. Geomagnetically induced currents in the Finnish high-voltage power system. Surv. Geophys. 15 (4), 383–408. https://doi.org/10.1007/BF00665999.

Viljanen, A., Nevanlinna, H., Pajunpaa, K., Pulkkinen, A., 2001. Time derivative of the horizontal geomagnetic field as an activity indicator. Ann. Geophys. 19, 1107–1118.

Viljanen, A., Pulkkinen, A., Amm, O., Pirjola, R., Korja, T., 2004. Fast computation of the geoelectric field using the method of elementary current systems and planar Earth models. Ann. Geophys. 22 (1), 101–113.

Viljanen, A., Pulkkinen, A., Pirjola, R., Pajunpää, K., Posio, P., Koistinen, A., 2006. Recordings of geomagnetically induced currents and a nowcasting service of the Finnish natural gas pipeline system Space Weather 4(10). https://doi.org/10.1029/2006SW000234.

Villante, U., Piersanti, M., 2008. An analysis of sudden impulses at geosynchronous orbit. J. Geophys. Res. Space Phys. 113(A8). https://doi.org/10.1029/2008JA013028.

Villante, U., Piersanti, M., 2009. Analysis of geomagnetic sudden impulses at low latitudes. J. Geophys. Res. Space Phys 114(A6). https://doi.org/10.1029/2008 JA013920.

Villante, U., Piersanti, M., 2011. Sudden impulses at geosynchronous orbit and at ground. J. Atm. Sol. Terr. Phys. 73 (1), 61–76. https://doi.org/10.1016/j.jastp.2010.01.008.

Zhang, J.J., Wang, C., Sun, T.R., Liu, C.M., Wang, K.R., 2015. GIC due to storm sudden commencement in low-latitude high-voltage power network in China: observation and simulation. Space Weather 13 (10), 643–655. https://doi.org/10.1002/2015SW001263.

Further reading

Inc, Wikimedia Foundation, 2018. Geomagnetically induced current. October 28, 2018, 19:08 UTC, Wikipedia: The Free Encyclopedia. Wikimedia Foundation Inc. Encyclopedia online. Available from: https://en.wikipedia.org/wiki/Geomagnetically_induced_current (Retrieved 10 April 2019).

PART III

Local irregularities

From instability to irregularities

D.L. Hysell

Earth and Atmospheric Sciences, Cornell University, Ithaca, NY, United States

1 Introduction

Plasma physics has been preoccupied with the study of instability since the first experiments in controlled thermonuclear fusion in the 1950s (Bishop, 1958). In laboratory plasmas, instabilities tend to promote transport, undermining magnetic confinement strategies and forestalling the development of practical fusion reactors. Instabilities have also been found to be widespread in space plasmas. As in laboratory plasmas, instabilities and attendant irregularities in space pose engineering problems, mainly by interfering with radio communication, navigation, and imaging systems. More fundamentally, they are crucial regulators of the space environment, shaping the equilibrium state of the heliosphere and the cosmos. Moreover, irregularities in space plasmas can be incisive telltales of natural phenomena that could otherwise go undetected. Finally, ionospheric irregularities facilitate some kinds of remote sensing by providing natural targets for ground-based radar and other instruments to observe.

Plasma instabilities can be classified by the source of the free energy in the environment that fuels them. Instabilities that arise from plasma inhomogeneity in physical space are termed configuration-space instabilities. These include plasma interchange instabilities where the weight of the plasma is supported by pressure. Also included are situations where the background macroscopic flow is inhomogeneous, that is, sheared, as in turning-point instabilities. In contrast, velocity-space instabilities arise from thermal distributions which are something other than simple Maxwellians. Examples here include two-stream and modified two-stream instabilities. Configuration-space instabilities are normally studied using fluid theory. Whereas fluid theory can give indications of velocity-space instabilities, they are normally studied using kinetic theory which can give more insight into the wave-particle interactions at work.

Plasma instabilities can also be classified as being essentially electrostatic or electromagnetic. The electric field in electrostatic instabilities arises primarily from induced dipole moments in inhomogeneous plasmas which behave like dielectrics. Instability results when the dynamics driven by the electric field increase the polarization. The electric field can be expressed as the gradient of a scalar potential. In contrast, the electric field in electromagnetic instabilities arises mainly from induction and current filaments which pinch under the influence of and subsequently intensify the electromagnetic fields. In this case, Faraday's and Ampere's laws are required for analysis, although the displacement current can often be neglected. In the low-beta ionospheric plasmas, electrostatic instabilities dominate.

Finally, instabilities can be classified in terms of the relative importance if inertia. Inertia is an essential agent in familiar hydrodynamic fluid instabilities such as Rayleigh-Taylor and Kelvin-Helmholtz. It is likewise essential in the closely related plasma instabilities (with both electromagnetic and electrostatic variants) which bear the same names. In their fully developed stages, inertial instabilities might be expected to exhibit some of the characteristics of inertial-range turbulence. In the partially ionized ionosphere, meanwhile, ion-neutral collisions and associated Pedersen currents can overwhelm the polarization currents associated with ion inertia. In such cases, the instabilities become collisional. Examples of collisional instabilities in the ionosphere include $\mathbf{E} \times \mathbf{B}$ instabilities in the F-region and gradient-drift instabilities in the E-region. While collisional instabilities can produce complex and self-similar irregularities and flows, this should not be mistaken for inertial-range turbulence.

The study of ionospheric plasma instabilities and irregularities is enormous and sprawling, and a review is beyond the scope of a single book chapter. The object of this work is to discuss the ways that the instabilities have been approached theoretically. Contemporary analysis methods reflect some historic choices. By reexamining these choices and their limitations, solutions to some stubborn problems may be found.

2 Theoretical background

An ideal gas of noninteracting molecules will approach a state of thermodynamic equilibrium characterized by minimum energy and a Maxwellian thermal distribution consistent with the temperature of the vessel containing it (Landau and Lifshitz, 1958). Binary collisions between gas molecules or with the container walls are all that is required to reach this state. The approach to thermodynamic equilibrium can be accelerated by forced mixing or by fluid turbulence, but the time to reach thermodynamic equilibrium will be bound by the reciprocal of the collision frequency.

In space, there are no walls, and in a plasma, long-range interactions between charged particles render the effects of binary Coulomb collisions essentially negligible compared to plasma collective effects, with relevant timescales related to the plasma frequency rather than the collision frequency. This might suggest that space plasmas in certain parameter regimes may have no means of relaxing to thermodynamic equilibrium and that higher-energy steady-state macroscopic configurations could be very long lived. The important question is whether or not these configurations are stable. If not, instability becomes the path to thermodynamic equilibrium.

Stability is usually analyzed in one of two ways—energy analysis and eigenvalue analysis. With energy analysis, the change in energy associated with all possible perturbations to a steady-state plasma configuration is calculated. If any perturbation leads to a reduction in energy, the configuration is unstable. Energy analysis can be intuitive and incisive but does not yield a growth-rate estimate for instability.

Much more common than energy analysis is eigenvalue analysis which involves two steps. First, the equations of motion for the plasma are linearized about a steady-state configuration. Next, unstable eigenvalues in the complex plane are sought. An unstable eigenvalue signifies the exponential growth of small perturbations in the equilibrium configuration. Only one eigenvalue need to be unstable. In the event of multiple unstable eigenvalues, the one associated with the fastest growing mode is usually considered. The analysis yields the linear growth rate and can be used to determine the threshold conditions for instability. The nature of the initial perturbations is not considered to be important in eigenvalue analysis; we presume that perturbations are present in nature.

When the medium is homogeneous so that the state equation derived from the equations of motion has constant coefficients, it is usually possible to guess the eigenfunctions or mode shapes which are often plane waves described by a wavevector **k**. Otherwise, the mode shapes may have to be found by solving a boundary-value problem. This is sometimes referred to as a nonlocal treatment. A common method for solving linear boundary-value problems is the WKB method (e.g., Bender and Orszag, 1978). If the medium is not too inhomogeneous, it may be possible to define a spatially varying wavevector **k(x)** and to develop equations that govern how it varies. In the method of geometric optics, those equations are Hamilton's equations, and the characteristics or ray paths of the waves are like ballistic trajectories in an appropriate potential field (Landau and Lifshitz, 1971). Using the eikonal method, a closely related analysis approach, the space in which the rays exist has axes in the position and wave number directions. Waves may be seen to become alternately stable or unstable as they propagate through parameter space along their characteristics. Such waves are termed convectively stable and unstable, respectively. A pedagogical treatment of the application of the eikonal method to gradient drift instability in the equatorial electrojet was given by Ronchi et al. (1989).

In the ionosphere, the conductivity parallel to the magnetic field lines is generally much larger than in the transverse directions and the magnetic field lines are approximately equipotentials. Consequently, many analysts reduce the dimensionality of their electrostatic ionospheric instability models by integrating the equations of motion along magnetic field lines and modeling the resulting flux-tube-integrated state variables. This is also sometimes called nonlocal treatment. The rationale for doing so was given by Farley (1959), and the necessary formalism by Haerendel (1973). The simplification afforded by this procedure is considerable, but it comes at a cost. The equipotential argument applies to cold plasmas. In a warm plasma, the parallel and perpendicular ambipolar electric fields around a plasma density irregularity are complicated and quadrupolar rather than dipolar (Drake and Huba, 1987). This can fundamentally alter the nature of configuration-space instabilities. Furthermore, flux-tube-integration neglects plasma instabilities that rely on finite parallel wave numbers explicitly. Allowing for nonequipotential magnetic field lines in instability analysis can be important.

While eigenvalue analysis yields accurate predictions of linear stability in some instances, it fails badly in others. In particular, eigenvalue analysis can fail to predict instability where there is significant shear flow (e.g., Farrell and Ioannou, 1993 and references therein). In such situations, the equations of motion may not satisfy the Sturm Liouville condition and the eigenfunctions of the state equation are not guaranteed to be normal. It has been shown that the amplitudes of initial perturbations in nonnormal systems can grow by factors of many thousands even when all the eigenvalues are stable (Boberg and Brosa, 1988; Gustavsson, 1991; Butler and Farrell, 1992; Reddy and Henningson, 1993). This result has profound overtones for instability in the ionosphere where sheared background flows are the rule rather than the exception. Eigenvalue analysis by itself is not guaranteed to predict ionospheric stability.

The anomalous growth in nonnormal systems can be viewed as a transient response (Trefethen et al., 1993). Even when the individual eigenmodes of the system are all decaying, their superposition can grow substantially if the modes are nonnormal and if the initial conditions are favorable. The ultimate size of the transient response therefore depends on the particular initial perturbations in a way that the size of the asymptotic response does not. While analytic methods exist for predicting the behavior of nonnormal systems, a common approach is to construct and solve the appropriate initial boundary value problem.

Finally, the state equations for plasma instabilities are nonlinear as a rule, and linear theory can be expected at best to govern the conditions for instability onset. After a few e-folding times (the time it takes for the amplitude to grow by a factor of e), linear perturbation theory may be expected to fail badly and for a number of reasons. For one, the instability may alter the average plasma parameters, notably the temperature (or the plasma distribution function in the case of kinetic instabilities). For another, large-amplitude plasma waves can be expected to interact or beat. In simple cases, a few discrete waves may interact in triads in such a way that each wave behaves approximately according to linear theory but with the total wave momentum and energy conserved. This is the domain of weak turbulence. In more complicated cases, a continuum of waves could interact and produce phenomena completely outside the domain of linear theory such as inertial-range turbulence. This is the domain of strong turbulence. That wave energy depends quadratically on wave amplitude tells us that something as fundamental as energy conservation is inherently a nonlinear problem.

Closed-form solutions for nonlinear wave equations are obtainable in some circumstances and can be used to spotlight phenomena like wave steepening and wave and particle trapping, secondary-wave formation, solitary-wave propagation, and anomalous transport. Most often, nonlinear theoretical treatments of plasma instabilities are computational. Numerical simulations of initial boundary value problems have the added advantage of capturing nonlocal and nonnormal instability behavior automatically.

The remainder of this chapter examines a few well-known ionospheric instabilities. The focus is on electrostatic instabilities which produce plasma density irregularities detectable from the ground by radar.

3 Interchange instabilities and equatorial spread F (ESF)

The equatorial F-region ionosphere was found to be unstable and to produce intense plasma density irregularities in the earliest days of radio and radar science (Booker and Wells, 1938). Plasma interchange instability operating on the steep postsunset bottomside F-region was suspected, but it was unclear how this mechanism could be responsible for irregularities detected by VHF radar in the topside as well as the bottomside F-region (Farley et al., 1970). By plotting VHF coherent scatter data as two-dimensional (2D) images, Woodman and La Hoz (1976) surmised that topside irregularities were produced by highly developed interchange instabilities and by plumes of depleted plasma that ascend from the bottomside through the F peak to the topside. The basic premise was supported by numerical simulations by Ossakow (1981) and many others subsequently (e.g., Huba et al., 2008; Retterer, 2010; Yokoyama et al., 2014).

Coherent radar backscatter from a typical ESF event like some of those described by Woodman and La Hoz (1976) is shown in Fig. 1. As is very typical, a thin "bottom-type" scattering layer appeared at the base of the F-region (around 350 km altitude) after sunset. Later, a series of ESF plumes complexes passed over the radar. The plumes signify the presence of intense meter-scale plasma density irregularities within deep plasma depletions extending through the F peak into the topside. Broadband irregularities act like a diffraction screen, causing scintillations in radio signals passing through them.

More detailed radar diagnostics are available for interrogating coherent backscatter today. At Jicamarca, imaging techniques adapted from

Wed Nov 26 02:13:45 2014

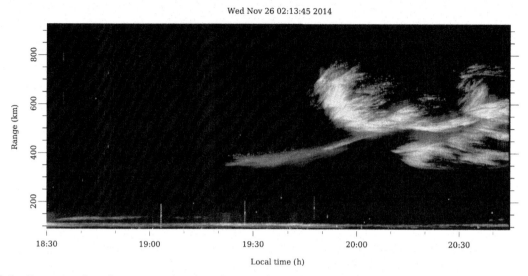

FIG. 1 Range-time-Doppler-intensity plot of an ESF event observed at the Jicamarca Radio Observatory on November 26, 2014. The brightness, hue, and saturation of the pixels reflect the signal-to-noise ratio, Doppler shift, and spectral width, respectively.

radio astronomy can be used to distinguish between spatial and temporal variations in the backscatter and produce true 2D images of the field-aligned radar plumes (Hysell and Chau, 2006). Fig. 2 shows some of the plumes comprising the complex that passed overhead around 20 LT in Fig. 1. The images depict multiple, closely spaced plumes with reverse "C" shapes drifting eastward with the background flow. As time progresses, the plumes undergo pinching and bifurcation. Comparisons with in situ data from the C/NOFS satellite indicate that the bright patches in the images correspond to compact regions of deeply depleted plasma adjacent to steep edges.

The theory of plasma instabilities and irregularities associated with ESF has undergone extensive refinement over the years. Zargham and Seyler (1989) distinguished between interchange instability in the inertial regime, which is the electrostatic form of Rayleigh-Taylor instability, and instability in the collisional regime, which is essentially $\mathbf{E} \times \mathbf{B}$ instability with gravity and vertical winds augmenting the

background, driving current. Zargham and Seyler (1987) performed an eigenvalue analysis of the collisional interchange instability, taking into account the effect of the finite depth of the bottomside F-region. This introduced a long-wavelength cutoff. The short-wavelength cutoff is due to diffusive dissipation. Sultan (1996) reformulated the collisional instability problem using flux tube-integrated quantities. Krall et al. (2010) determined the conditions that control the terminal altitude of spread F plumes. Dynamic and thermodynamic aspects of the instability in three dimensions were investigated systematically by Huba et al. (2011). Dao (2012) investigated the electromagnetic properties of the instability, demonstrating how quasiequipotential magnetic field lines are maintained by reflecting shear Alfven waves. The main nonlinear effect in the collisional interchange instability is wave steepening whereas the inertial interchange instability supports inertial-range turbulence (Costa and Kelley, 1978; Hysell et al., 1994; Zargham and Seyler, 1989; Hysell and Shume, 2002).

III. Local irregularities

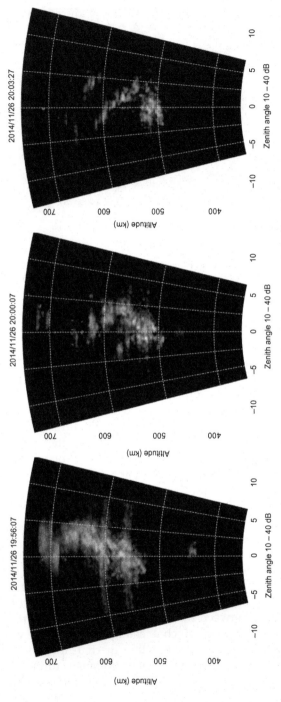

FIG. 2 In-beam radar images of some of the plumes in the complex shown in Fig. 1 around 20 LT.

III. Local irregularities

A number of investigators analyzed how persistent shear flow in the bottomside F-region around twilight might reduce the growth rate of the interchange instability and shift the fastest-growing eigenmode to longer wavelengths (Perkins and Doles III, 1975; Satyanarayana et al., 1984, 1987; Rappaport, 1998; Sekar and Kelley, 1998; Shukla and Rahman, 1998). However, first (Fu et al., 1986) and later (Flaherty et al., 1999) challenged those findings, pointing out the deficiencies of boundary-value analysis in the context of sheared flows and arguing that large-amplitude, fast-growing, broadband transients should predominate in sheared bottomside flows for a time.

Hysell and Kudeki (2004) then considered explicitly whether the shear flow itself could be destabilizing. Reworking an analysis of electrostatic Kelvin-Helmholtz instabilities in polar cap arcs (Keskinen et al., 1988), they found a robust, collisional branch of the instability that could function in the bottomside. The fastest-growing eigenmode had a growth rate several times faster than the conventional collisional interchange instability and a wavelength given by $\lambda \approx 4\pi L$ where L is the vertical length scale of the shear. In practice, this implies a preference for waves with horizontal wavelengths of about 150–200 km. This is the asymptotic behavior of the instability. The mode shapes are like plane waves but confined in a narrow altitude layer and propagating in a direction intermediate between vertical and horizontal. In numerical simulations, an even faster-growing transient response was found to have a preferred wavelength of the order of L. The mechanism, termed "collisional shear instability," does not rely on ion inertia and is really just another variant of $\mathbf{E} \times \mathbf{B}$ instability driven by the vertical currents that inevitably arise in the bottomside when shear flow is present.

Hysell and Kudeki (2004) focused on the asymptotic form of the instability. Here, the transient response will be examined. A simplified version of the model they considered for ionospheric perturbations in a vertical shear flow including the effects of Pedersen and wind-driven currents but with polarization currents neglected is

$$(\omega - kv)n_1 = \frac{k}{B}\frac{dn_0}{dz}\phi_1 \tag{1}$$

$$\frac{d}{dz}\left(n_0\nu_{\text{in}}\frac{d\phi_1}{dz}\right) - n_0\nu_{\text{in}}k^2\phi_1 = -B\frac{d}{dz}(\nu_{\text{in}}(u-v)n_1) \tag{2}$$

where $n(z,\omega)$ and $\phi(z,\omega)$ are the electron number density and electrostatic potential, respectively, $\nu_{\text{in}}(z)$ is the ion-neutral collision frequency, $B(z)$ is the magnetic induction, $u(z)$ is the zonal wind speed, and $v(z)$ is the zonal plasma drift speed. Perturbations are denoted by subscripts, and the perturbed state variables have been taken to be of the form $\{n_1(x,z,t),\phi_1(x,z,t)\} = \{n_1(z),\phi_1(z)\}\exp(ikx - i\omega t)$. Since inertia has been neglected in this model, there is no frequency dependence in Eq. (2) which derives from the equation for ion momentum. The frequency dependence in Eq. (1) can be identified with the time derivative in the fluid continuity equation.

A discretized version of the model equations can be expressed as

$$\dot{n}_1 = Bn_1 + C\phi_1 \tag{3}$$

$$D\phi_1 = En_1 \tag{4}$$

where n_1 and ϕ_1 are now column vectors and where B, C, D, and E are appropriately defined linear operators built from first- and second-derivative finite-difference operators (using Neumann boundary conditions in this case). Combining Eqs. (3), (4) yields a first-order linear system

$$\dot{n}_1 = (B + CD^{-1}E)n_1 = An_1 \tag{5}$$

Finding the eigenvalues and eigenvectors of A is routine. To gage the behavior of the transient response and the effect of nonnormality, the

ϵ-pseudospectrum of A can instead be calculated (Trefethen and Embree, 2005). For a detailed discussion of the application of pseudospectral analysis to instabilities in space plasmas, see Flaherty et al. (1999).

The ϵ-pseudospectrum of A can be defined as

$$\Lambda_\epsilon(A) = \left\{ z \in \mathbb{C} : \| (zI - A)^{-1} \| \geq \epsilon^{-1} \right\} \quad (6)$$

which is the space of complex numbers z for which the norm of $(zI-A)^{-1}$ exceeds the reciprocal of a small parameter ϵ. If z is an eigenvalue of A, then $\|(zI-A)^{-1}\|$ is defined as infinity. In the event A is normal, the pseudospectrum will appear as small closed contours of radius ϵ surrounding the eigenvalues. For nonnormal matrices, however, the pseudospectrum can extend well outside of the eigenvalues. This is an indication of nonnormal, "overlapping" eigenfunctions.

Consider the matrix B whose columns are the eigenvectors of A. If the initial perturbations to the system are n_1, then the amplitudes of the eigenvectors b satisfy $Bb = n_1$ or $b = B^{-1}n_1$ and are bound by $\|b\| \leq \|B^{-1}\| \|n\|$. Nonnormality implies that B will be poorly conditioned implies that the eigenmode amplitudes may have to be very large to satisfy the initial conditions even though the n_1 themselves are small. The eigenmodes at first interfere destructively. As the system evolves, however, the interference can become constructive and dominate the overall behavior of the system rather than the growth or decay of any one mode. Only when the transient decays will the asymptotic response found through boundary-value analysis dominate.

Another interpretation of the ϵ-pseudospectrum is that it indicates the set of eigenvalues corresponding to all perturbed systems $A + A_1$ such that $\|A_1\| \leq \epsilon$. Contours extending far from the eigenvalues denote a system with eigenvalues that are highly sensitive to small perturbations. Such sensitivity would make it difficult to predict the behavior of a system on the basis of a single eigenvalue.

For simplicity, we consider the altitude range between 300–400 km and take $v(z) = 90 \tanh((z-350)/L)$ m s^{-1}, $\nu_{in}(z) = 0.1 \exp(-(z-300)/L)$ s^{-1}, and $\log(n(z)) = 11 + \tanh((z-300)/L)$ in MKS units, m^{-3} and also $u = 100$ m s^{-1}. Here, L is the vertical length scale of the problem which is taken to be 25 km. The system is discretized with dimension 100.

The ϵ-pseudospectrum for the case of a 20-km horizontal wavelength perturbation is shown in Fig. 3. A number of unstable eigenvalues are present for this case. The real part of the eigenvalues are significant although about an order of magnitude smaller than the case of $\lambda = 4\pi L$. The figure also shows clear evidence of nonnormality where the ϵ contours extend several times ϵ farther away from the unstable eigenvalues. A rough estimate of the transient amplitude amplification factor is given by the maximum abscissa of an ϵ contour divided by ϵ. In this case, the $\epsilon = 2 \times 10^{-4}$ contour crosses 0.01 implying an amplification factor of more than 50. This can be large enough for the transient response to dominate during the initiation of ESF.

The most comprehensive theoretical analysis of the instability comes through numerical simulation of the full initial boundary value problem. A regional, three-dimensional (3D) numerical model designed for this purpose was described by Hysell et al. (2015). The finite-volume model consists of conservation equations for the number density and momentum of O$_2^+$, NO$^+$, O$^+$, and H$^+$ ions and electrons together with the quasineutrality condition. The potential is solved completely in three spatial dimensions. Neutral atmospheric parameters are taken from the NRLMSISE-00 and HWM14 models (Picone et al., 2002; Drob et al., 2015).

The finite-volume model can be initialized and forced using incoherent scatter measurements from Jicamarca. It is initialized at time t_0 with plasma number densities taken from the SAMI2 model (Huba et al., 2000). SAMI2

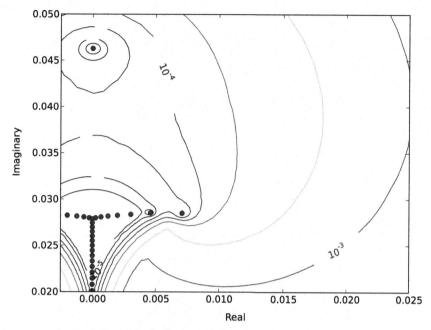

FIG. 3 Pseudospectrum for instability in sheared bottomside F-region flow. The *horizontal and vertical axes* represent the real and imaginary parts of z, respectively. Contours for ϵ follow a 1:2:5 progression with the outermost contour being 1×10^{-3}. *Plotter symbols* denote eigenvalues. The right-half plane of the pseudospectrum corresponds to growing solutions.

incorporates the Scherliess and Fejer (1999) ionospheric electric field model. The electric field model is scaled to maximize congruity between the SAMI2 results and Jicamarca density profiles measured at t_0. Likewise, the HWM-14 model winds are scaled for maximum congruity between the zonal plasma drift profiles predicted by the finite-volume model and measured at Jicamarca at t_0. Finally, for times greater than t_0, zonal electric fields measured at Jicamarca are continuously imported and imposed in the finite-volume model. Finally, the model output can be compared with coherent scatter observations of ESF from Jicamarca for validation.

Results for a model run for November 26, 2014 and for $t_0 = 1830$ LT are shown in Fig. 4. The left (right) panel shows conditions for 1910 LT (1950 LT). At 1910 LT, a bottom-type layer was on the verge of appearing at about 350-km altitude over Jicamarca. The simulation shows a vortex in the bottomside accompanied by strong vertical current density. Vertical current must be supplied to the F-region due to the quasineutrality condition and the imperfectly efficient F-region dynamo. The vertical currents cause intermediate-scale irregularities to form where the valley region meets the bottomside, where the plasma drifts and neutral winds differ the most. The irregularities have westward-tilting wavefronts and a horizontal wavelength of about 30 km. They represent the transient response of the system to the shear flow and are the first irregularities to emerge. Bottom-type scattering layers are the signature of the penetration of the transient response upward into denser bottomside strata.

Bottom-type layers are a necessary but not sufficient condition for ESF plumes. The transient response is not long lived and cannot extend far outside the shear-flow region. The transient can evolve into conventional

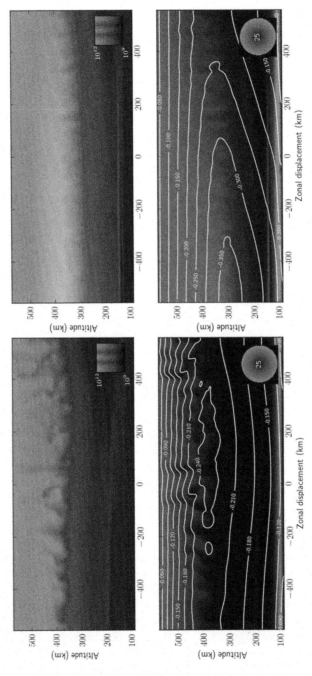

FIG. 4 Numerical simulations of the November 26, 2014 ESF event. The *left and right columns* show conditions at 1910 LT and 1950 LT, respectively. The *upper panels* show plasma number density and composition, with molecular ion, atomic ion, and protons dominating in the valley region, F region, and topside respectively. The *lower panels* show current density in the plane perpendicular to B with a maximum scale of 25 nA m^{-2}. Contours are equipotential curves.

III. Local irregularities

collisional interchange instability, however. If the zonal electric field is sufficiently strong, the irregularities can grow and expand until zonal gravitational currents become important and sustain further growth. Only in this case do topside ESF plumes emerge. By 1950 LT, the plumes in the simulation had reached the upper boundary and could grow no further. This is when plumes in the topside began passing over Jicamarca.

Overall, the depletions in Fig. 4 resemble the radar plumes in Fig. 2. Both have reverse "C" shapes, exhibit multiple bifurcations, and tend to become narrow and pinched in their midsections. The widths and separation between depletions and overall rates of development are also comparable to what is seen in nature. Realistic initialization and forcing of the finite-volume model together with a 3D treatment of the electrostatic potential, allowances for vertical, wind-driven current, and incorporation of the transient response are key for reproducing accurate bottomside flows and recovering essential ESF phenomenology.

4 Instability in sporadic-E-layers

In ESF, nonequipotential magnetic field lines contributed to enhanced bottomside shear flow and the tendency for instability. Finite electrostatic potential variations along the magnetic field may have even more important effects at middle latitudes where B-field lines cut directly through ionospheric strata. There, a class of drift-wave instability is likely responsible for kilometer-scale plasma density irregularities in patchy sporadic E-layers (E_s), the most prevalent irregularities found at middle latitudes.

Thin, dense, sporadic layers of metallic ions in the mid-latitude E-region have been known about since the pioneering days of radio. It is now well known that the layers often present as patchy rolls or fronts (Pan et al., 1994; Hysell et al., 2002; Saito et al., 2006). Images from coherent-scatter radars show that the fronts propagate mainly equatorward and westward and have wavelengths of tens of kilometers and periods of 5–10 min. Electric fields entrained in the fronts can be strong enough to drive Farley-Buneman (FB) instability (Haldoupis and Schlegel, 1994; Schlegel and Haldoupis, 1994). Even absent FB instability, however, the layers are replete with meter-scale irregularities visible to radar. The fronts themselves have been attributed to neutral dynamic instability (Larsen, 2000; Bernhardt, 2002; Larsen et al., 2007; Hysell et al., 2012), gravity waves (Woodman et al., 1991; Didebulidze and Lomidze, 2010; Chu et al., 2011), and an E-region variant of the plasma instability described by Perkins (1973) and Cosgrove and Tsunoda (2002, 2004).

The patchy E_s-layers are moreover home to irregularities with scale sizes of the order of 1 km. Barnes (1992) first detected these irregularities in HF soundings. They were observed subsequently in data from a sounding rocket that passed through a patchy E_s layer (Kelley et al., 1995) and in VHF radio scintillations (Maruyama et al., 2000). Bernhardt et al. (2003) observed kilometric structures in airglow imagery of sporadic E-layers in ionospheric heating experiments at Arecibo.

Finally, kilometer-scale irregularities were clearly depicted in ISR images of E_s layers over Arecibo (Hysell et al., 2013b). Imagery of that event is reproduced in Fig. 5. The irregularities in question are most evident in the topmost layer. The horizontal drift speed of the layer was determined to be about $70 \, \text{m s}^{-1}$ on the basis of Doppler beam-swinging information. The 30–40 s periodicity of the structuring therefore corresponds to a wavelength of about 2–3 km.

Kilometer-scale structuring is too fine to be explained by the candidate mechanisms for the roll-like fronts listed earlier but can be produced by ionospheric drift-wave instabilities. Drift-waves rely on finite variations along B.

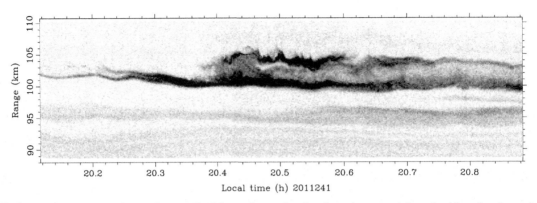

FIG. 5 High-resolution imagery of a sporadic E-layer observed at Arecibo using a special maximal-length pulse code.

Electrons stream along magnetic field lines to preserve quasineutrality but overshoot, causing instability. The overshoot can occur in the inertial or the collisional regime. In the collisional regime in the mid- or high-latitude F-region, the mechanism is a generalized form of $\mathbf{E} \times \mathbf{B}$ instability called current convective instability. In the E-region, it is like a generalized form of gradient drift instability called resistive drift instability (Hysell et al., 2013b).

The E-region instability can be understood using a two-fluid model. The linearized continuity equation for the streaming electrons is

$$\frac{\partial n_1}{\partial t} = -\mathbf{v}_{e1} \cdot \nabla_\perp n_0 - \mathbf{v}_{e0} \cdot \nabla_\perp n_1 - \nabla_\perp \cdot \mathbf{v}_{e1} n_0$$

$$+ \nabla_\parallel (n_0 \mu_{e\parallel} \nabla_\parallel \phi_1) \tag{7}$$

where n is number density, $\mathbf{v}$ is velocity, ϕ is the electrostatic potential, μ is the mobility, and where the subscripts 0 and 1 denote background and perturbed quantities, respectively, such that $n = n_0 + n_1 + \cdots$, $\mathbf{v}_e = \mathbf{v}_{eo} + \mathbf{v}_{e1} + \cdots$, and $\mathbf{E} \approx \mathbf{E}_o - \nabla\phi_1 + \cdots$ to first order in the perturbations. The perpendicular ($\perp$) and parallel ($\parallel$) subscripts are with respect to the background magnetic field.

The ingredients for all the familiar E-region plasma instabilities are contained in Eq. (7). The first term to the right side of the equation is critical for gradient drift instability, and the

second for FB instability. The last term can be stabilizing when streaming electrons merely "short out" electrostatic fluctuations with finite parallel wave number components. However, it can also be destabilizing when the density and potential fluctuations are in phase (note that $\mu_{e\parallel} < 0$).

The current density carried by electrons and ions is

$$\mathbf{J} = ne \Big(\mu_\perp \mathbf{E}_\perp + \mu_\parallel \mathbf{E}_\parallel + \mu_\times \hat{b} \times \mathbf{E} - D_\perp \nabla_\perp$$

$$\ln n - D_\parallel \nabla_\parallel \ln n \Big) \tag{8}$$

Here, D and μ are the plasma diffusivity and mobility and $\mu_\times$ is the plasma Hall mobility. Taking $\nabla \cdot \mathbf{J} = 0$ (the quasineutrality condition) and linearizing then gives

$$\phi_1 = \frac{-i\mathbf{k}_\perp \cdot \mathbf{E}_0 \mu_\perp + i(\mathbf{k}_\perp \times \mathbf{E}_0) \cdot \hat{b}\mu_\times - D_\perp k_\perp^2 - D_\parallel k_\parallel^2}{\mu_\perp k_\perp^2 + \mu_\parallel k_\parallel^2 - ik_\parallel \mu_\parallel / L} \frac{n_1}{n_0} \tag{9}$$

where L is the parallel conductivity parallel gradient length scale and where all other spatial variations in the transport coefficients and in n_0 have been neglected. If $L \to \infty$, then the diffusion terms in Eq. (9) cause the density and potential fluctuations to be out of phase, leading to damping and stability. The electric-field terms push the phase relationship toward ± 90 degrees and so do not directly affect stability through the rightmost term in Eq. (7). For finite L, however, the density and potential fluctuations can have an in-phase component. The effect is greatest when the two real terms in the denominator of Eq. (9) match, that is, $\mu_\perp k_\perp^2 = \mu_\parallel k_\parallel^2$.

To develop a dispersion relation for drift waves, the equation for ion momentum is required. Assuming plane-wave solutions for the waves, the ions obey

$$\left(-i\omega + i\mathbf{k}_\perp \cdot \mathbf{E}_o \mu_{\perp i} - i(\mathbf{k}_\perp \times \mathbf{E}_o) \cdot \hat{b}\mu_{\times i} + D_{\perp i} k_\perp^2 \right.$$
$$\left. + D_{\parallel i} k_\parallel^2 \right) n_1 + n_o \mu_{\perp i} k_\perp^2 \phi_1 = 0 \qquad (10)$$

which neglects parallel ion mobility but not parallel ion diffusivity. The unperturbed electric field $\mathbf{E}_o$ represents the sum of the background field in the ionosphere and the polarization field in the patchy E_s layer. Terms associated with transverse density gradients in the patchy layers have been neglected for simplicity.

Combining Eq. (10) with Eq. (9) yields a dispersion relation for collisional drift waves

$$-i\omega + i\mathbf{k}_\perp \cdot \mathbf{E}_o\mu_{\perp i} - i(\mathbf{k}_\perp \times \mathbf{E}_o) \cdot \hat{b}\mu_{\times i} + D_{\perp i}k_\perp^2 + D_{\parallel i}k_\parallel^2$$
$$+ \frac{\mu_{\perp i}}{\mu_\perp} \frac{-i\mathbf{k}_\perp \cdot \mathbf{E}_o\mu_\perp + i(\mathbf{k}_\perp \times \mathbf{E}_o) \cdot \hat{b}\mu_\times - D_\perp k_\perp^2 - D_\parallel k_\parallel^2}{2 - i/k_\parallel L} = 0$$
$$(11)$$

The growth rate for the instability is controlled by the real part of the quotient in Eq. (11) which is maximized when $k_\parallel L = 1/2$, setting the preferred scale size for instability. Finally, separating the real from the imaginary parts of Eq. (11) gives expressions for the frequency and growth rate of collisional drift waves

$$\omega_r = \frac{3}{4}\mu_{\perp i}\mathbf{k}_\perp \cdot \mathbf{E}_o + \left(\frac{1}{4}\mu_{\perp i}\frac{\mu_\times}{\mu_\perp} - \mu_{\times i} \right)$$
$$(\mathbf{k}_\perp \times \mathbf{E}_o) \cdot \hat{b} - \frac{1}{4}\frac{\mu_{\perp i}}{\mu_\perp}\left(k_\perp^2 D_\perp + k_\parallel^2 D_\parallel \right) \qquad (12)$$

$$\gamma = \frac{1}{4}\mu_{\perp i}\left[-\mathbf{k}_\perp \cdot \mathbf{E}_o + \frac{\mu_\times}{\mu_\perp}(\mathbf{k}_\perp \times \mathbf{E}_o) \cdot \hat{b} \right]$$
$$-D_{\perp a}k_\perp^2 - D_{\parallel a}k_\parallel^2 \qquad (13)$$

where it should be emphasized that $k_\parallel$ and $k_\perp$ are fixed here by L and the related conditions imposed earlier. According to Eq. (13), the fastest-growing waves propagate in nearly the $\mathbf{E} \times \mathbf{B}$ direction in ionized layers at altitudes where the Hall mobility is significant. The ratio $\mu_{\perp i}\mu_\times/\mu_\perp$ has a wide maximum between 100 and 110 km at mid-latitudes. The phase velocity will be approximately in that direction of but much smaller than the $\mathbf{E} \times \mathbf{B}$ drift speed. Meanwhile, the dominant wavelength will be determined by L and of the order of a few kilometers, depending on the layer altitude and vertical depth. The growth time for the waves will be of the order of 1 min, depending on the wave number and magnitude of $\mathbf{E}_o$.

The results of a 3D numerical simulation of a mid-latitude sporadic E-layer blob are shown in Fig. 6. The simulation was described in detail by Hysell et al. (2013b). The figure shows three planar cuts through the simulation volume 3 min after initialization at 1915 LT. Longitudinal waves can be seen throughout the layer, propagating in the direction nearly opposite the Hall current, as anticipated. The waves follow the S-shaped bend in the transverse current. The wavelengths of the waves vary spatially but are nearly 1 km in most instances. The wavelength, propagation direction, and growth rate are consistent with the linear analysis performed previously. Overall, the simulation replicated the irregularities observed over Arecibo reasonably closely (Hysell et al., 2013b).

5 Farley-Buneman waves

Modified two-stream waves were among the first unstable plasma waves detected in the ionosphere (Eckersley, 1937; Bowles, 1954). The small-scale waves extract free energy from Hall-drifting E-region electrons moving at supersonic speeds. The waves are electrostatic and strongly field aligned. Farley (1963) derived an implicit dispersion relation for the waves

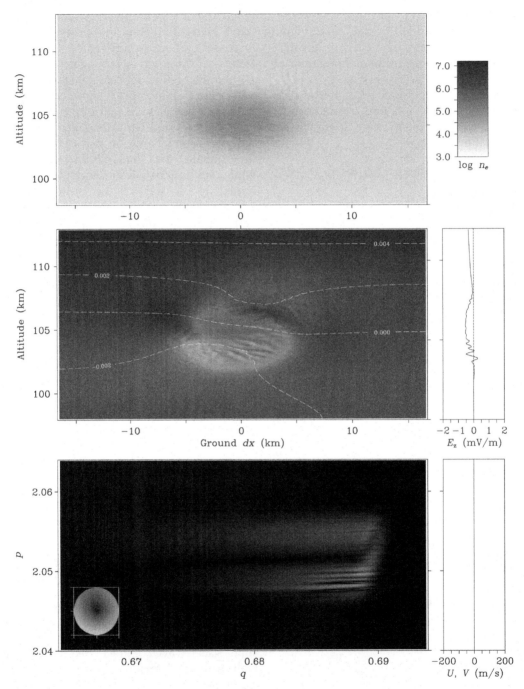

FIG. 6 Numerical simulation of instability in mid-latitude sporadic-E-layer cloud. *Top panel*: plasma number density in a plane perpendicular-to-B through the layer. *Middle panel*: transverse current density in the plane perpendicular-to-B with superimposed, perturbation equipotential contours. The *line plot to the right* shows the vertical/meridional electric field component. *Bottom panel*: meridional current density in the meridional plane. The legend for the current densities is given by the color wheel. Maximum scale for the transverse (meridional) current densities is 20 (200) nA m^{-2}. The *line plot to the right* shows the assumed zonal wind profiles (zero here).

using plasma kinetic theory, showing how the short-wavelength cutoff is established by Landau damping. Buneman (1963) derived an approximate, explicit dispersion relation for the waves using fluid theory. The waves are now normally referred to as FB waves. They have been studied in the intervening years with coherent scatter radars, sounding rockets, and recently with the RAX CubeSats (Bahcivan et al., 2014). They have been found in the auroral, equatorial, and mid-latitude E-region ionospheres and are also believed to exist in the Sun's partially ionized chromosphere where they contribute to heating in the transition region (Madsen et al., 2013). FB waves are important because they alter plasma temperature and conductivity. They also facilitate incisive ionospheric diagnostics. For reviews of experimental work on FB waves, see Haldoupis (1989), Sahr and Fejer (1996), Bahcivan et al. (2005), and Hysell (2016).

FB waves are meter-scale electrostatic waves excited by strong Hall currents in weakly ionized plasmas. When perturbations in the electron density convert past unmagnetized ions in the E-region, the ions see the associated perturbations in the electrostatic field. Ion inertia causes them to drift into (out of) locally enhanced (depleted) plasma regions. Where inertia is able to overcome diffusion, instability produces field-aligned waves which are not shorted out by field-aligned currents. Instability requires the convection speed to exceed the ion-acoustic speed.

Fig. 7 shows radar imagery of 30-MHz coherent radar backscatter from FB waves in the auroral zone. The colored pixels in the images convey information about the signal-to-noise ratio,

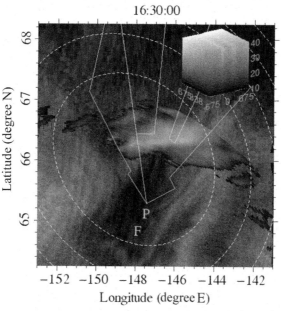

16:30:00

FIG. 7   Representative imagery of FB waves observed by a coherent scatter radar in Homer, Alaska. The brightness of the pixels in the image represent the signal-to-noise ratio from 10 to 40 dB. The hue represents the Doppler shift on a scale from $\pm 675 \text{ m s}^{-1}$. The saturation represents the spectral width on a scale from 0 to 675 m s^{-1} (see the color cube for legend). "P" and "F" show the locations of Poker Flat and Fairbanks. The incoherent integration time for the radar echoes was 3 s. *Gray scales* depict white-light auroral imagery from Poker Flat. *Courtesy Robert Michell.*

Doppler shift, and spectral width of the echoes. The radar echoes are superimposed on white-light auroral imagery. Both the radar and the optical aurora evolve rapidly in time. In this case, gross features in both drifted from west to east. As a rule, radar echoes and optical auroral forms tend to occur exclusively, with the strongest radar echoes coming from voids in the optical emissions. Echoes are only observed during periods of geomagnetic activity. Because the echoes are finely resolved telltales of strong convection, they can serve as incisive diagnostics during substorms. However, precipitation and absorption associated with very strong activity can also suppress the radar echoes.

The gap between theory and observations of FB waves remains wide, however. Most theoretical analyses in the literature have been based on linear, local fluid theory which is tractable but limited. Fluid theory neglects potentially important kinetic physics, although it can be extended to capture approximately the effects of nonisothermal fluids as well as certain destabilizing thermal effects (St.-Maurice et al., 2003; Kagan and St.-Maurice, 2004; Dimant and Oppenheim, 2004). While some work has been done on the theory of FB waves incorporating the effects of finite variations of background state variables along the magnetic field (Bahcivan and Cosgrove, 2010), the transient behavior has not been considered.

Most importantly, the FB waves observed in the auroral zone especially are usually driven well beyond threshold (i.e., by convection speeds much greater than the ion-acoustic speed). They exist in a quasiequilibrium saturated state, calling into question the validity of any linear theory. Quasilinear theories introducing anomalous corrections to linear theory have been developed with some degree of success (Sudan, 1983; Dimant and Oppenheim, 2011; Rojas et al., 2016). Attempts to salvage linear theory with the incorporation of some heuristics have also been made with some degree of

success (Dimant and Milikh, 2003; Milikh and Dimant, 2003; Hysell et al., 2013a). The "marginal stability" concept in particular has been useful in reconciling linear theory with observations. This assumes that conditions are maintained such that the linear growth rate for instability is always close to zero. An accurate prediction of marginal stability theory is that the phase speeds of FB waves should always be close to the ion-acoustic speed.

The state of the art in numerical work on FB waves is represented by the massively parallel initial-value PIC simulations from Oppenheim and Dimant (2013). These simulations predict that FB waves propagate at speeds close to the nonisothermal ion-acoustic speed, propagate at angles offset from the main convection direction due to certain thermal effects, and cause significant electron heating.

Another prediction of the numerical simulations is that both the Doppler shift and spectral width of radar echoes should be monotonically increasing functions of the convection speed. Moreover, the Doppler shift (spectral width) should be monotonically decreasing (increasing) functions of the flow angle, the angle between the convection velocity and the radar wavevector, with a small offset due to thermodynamic effects. The functional dependencies of Doppler shift and spectral width on flow angle are approximately cosinusoidal and sinusoidal, respectively. Experimental support for these theoretical predictions came from radar observations made during the JOULE I and JOULE II sounding rocket flights from Poker Flat, Alaska (Hysell et al., 2008).

The behavior of the Doppler spectrum of FB waves can be understood with the help of two-fluid analysis. While the fluid equations used at work here are linear, the energy analysis which follows is necessarily nonlinear (and yet still tractable with some simplifying assumptions). The analysis will also invoke the coexistence of primary and secondary waves, another heuristic concept.

Combining the linearized momentum and continuity equations for magnetized electrons and unmagnetized ions and assuming perturbations of the form $\exp(-i\mathbf{k} \cdot \mathbf{x})$ leads to an equation for the linear evolution of the FB-wave modal amplitude $n_\mathbf{k} \equiv n(\mathbf{k}, t)$

$$\left(\frac{\partial}{\partial t} + i \frac{\mathbf{k} \cdot \mathbf{V}_d}{1 + \psi} \right) n_\mathbf{k} = -\frac{1}{1 + \psi} \left\{ \frac{\psi}{\nu_i} \left(\frac{\partial^2}{\partial t^2} + k^2 C_s^2 \right) \right. $$

$$\left. + i \frac{\mathbf{k} \cdot \boldsymbol{\kappa} \, \nu_i}{k^2} \frac{\partial}{\Omega_i \, \partial t} \right\} n_\mathbf{k} \tag{14}$$

$$\boldsymbol{\kappa} \equiv \frac{\nabla n_o}{n_o} \times \hat{b} \tag{15}$$

where $\mathbf{V}_d$ is the background electron convection velocity, ψ is the anisotropy factor, C_s is the ion-acoustic speed, ν_i is the ion-neutral collision frequency, and n_o is the background plasma number density. The term involving κ defined in Eq. (15) is associated with destabilization due to gradient-drift effects. For details regarding the derivation of Eq. (14), see Hamza and St-Maurice (1993).

An equation for the modal energy can be derived next by multiplying Eq. (14) by $n_\mathbf{k}^*$ and retaining the real part:

$$\frac{\partial}{\partial t} |n_\mathbf{k}|^2 = -\frac{2}{1 + \psi} \left\{ \frac{\psi}{\nu_i} \Re \left(n_\mathbf{k}^* \frac{\partial^2}{\partial t^2} n_\mathbf{k} + k^2 C_s^2 |n_\mathbf{k}|^2 \right) \right.$$

$$\left. - \Im \left(\frac{\mathbf{k} \cdot \boldsymbol{\kappa} \, \nu_i}{k^2} \frac{1}{\Omega_i} n_\mathbf{k}^* \frac{\partial}{\partial t} n_\mathbf{k} \right) \right\} \tag{16}$$

where Ω_i is the ion gyrofrequency. The equation describes how wave energy grows from free energy in the background. The free energy resides both in the streaming electrons, traceable to the second time derivative term, and in the plasma inhomogeneity, represented by the κ term. At this point, and on the basis of the empirical success of the model, the condition of marginal stability may be imposed for purposes of simplification. This means that the terms on the right side of Eq. (16) are taken to cancel approximately so that the left side may be set to zero.

A given wave mode can be decomposed into its Doppler spectral components according to $n_\mathbf{k}(t) = \sum_\omega n_{\mathbf{k},\omega} e^{-i\omega t}$ with $|n_{\mathbf{k},\omega}|^2$ corresponding to the power in the frequency bin ω for the scattering wavevector $\mathbf{k}$. This is identically what a backscatter radar measures and the quantity of interest here. Substituting this into Eq. (16) (with the left side set to zero) produces a balance law for the individual spectral components

$$\left\{ \frac{\psi}{\nu_i} (\omega^2 - k^2 C_s^2) + \frac{\mathbf{k} \cdot \boldsymbol{\kappa} \, \nu_i}{k^2} \frac{\omega}{\Omega_i} \right\} |n_{\mathbf{k},\omega}|^2 = 0 \tag{17}$$

Of greatest experimental interest are the gross spectral characteristics—the Doppler shift $\overline{\omega}$ and spectral width (squared) $\overline{\delta\omega}^2 = \overline{\omega^2} - \overline{\omega}^2$. Explicitly, these are defined by the moments

$$\overline{\omega^n} = \sum_\omega \omega^n |n_{\mathbf{k},\omega}|^2 / \sum_\omega |n_{\mathbf{k},\omega}|^2$$

If the gradient drift term is neglected for the moment, then Eq. (17) becomes (after summing over all frequencies)

$$\overline{\delta\omega}^2 + \overline{\omega}^2 = k^2 C_s^2$$

showing how the Doppler shift and spectral width of the echoes might be expected to combine in the Pythagorean sense to form C_s and therefore to be bound by C_s (Hamza and St-Maurice, 1993). This is not surprising; the marginal stability condition inevitably ties the spectral moments to the ion-acoustic speed.

It is possible to proceed further with this analysis by considering that coherent scatter is observed along all flow angles and so is best interpreted as coming from secondary FB waves coexisting with a primary wave (which can only be observed at zero flow angle). Secondary waves can been seen propagating in opposing channels in the crests and troughs of primary FB waves in the Oppenheim and Dimant (2013) PIC simulations. One might expect spectral broadening to accompany FB-wave observations as flow angle increases. Density variations due to the primary wave would affect the secondary waves through the heretofore neglected gradient drift term. This can allow secondary-wave phases speeds to saturate at something less than C_s.

In this context, $\mathbf{k}$ can be interpreted as the wavevector of a secondary wave propagating at a flow angle θ so that $\mathbf{k} \cdot \boldsymbol{\kappa} = k\kappa \sin\theta$. Next, consider the moment

$$\overline{\kappa\omega} = \sum_\omega \kappa\omega |n_{\mathbf{k},\omega}|^2 / \sum_\omega |n_{\mathbf{k},\omega}|^2$$

One can expect $\kappa\omega$ and $|n_{\mathbf{k},\omega}|^2$ to be correlated since the gradient drift mechanism excites (damps) modes for which $\boldsymbol{\kappa} \cdot \mathbf{k}\omega$ is positive (negative). A priori, the degree of correlation is unknown, but the correlation coefficient may be defined to be $\mathcal{C} \equiv \overline{\kappa\omega}/\kappa_{\mathrm{rms}}\delta\omega_{\mathrm{rms}}$. Making use of this definition and retaining the gradient drift term in Eq. (17) then gives

$$\delta\omega_{\mathrm{rms}}^2 + \overline{\omega}^2 + \frac{\nu_i \Omega_e}{\nu_e} \frac{\mathcal{C}}{kL_{\mathrm{rms}}} \sin\theta \delta\omega_{\mathrm{rms}} = k^2 C_s^2$$

Finally, assuming a cosinusoidal dependence of the Doppler shift on flow angle, that is, $\overline{\omega} = kC_s \cos\theta$ in accordance with the Oppenheim and Dimant (2013) PIC simulations and substituting this into Eq. (18) gives an expression for the spectral width of the echoes

$$\delta\omega_{\mathrm{rms}} = |\sin\theta| \left(\sqrt{k^2 C_s^2 + \Gamma^2} - \Gamma \right)$$

$$\Gamma \equiv \frac{\nu_i \Omega_e}{2\nu_e} \frac{\mathcal{C}}{kL_{\mathrm{rms}}}$$

(18)

where $L_{\mathrm{rms}} \equiv \kappa_{\mathrm{rms}}^{-1}$. This predicts that the spectral width should have a sinusoidal dependence on flow angle and a magnitude that is related to but less than the ion-acoustic speed. In the JOULE I and JOULE II sounding rocket experiments, the relationship was found to be $\delta\omega_{\mathrm{rms}} \approx (1/2)kC_s|\sin\theta|$.

The value of C_s meanwhile depends on the temperature and on the electron and ion ratios of specific heat. The temperature is elevated by FB-wave heating, and the ratios of specific heat

depend on wave frequency. Consequently, the ion-acoustic speed depends on the convection speed. It also varies with altitude, and so radar measurements of FB waves need to be interpreted as an average over an altitude kernel. Hysell et al. (2013a) estimated how the average ion-acoustic speed varies with convection speed, recovering the empirical formula reported many years earlier by Nielsen and Schlegel (1983, 1985).

The preceding discussion implies a means of estimating the convection electric field from coherent backscatter from FB waves in the auroral zone. Since the Doppler shift and spectral width are determined by the convection speed and direction, the latter can be estimated from the former wherever a Doppler spectrum can be measured. Using radar imaging, finely resolved convection velocity estimates can be derived over a wide field of view. The short vectors in Fig. 8 show the results of this analysis for the data in Fig. 7.

The contours in Fig. 8 are equipotential curves which have been found through a global optimization process. The residuals to the fit are very small in a relative sense, and the fit is good. This means that the convection field inferred from the FB echoes is very nearly incompressible. This would be unlikely to occur by chance and argues that the interpretation of the spectral moments is reliable.

6 Artificial irregularities and ionospheric modification

Ionospheric irregularities can also be created artificially using high-power radio waves. The irregularities are generated by thermal parametric instabilities (Grach et al., 1978a; Das and Fejer, 1979; Fejer, 1979; Kuo and Lee, 1982; Dysthe et al., 1983; Mjølhus, 1990) and, upon entering the nonlinear regime, by resonance instability (Vas'kov and Gurevich, 1977;

16:30:00

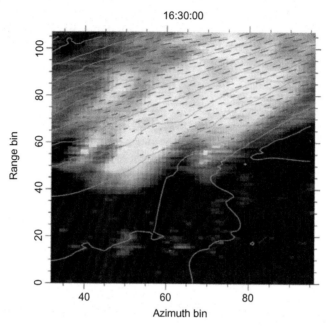

FIG. 8 Convection pattern derived from radar imagery in Fig. 7. *Short lines* are a sampling of vector velocities derived from individual spectra. *Contours* are equipotentials derived using a global optimization method. Results are shown in radar coordinates. Very fine spatiotemporal resolution is required for this to work since the relationship between the convection velocity and the spectral moments is nonlinear and so superposition does not hold.

Inhester et al., 1981; Grach et al., 1981; Dysthe et al., 1982; Lee and Kuo, 1983; Mjølhus, 1993). Thermal parametric instability (TPI) occurs when the electromagnetic O-mode pump wave undergoes linear mode conversion in the vicinity of nascent field-aligned plasma density irregularities into an electrostatic mode (mainly upper-hybrid waves) at the altitude where the pump frequency equals the upper-hybrid frequency. Inhomogeneous wave heating and thermal forcing cause the density irregularities to grow in amplitude, leading to instability. In the nonlinear regime, wave trapping occurs, leading to resonance instability. Whereas TPI only occurs when the pump-mode amplitude exceeds a specific threshold, resonance instability can be sustained with very low pump-mode power.

Like their natural counterparts, artificial field-aligned irregularities (AFAIs) cause coherent radar backscatter that can be detected by appropriately situated HF and VHF radars (e.g., Senior et al., 2004). The first coherent scatter observations were made at Platteville in Colorado using a range of probe wavelengths between 1 and 10 m (e.g., Minkoff and Kreppel, 1976; Haslett and Megill, 1974). Subsequent observations of AFAIs have been made in concert with heating experiments at Sura in Russia (e.g., Belenov et al., 1977), Arecibo (e.g., Coster et al., 1985), the EISCAT heater near Tromsø (e.g., Stubbe et al., 1982; Hedberg et al., 1983), SPEAR (e.g., Robinson et al., 2006), and at HAARP near Gakona, Alaska (e.g., Hughes et al., 2003). Radar observations of AFAIs have been used to estimate ionospheric convection, diffusivity, and conductivity and to monitor the propagation of MHD waves (e.g., Hysell et al., 1996; Sinitsin et al., 1999; Yampolski et al., 1997).

III. Local irregularities

The relative absence of day-to-day and quiet-time variation in the daytime E-region compared to the F-region together with the minimal impact of propagation effects make E-region AFAIs ideal for radar studies, and so they will be the focus here. Generating AFAIs in the E-region requires either very low pump frequencies or sporadic E-layers or auroral precipitation. Planning experiments around these phenomena is impractical, and the latter can attenuate both the pump and probe signals. Both phenomena furthermore negate the aforementioned consistency advantage of E-region experiments. The Platteville and Arecibo heaters operated low enough frequencies for generating E-region AFAIs under normal daytime conditions, as did the EISCAT facility in the early 1980s (Frey, 1986). HAARP can operate at frequencies comfortably below the E-region critical frequency in the summer daytime.

Below, the salient characteristics of E-region AFAIs are reviewed. These are demonstrated with data from a 30-MHz coherent scatter radar imager located in Homer, Alaska, which can observe AFAIs over HAARP.

6.1 Excitation threshold

The pump power necessary to excite TPI was calculated by Dysthe et al. (1983) who included magnetoionic, thermal, transport, and dissipative effects in their balance calculations (see also Grach et al., 1977; Das and Fejer, 1979; Dysthe et al., 1982). The result was intended for F-region applications and neglected two potentially important E-region effects. One is cooling due to inelastic electron-neutral collisions. The other is a correction accounting for the finite altitude extent of the ionospheric interaction region compared to the pump-mode wavelength, which is wider in the E-region where dissipation is greater. Like inelastic collisions, this effect is stabilizing.

Hysell et al. (2019) recently reformulated the Dysthe et al. (1983) calculations for application in the E-region. Whereas Dysthe et al. (1983) regarded the heated region as a point-source in altitude, Hysell et al. (2019) solved the appropriate boundary-value problem, finding the mode shapes of the perturbed density and temperature as a function of altitude and identifying the threshold electric field for instability which plays the role of the eigenvalue. The most important parameter controlling the threshold is the inelastic electron-neutral collision frequency which varies strongly with altitude.

Experimental validation of the threshold theory has been sought at HAARP where the pump-mode power can be varied gradually over time. When considering the threshold for instability, when pump power levels are small by definition, the linear theory outlined earlier is applicable. Experiments were performed where the pump-mode power increased in gradual, discrete steps lasting 10 s each up to a maximum and then decreased again back to zero. The pump frequency was 2.7 MHz. At this frequency, the maximum effective radiated power for O-mode zenith heating is 390 MW.

Representative results are presented for two heating cycles in Fig. 9, which shows the coherent backscatter signal-to-noise ratio measured at Homer. In this and several other examples recorded on the same day, following long heater-off intervals, coherent scatter appeared first 30 s into the heating cycle (here at 2040 UT) when the pump-mode power reached just 4% of full rated power. Using full-wave theory to account for magnetoionic effects and the effects of D-region absorption on the pump mode, Hysell and Nossa (2009) estimated the corresponding electric field at the upper-hybrid interaction height to be less than $200 \, \text{mV m}^{-1}$. In more precise experiments conducted in 2018, Hysell et al. (2019) measured the threshold electric field for instability as a function of pump-mode frequency (and therefore altitude), reporting a decrease from $83 \, \text{mV m}^{-1}$ at $2.75 \, \text{MHz}$ to $51 \, \text{mV m}^{-1}$ at $3.026 \, \text{MHz}$. These values were comparable to if somewhat greater

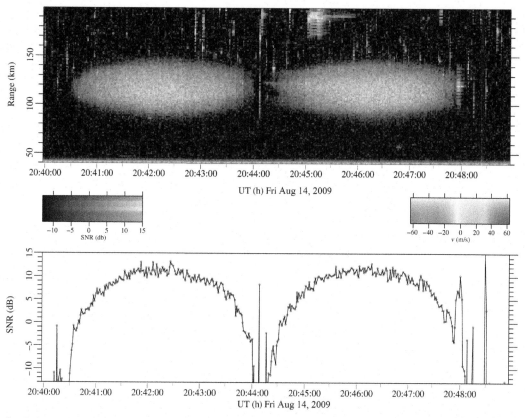

FIG. 9 Coherent backscatter from E-region AFAIs generated over HAARP observed near Homer, Alaska versus apparent range and time. The brightness, hue, and saturation of the pixels represent signal-to-noise ratio, Doppler shift, and spectral width. Pump-mode power was ramped gradually up and down during the experiment. A long pause in heating preceded the first heating cycle shown. The *bottom panel* of the figure shows the average signal-to-noise ratio in range gates between 90 and 140 km. Note that the echoes shown are range aliased and that the true range is the apparent range plus 370 km.

than the values predicted by the theoretical analysis described earlier. The discrepancy was attributed mainly to absorption which is difficult to account for fully.

6.2 Hysteresis and preconditioning

Fig. 9 illustrates two more remarkable features of E-region AFAIs. The first is hysteresis. The first heating cycle in the figure ended at 2044 UT. During the last 10 s of the cycle, the pump-mode power was just 1% of available power, the corresponding electric field being just 100 mV m^{-1}.

This was evidently enough to sustain the AFAIs once generated. The ability to sustain irregularities with pump power levels well below excitation threshold is evidence of nonlinear behavior and the resonance instability, which involves the trapping of upper-hybrid waves in density striations created by TPI. The time history of the radar signal-to-noise ratio furthermore is asymmetric, exhibiting higher values and a flatter shape from 2042 to 2044 during ramp-down than 2040 to 2042 during ramp-up.

Djuth et al. (1985) also reported that E-region irregularities could be sustained by a pump-mode electric field amplitude of only about $100\,\mathrm{mV\,m^{-1}}$. While their experiments were run with an ERP comparable to the HAARP experiment, they assumed greater classical absorption and also invoked anomalous absorption in evaluating the steady-state pump electric field amplitude at the upper-hybrid resonance height. Greater classical absorption was needed since theirs were postsunset experiments, relying on particle precipitation for E-layer ionization. Anomalous absorption was included using the formalism of Graham and Fejer (1976) with the assumption of fully developed irregularities.

Grach et al. (1978b) and Dysthe et al. (1982) anticipated such hysteresis effects associated with resonance instability and wave trapping. Hysteresis has been reported by Erukhimov et al. (1978), Jones et al. (1983), and Wright et al. (2006), although in an F-region context.

A related phenomenon illustrated by Fig. 9 is preconditioning. During the second heating cycle shown in the figure, irregularities emerged just 10 s after 2044, at the moment the heater power rose to the 1% level. The plasma striations necessary for resonance instability were therefore still present after a 10-s heating interruption. They were not present during the first heating cycle shown in Fig. 9, which followed a 2-min heating interruption.

6.3 Echo timescales

Homer radar AFAI observations made with HAARP operating at full power and processed with fine time resolution show that the AFAIs exhibit behavior on several characteristic timescales. The rise of backscatter power after heater turn-on and the fall after turn-off usually exhibit e-folding times close to 100 ms. Noble et al. (1987) reported echo rise and fall times of hundreds of milliseconds for so-called "type A" E-region AFAIs. Comparing observations at 21.4, 46.9, and 143.8 MHz, they proposed that the rise and fall times scaled with the probe

wavelength and the square of the probe wavelength, respectively. The former is expected for the TPI linear growth rate (Grach et al., 1978a) and the latter for transverse ambipolar diffusion. The $\sim\!100\,\mathrm{m\,s^{-1}}$ figure for the 30 MHz echoes falls between those reported for 46.9 and 143.8 MHz probe signals by Noble et al. (1987) and are also shorter than the decay times reported by Coster et al. (1985) for 50 MHz echoes at Arecibo, partially defying the aforementioned scaling laws. Variations may be due to the steep gradient in electron-neutral collisions with altitude and a strong sensitivity to heating frequency/altitude.

Noble et al. (1987) also observed AFAIs with rise times 10 or more times longer than the type A irregularities. These so-called "type B" events have also been observed with the Homer radar. Noble et al. (1987) associated type B events with gradient drift instability. If so, we would expect the Doppler shifts of the irregularities to vary from one side of the modified region to the other, behavior which is often observed over HAARP.

Finally, a third timescale associated with a systematic intensification and broadening of AFAIs over intervals of tens of seconds has also been seen in E-region modification experiments at HAARP. This phenomenon was first reported in experiments at Platteville, where it was tied to the "cold turn-on" of the heater (Frank, 1974) (see also Djuth et al., 1985). The cold turn on has been observed at Arecibo but only rarely (Coster et al., 1985). The long timescale of the cold turn-on suggests self-focusing, where irregularity formation permits the upper-hybrid matching condition to be met more easily over larger volumes.

6.4 Gyroharmonic effects

The key role of upper-hybrid waves in AFAI generation implies that unusual phenomena might occur close to electron gyroharmonic double-resonance frequencies, that is, where $\omega \sim \omega_{\mathrm{uh}} \sim n\Omega$. For E-region AFAIs, the only routinely accessible electron gyroresonance is the second gyroharmonic ($n = 2$). Ionospheric modifications at $n = 2$ differ fundamentally from the $n \geq 3$ cases (see, e.g., Grach, 1979 for basic theory). Mjølhus (1993) pointed out that wave trapping might be prohibited entirely at pump frequencies below the second electron gyroharmonic frequency, although the excitation threshold for TPI may also be reduced there (Grach, 1979). Experimentally, coherent scatter has been observed at pump frequencies near $2\Omega_e$, and enhancements have been reported at frequencies just above $2\Omega_e$ (Fialer, 1974; Minkoff et al., 1974; Kosch et al., 2007). Airglow intensification

also occurs at pump frequencies slightly above $2\Omega_e$ (Haslett and Megill, 1974; Djuth et al., 2005; Kosch et al., 2005, 2007).

Fig. 10 shows Homer radar coherent backscatter data when the O-mode pump frequency was varied between 2.9 and 3.1 MHz and back, a span that encompasses the second electron gyroharmonic frequency of about 3.025 MHz at 100 km altitude. (Note that the pump frequency approached or exceeded FoE at the upper extreme of the sweep, explaining why the echo intensity diminishes at the midpoints of the four heater cycles shown.) The figure illustrates three important aspects of E-region AFAIs: (1) Strong

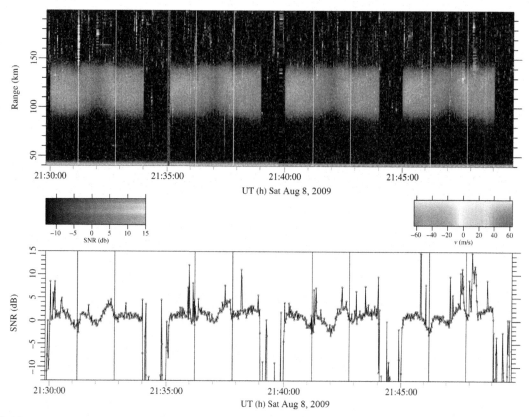

FIG. 10 Same as Fig. 2, only for experiments in which the pump power was constant but the pump frequency was varied. Each heating cycle depicted represents a pump-mode frequency sweep from 2.9 to 3.1 MHz and back to 2.9 MHz, with sweeping occurring at a uniform rate. *Vertical lines* depict instances where the pump frequency matched the second electron gyroharmonic frequency.

III. Local irregularities

irregularities are produced at frequencies both above and below the second electron gyroharmonic frequency. (2) There is a broad depression in echo intensity surrounding the second electron gyroharmonic frequency 10 s of kHz wide. (3) The depression is asymmetric, exhibiting a steeper shoulder on the high-frequency than the low-frequency side. The depression is due to the extinction of scatterers at the physical periphery of the modified region. Only marginally unstable regions of space appear to be affected by heating at the double-resonance frequency.

Mjølhus (1993) pointed out the necessity of both lower and upper cutoff frequencies in the dispersion relation for upper-hybrid waves for wave trapping to occur. The analysis predicted AFAI suppression at frequencies just below gyroharmonic frequencies and below $2\Omega_e$ entirely due to the absence of a lower cutoff frequency under those conditions. However, Hysell and Nossa (2009) showed that allowing for finite parallel wave numbers in the dispersion relation introduces a lower cutoff frequency below every gyroharmonic frequency including $2\Omega_e$. Consequently, there seems to be no theoretical prohibition against wave trapping below $2\Omega_e$ or near any gyroharmonic frequency. This is another example of the crucial destabilizing role played by finite parallel wave numbers in some ionospheric plasma instabilities.

What is the explanation for irregularity suppression near gyroharmonic frequencies? Hysell et al. (2010) argued that cyclotron damping should only occur in a neighborhood about 1-kHz wide around gyroharmonic frequencies, much narrower than what is actually observed even when certain beam broadening effects are taken into account. Another possibility is the mechanism suggested by Rao and Kaup (1990) whereby upper-hybrid waves are damped near gyroharmonic frequencies by mode conversion into electron Bernstein waves. Rao and Kaup (1990) discounted the viability of this mechanism near the second electron gyroharmonic

frequency, but the mechanism can be shown to be viable there once again by including the effects of finite parallel wave numbers.

6.5 Fine structure

During moderately disturbed conditions, the coherent echoes from E-region AFAIs often exhibit fine structure in intensity and Doppler shift. The fine structure can last for hours and does not appear to be a direct result of ionospheric modification, patterns in the fine structure seeming to persist across long gaps in the heating. Ionospheric modification can consequently be used to diagnose natural background ionospheric structuring.

Radar imagery constructed using aperture synthesis methods is shown in Fig. 11 for a heating experiment conducted under moderately disturbed conditions. The image shows a plan view of the AFAIs. Note how the echoes in the image are red-shifted (blue-shifted) on the northern (southern) side of the heater-modified volume. The line between red- and blue-shifted echoes is irregular and evolves over time. In the image shown, two intense, compact scattering regions with opposite Doppler shift occupy the northwest quadrant of the volume. Over time, these move southward, remaining intact throughout multiple on-off heating cycles. Features in successive images evolve gradually and systematically from image frame to image frame.

The imagery suggests clockwise circulation concentric with the heater-modified E-region volume. Such vorticity is inconsistent with the polarization of the thermally depleted volume by a background electric field. Such polarization would divert the flow within the volume but not cause vorticity (Hysell and Drexler, 2006). The circulation is consistent with electron diamagnetic drift around the thermally depleted volume. However, diamagnetic drifts should be present all the time and not just during disturbed periods. Another possibility is polarization associated with upward field-aligned current. The direction of the attendant polarization electric field in a depleted E-region volume would be inward, causing counterclockwise electron convection. This scenario is also consistent with precipitation-induced absorption.

The imagery furthermore suggests plasma irregularities created by natural plasma instabilities and highlighted by HAARP. One candidate is gradient drift instability operating on precipitation-induced plasma density gradients in the E-region in the presence of a background convection electric field. The instability is evidently not strong enough to produce meter-scale irregularities and VHF backscatter but strong enough to drive kilometer-scale irregularities. The features in Fig. 11 bear some resemblance to gradient drift waves and instabilities in the equatorial electrojet, only operating at longer scales (Hysell and Chau, 2006).

7 Summary and conclusions

This chapter has examined some of the most important plasma instabilities in the Earth's ionosphere. The review has not been comprehensive, and some key instabilities have been neglected. Among these are $\mathbf{E} \times \mathbf{B}$ and current-convective instabilities in the high-latitude F-region, instabilities related to medium-scale traveling ionospheric disturbances (MSTIDs) and spread F at middle latitudes, gradient-drift instabilities in the equatorial electrojet, and streaming instabilities underlying so-called 150-km echoes (Oppenheim and Dimant, 2016). Numerous instabilities occurring during ionospheric modification experiments are also absent here. Many of the themes touched upon here apply equally well in those contexts, however.

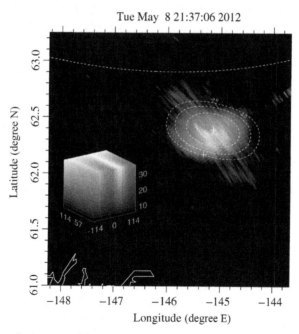

Tue May 8 21:37:06 2012

FIG. 11 Aperture synthesis radar imagery of E-region AFAIs created over HAARP. Contours show the HAARP radiation pattern, projected to an altitude of 100 km. The brightness, hue, and saturation of the image pixels represent the signal-to-noise ratio, Doppler shift, and spectral width according to the legend shown. The incoherent integration time for the image is approximately 3 s.

One of those themes is the shortcomings of eigenvalue analysis, the tool of choice of the ionospheric analyst. Boundary-value analysis is incomplete wherever the eigenmodes are non-normal, including anywhere where shear flow exists. As strong plasma shear flow is present in the E and F-regions at all latitudes and in the electrojets, the bottomside equatorial F-region, and throughout the auroral zone in particular, the limitations of eigenvalue analysis are crucial. Ronchi et al. (1991) encountered the problem analyzing gradient drift waves in the equatorial electrojet. In that case, neither local nor boundary-value nor eikonal analysis correctly predicts the predominance of kilometer-scale waves in the daytime. Using initial boundary-value analysis and full numerical simulation, Ronchi et al. (1991) determined that kilometer-scale waves are transients which are continuously reintroduced by nonlinear mode coupling which maintains a dynamic equilibrium. This was a case where nothing short of a complete direct numerical simulation could give reliable theoretical predictions.

Another theme is the problem with the widespread assumption of equipotential magnetic field lines. This is a blunt approximation meant to simplify electrodynamic calculations but neglects important properties of warm plasmas. A 3D potential calculation is necessary to recover the background flow in the equatorial ionosphere, for example. As that flow is inherently unstable, capturing it correctly is important for stability analysis. Moreover, a number of instabilities depend on nonequipotential field lines including current-driven instabilities and drift-wave instabilities. The middle- and low-latitude ionosphere appears to be prone to collisional drift-wave instability which appear to be important in sporadic E-layers and related phenomena. Finite $k_\parallel$ effects are also necessary for resonance instability to occur at frequencies below the second electron gyroharmonic frequency.

Neither natural FB waves and heater-induced thermal parametric instabilities can be characterized fully by linear analysis. Nonetheless, certain aspects of their behavior can be understood with the help of intuitive heuristics. For FB waves, the only reliable result of linear analysis is that growth should only occur when the convection speed exceeds the ion-acoustic speed. A useful heuristic is the assumption that the waves exist in a dynamic equilibrium state of marginal stability wherein the linear growth rate is always small. This assumption yields reasonable predictions about the Doppler shift and spectral width of the echoes observed with coherent scatter. It also leads to accurate estimates of wave heating and electron temperatures in the auroral electrojet (Hysell et al., 2013a).

In ionospheric modifications, TPI evolves rapidly into resonance instability which is not accessible to linear analysis. However, linear theory can be applied to the analysis of the threshold conditions for instability onset. The assumption here is that linear theory applies up to the point where heater-induced irregularities are detected and not beyond. The theory for onset of TPI is among the best quantitative predictions to emerge from ionospheric modifications.

The value of ionospheric irregularities as ionospheric diagnostics is difficult to overstate. In the case of ESF events, bottom-type scattering layers are reliable precursors of radar plumes. Being confined to thin layers at the base of the F-region, they are not evidence of conventional collisional interchange instability in the bottomside. They are instead telltales of a variant of the instability associated with shear flow and vertical winds. The dominant wavelength of the instability is either about 30 km when the transient response dominates or about 150–200 km when the asymptotic response is dominant. The distance between bottom-type layer patches distinguishes one case from the other.

We would not know about plasma instabilities in E_s layers or in the auroral electrojet were it not for the coherent scatter observed there. In the case of auroral FB waves, the characteristics of the Doppler spectra are related to the vector convection velocity. Since the relationship is invertible, the spectra provide a means of inferring the velocity, at least where the convection speed is above threshold. In the case of thermal parametric instabilities, coherent scatter provides an incisive test of the theory for instability threshold, the transition to resonance instability, and the plasma wave interactions that occur near the double-resonance condition where the RF pump frequency equals the upper-hybrid frequency equals a gyroharmonic frequency.

Finally, by illuminating the background ionosphere for coherent scatter radars, ionospheric modifications and the associated plasma instabilities create a screen on which natural ionospheric irregularities can be projected. This method of observation is more reliable than simply waiting for nature to produce its own radar echoes and more sensitive than can be accomplished through incoherent scatter or other remote-sensing means. It can be exploited in the future in pursuit of long-standing problems in ionospheric plasma physics and aeronomy.

Acknowledgments

This work was supported by award FA9550-12-1-0462 from the Air Force Office of Scientific Research by awards AGS-1634014 and AGS-1818216 from the National Science Foundation to Cornell University. The Jicamarca Radio Observatory is a facility of the Instituto Geofisíco del Perú operated with support from NSF award AGS-1732209 through Cornell. The help of the staff is much appreciated. Help from Dr. Joseph Huba at the Naval Research Laboratory in using the SAMI2 model is also greatly appreciated.

References

Bahcivan, H., Cosgrove, R., 2010. On the generation of large wave parallel electric fields responsible for electron heating in the high-latitude E region. J. Geophys. Res. 115, A10304. https://doi.org/10.1029/2010JA015424.

Bahcivan, H., Hysell, D.L., Larsen, M.F., Pfaff, R.F., 2005. 30 MHz imaging radar observations of auroral irregularities during the JOULE campaign. J. Geophys. Res. 110, A05307. https://doi.org/10.1029/2004JA010975.

Bahcivan, H., Cutler, J.W., Springmann, J.C., Doe, R., Nicolls, M.J., 2014. Magnetic aspect sensitivity of high-latitude E region irregularities measured by the RAX Cubesat. J. Geophys. Res. 119 (2), 1233–1249. https://doi.org/10.1002/2013JA019547.

Barnes, R.I., 1992. An investigation into the horizontal structure of spread-E. J. Atmos. Terr. Phys. 54 (3/4), 391–399.

Belenov, A.F., Bubnov, V.A., Erukhimov, L.M., Kiselev, Y.V., Kokrakov, G.P., Mityakova, E.E., Rubstov, L.N., Uryadov, V.P., Frolov, V.L., Chagnuov, Y.V., Ykhmatov, B.V., 1977. Parameters of artificial small-scale ionospheric irregularities. Radiophys. Quantum Electron. 20, 1805. Engl. Transl.

Bender, C., Orszag, S., 1978. Advanced Mathematical Methods for Scientists and Engineers. McGraw-Hill, New York, NY.

Bernhardt, P.A., 2002. The modulation of sporadic-E layers by Kelvin-Helmholtz billows in the neutral atmosphere. J. Atmos. Sol. Terr. Phys. 64 (12–14), 1487–1504.

Bernhardt, P.A., Gondarenko, N.A., Guzdar, P.N., Djuth, F.T., Tepley, C.A., Sulzer, M.P., Ossakow, S.L., Newman, D.L., 2003. Using radio-induced aurora to measure the horizontal structure of ion layers in the lower thermosphere. J. Geophys. Res. 108 (A9), 1336. https://doi.org/10.1029/2002JA009712.

Bishop, A.S., 1958. Project Sherwood: The U.S. Program in Controlled Fusion. Addison-Wesley, Reading, MA.

Boberg, L., Brosa, U., 1988. Onset of turbulence in a pipe. Z. Naturforsch A43, 697.

Booker, H.G., Wells, H.W., 1938. Scattering of radio waves by the F region. Terres. Magn. 43, 249.

Bowles, K.L., 1954. Doppler-shifted radio echoes from the aurora. J. Geophys. Res. 59, 553–555.

Buneman, O., 1963. Excitation of field aligned sound waves by electron streams. Phys. Rev. Lett. 10, 285–287.

Butler, K.M., Farrell, B.F., 1992. Three-dimensional optimal perturbations in a viscous shear flow. Phys. Fluids A8, 1637–1650.

Chu, Y.H., Wang, C.Y., Su, S.L., Kuong, R.M., 2011. Coordinated sporadic E layer observations made with Chung-Li 30 MHz radar, ionosonde and FORMOSAT-3/COSMIC satellites. J. Atmos. Sol. Terr. Phys. 73 (9), 883–894.

Cosgrove, R.B., Tsunoda, R.T., 2002. A direction-dependent instability of sporadic-E layers in the nighttime midlatitude ionosphere. Geophys. Res. Lett. 29 (18), 1864. https://doi.org/10.1029/2002JA009728.

Cosgrove, R.B., Tsunoda, R.T., 2004. Instability of the E-F coupled nighttime midlatitude ionosphere. J. Geophys. Res. 109. https://doi.org/10.1029/2003JA010243.

Costa, E., Kelley, M.C., 1978. On the role of steepened structures and drift waves in equatorial spread F. J. Geophys. Res. 83, 4359.

Coster, A.J., Djuth, F.T., Jost, R.J., Gordon, W.E., 1985. The temporal evolution of 3-m striations in the modified ionosphere. J. Geophys. Res. 90, 2807–2818.

Dao, E.V., 2012. Electromagnetic Properties of Low-Latitude Plasma Irregularities (Ph.D. thesis), Cornell University.

Das, A.C., Fejer, J.A., 1979. Resonance instability of small-scale field-aligned irregularities. J. Geophys. Res. 84, 6701–6704.

Didebulidze, G.G., Lomidze, L.N., 2010. Double atmospheric gravity wave frequency oscillations of sporadic E formed in a horizontal shear flow. Phys. Lett. A 374 (7), 952–959.

Dimant, Y.S., Milikh, G.M., 2003. Model of anomalous electron heating in the E region, 1. Basic theory. J. Geophys. Res. 108 (A9), 1350. https://doi.org/10.1029/2011JA016648.

Dimant, Y.S., Oppenheim, M.M., 2004. Ion thermal effects on E-region instabilities: linear theory. J. Atmos. Sol. Terr. Phys. 66, 1639–1654.

Dimant, Y.S., Oppenheim, M.M., 2011. Magnetosphere ionosphere coupling through E region turbulence: 2. Anomalous conductivities and frictional heating. J. Geophys. Res. 116, A09304. https://doi.org/10.1029/2011JA016649.

Djuth, F.T., Jost, R.J., Noble, S.T., Gordon, W.E., Stubbe, P., Nielsen, H.K.E., Boström, R., Derblom, H., Hedberg, Å., Thidé, B., 1985. Observations of E region irregularities generated at auroral latitudes by a high-power radio wave. J. Geophys. Res. 90, 12293–13206.

Djuth, F.T., Pedersen, T.R., Gerken, E.A., Bernhardt, P.A., Selcher, C.A., Bristow, W.A., Kosch, M.J., 2005. Ionospheric modification at twice the electron cyclotron frequency. Phys. Rev. Lett. 94, 125001.

Drake, J.F., Huba, J.D., 1987. Dynamics of three-dimensional ionospheric plasma clouds. Geophys. Res. Lett. 58 (3), 278–281.

Drob, D.P., Emmert, J.T., Meriwether, J.W., Makela, J.J., Doornbos, E., Conde, M., Hernandez, G., Noto, J., Zawdie, K.A., McDonald, S.E., Huba, J.D., Klenzing, J.H., 2015. An update to the horizontal wind model (HWM): the quiet time thermosphere. Earth Space Sci. 2, 301–319. https://doi.org/10.1002/2014EA000089.

Dysthe, K., Mjølhus, E., Pécseli, H., Rypdal, K., 1982. Thermal cavitons. Phys. Scr. T. 2, 548–559.

Dysthe, K., Mjølhus, E., Pécseli, H., Rypdal, K., 1983. A thermal oscillating two-stream instability. Phys. Fluids 26, 146.

Eckersley, T.L., 1937. Irregular ionic clouds in the E layer of the ionosphere. Nature 140, 846.

Erukhimov, L.E., Metelev, S.A., Mityakov, N.A., Frolov, B.L., 1978. Hysteresis effect in the artificial excitation of inhomogeneities in the ionospheric plasma. Radiophys. Quantum Electron. 21, 1738–1740.

Farley, D.T., 1959. A theory of electrostatic fields in a horizontally stratified ionosphere subject to a vertical magnetic field. J. Geophys. Res. 64, 1225.

Farley, D.T., 1963. A plasma instability resulting in field-aligned irregularities in the ionosphere. J. Geophys. Res. 68, 6083.

Farley, D.T., Balsley, B.B., Woodman, R.F., McClure, J.P., 1970. Equatorial spread F: implications of VHF radar observations. J. Geophys. Res. 75, 7199.

Farrell, B.F., Ioannou, P.J., 1993. Perturbation growth in shear flow exhibits universality. Phys. Fluids A5 (9), 2298–2300.

Fejer, J.A., 1979. Ionospheric modification and parametric instabilities. Rev. Geophys. Space Phys. 17, 135–153.

Fialer, P.A., 1974. Field-aligned scattering from a heated region of the ionosphere—observations at HF and VHF. Radio Sci. 9, 923–940.

Flaherty, J.P., Seyler, C.E., Trefethen, L.N., 1999. Large-amplitude transient growth in the linear evolution of equatorial spread F with a sheared zonal flow. J. Geophys. Res. 104, 6843.

Frank, V.R., 1974. E-region scatter observed at Haswell, Colorado. In: Proceedings of the Prarie Smoke V RF Measurements Data WorkshopStanford Research Institute, Menlo Park, CA, p. 75.

Frey, A., 1986. The observation of HF-enhanced plasma waves with the EISCAT/UHF-radar in the presence of strong Landau-damping. Geophys. Res. Lett. 13, 438–441.

Fu, Z.F., Lee, L.C., Huba, J.D., 1986. A quasi-local theory of the $\mathbf{E} \times \mathbf{B}$ instability in the ionosphere. J. Geophys. Res. 91, 3263.

Grach, S.M., 1979. Thermal parametric instability in ionospheric plasma at frequencies close to ω_{He} and $2\omega_{He}$. Radiophys. Quantum Electron. 22, 357–363.

Grach, S.M., Karashtin, A.N., Mityzkov, N.A., Rapoport, V.O., Trakhtengerts, V.Y., 1977. Parametric interactions between electromagnetic radiation and ionospheric plasma. Radiophys. Quantum Electron. 20, 1254–1258.

Grach, S.M., Karashtin, A.N., Mityzkov, N.A., Rapoport, V.O., Trakhtengerts, V.Y., 1978a. Theory of thermal parametric instability in an inhomogenous plasma. Sov. J. Plasma Phys. 4, 737–741.

Grach, S.M., Karashtin, A.N., Mityzkov, N.A., Rapoport, V.O., Trakhtengerts, V.Y., 1978b. Theory of thermal parametric instability in an inhomogenous plasma (nonlinear theory). Sov. J. Plasma Phys. 4, 742–747.

Grach, S., Mityakov, N., Rapoport, V., Trakhtengertz, V., 1981. Thermal parametric turbulence in a plasma. Physica D 2, 102–106.

Graham, K.N., Fejer, J.A., 1976. Anomalous radio wave absorption due to ionospheric heating effects. Radio Sci. 11 (12), 1057–1063.

Gustavsson, L., 1991. Energy growth of three-dimensional disturbances in plane Poiseuille flow. J. Fluid Mech. 224, 241.

Haerendel, G., 1973. Theory of Equatorial Spread F. Max-Planck Institute für Physik und Astrophysik, Garching, West Germany.

Haldoupis, C., 1989. A review on radio studies of auroral E-region ionospheric irregularities. Ann. Geophys. 7, 239–258.

Haldoupis, C., Schlegel, K., 1994. Observation of the modified two-stream plasma instability in the midlatitude E region ionosphere. J. Geophys. Res. 99, 6219.

Hamza, A.M., St-Maurice, J.-P., 1993. A self consistent fully turbulent theory of auroral E region irregularities. J. Geophys. Res. 98, 11601–11613.

Haslett, J.C., Megill, L.R., 1974. A model of the enhanced airglow excited by RF radiation. Radio Sci. 9 (11), 1005–1019.

Hedberg, Å., Derblom, H., Thidé, B., 1983. Observations of HF backscatter associated with the heating experiment at Tromsø. Radio Sci. 18, 840–850.

Huba, J.D., Joyce, G., Fedder, J.A., 2000. Sami2 is another model of the ionosphere (SAMI2): a new low-latitude ionospheric model. J. Geophys. Res. 105, 23035–23054.

Huba, J.D., Joyce, G., Krall, J., 2008. Three-dimensional equatorial spread F modeling. Geophys. Res. Lett. 35, L10102. https://doi.org/10.1029/2008GL033509.

Huba, J.D., Joyce, G., Krall, J., 2011. Three-dimensional modeling of equatorial spread F. In: Abdu, M.A., Pancheva, D. (Eds.), Aeronomy of the Earth's Atmosphere and Ionosphere. In: vol. 2. IAGA Spec. Sopron Book Ser., pp. 211–218

Hughes, J.M., Bristow, W.A., Parries, R.T., Lundell, E., 2003. SuperDARN observations of ionospheric heater-induced upper hybrid waves. Geophys. Res. Lett. 30 (24), 2276. https://doi.org/10.1029/2003GL018772.

Hysell, D.L., 2016. The radar aurora. In: Zhang, Y., Paxton, L.J. (Eds.), Auroral Dynamics and Space Weather. In: vol. 215. American Geophysical Union, pp. 193–209.

Hysell, D.L., Chau, J.L., 2006. Optimal aperture synthesis radar imaging. Radio Sci. 41, RS2003. https://doi.org/10.1029/2005RS003383.

Hysell, D.L., Drexler, J., 2006. Polarization of E region plasma irregularities. Radio Sci. 41, RS4015. https://doi.org/10.1029/2005RS003424.

Hysell, D.L., Kudeki, E., 2004. Collisional shear instability in the eqautorial F region ionosphere. J. Geophys. Res. 109, A11301.

Hysell, D.L., Nossa, E., 2009. Artificial E region field-aligned plasma irregularities generated at pump frequencies near the second electron gyroharmonic. Ann. Geophys. 27, 2711–2720.

Hysell, D.L., Shume, E.B., 2002. Electrostatic plasma turbulence in the topside equatorial F region ionosphere. J. Geophys. Res. 107, 1269.

Hysell, D.L., Seyler, C.E., Kelley, M.C., 1994. Steepened structures in equatorial spread F. 2. Theory. J. Geophys. Res. 99, 8841.

Hysell, D.L., Kelley, M.C., Yampolski, Y.M., Beley, V.S., Koloskov, A.V., Ponomarenko, P.V., Tyrnov, O.F., 1996. HF radar observations of decaying artificial field aligned irregularities. J. Geophys. Res. 101, 26981.

Hysell, D.L., Yamamoto, M., Fukao, S., 2002. Imaging radar observations and theory of type I and type II quasi-periodic echoes. J. Geophys. Res. 107 (A11), 1360.

Hysell, D.L., Michhue, G., Larsen, M.F., Pfaff, R., Nicolls, M., Heinselman, C., Bahcivan, H., 2008. Imaging radar observations of Farley Buneman waves during the JOULE II experiment. Ann. Geophys. 26, 1837–1850.

Hysell, D.L., Nossa, E., McCarrick, M., 2010. Excitation threshold and gyroharmonic suppression of artificial E region field-aligned plasma density irregularities. Radio Sci. 45, 1–17. https://doi.org/10.1029/2010RS004360.

Hysell, D.L., Nossa, E., Larsen, M.F., Munro, J., Smith, S., Sulzer, M.P., González, S.A., 2012. Dynamic instability in the lower thermosphere inferred from irregular sporadic layers. J. Geophys. Res. 117 (A8), 13. https://doi.org/10.1029/2012JA017910.

Hysell, D.L., Miceli, R.J., Huba, J.D., 2013a. Implications of a heuristic model of auroral Farley Buneman waves and heating. Radio Sci. 48 (5), 527–534. https://doi.org/10.1002/rds.20061.

Hysell, D.L., Nossa, E., Aveiro, H.C., Larsen, M.F., Munro, J., Sulzer, M.P., González, S.A., 2013b. Fine structure in midlatitude sporadic E layers. J. Atmos. Sol. Terr. Phys. 103, 16–23. https://doi.org/10.1016/j.jastp.2012.12.005.

Hysell, D.L., Milla, M.A., Condori, L., Vierinen, J., 2015. Data-driven numerical simulations of equatorial spread F in the Peruvian sector 3: Solstice. J. Geophys. Res. 120, 10809–10822. https://doi.org/10.1002/2015JA021877.

Hysell, D.L., Munk, J., McCarrick, M., 2019. Investigating transport and dissipation in the subauroral E region with ionospheric modification experiments and VHF radar backscatter. Radio Sci 54. 245–253.

Inhester, B., Das, A.C., Fejer, J.A., 1981. Generation of small-scale field-aligned irregularities in ionospheric heating experiments. J. Geophys. Res. 86, 9101–9105.

Jones, T.B., Robinson, T., Stubbe, P., Kopka, H., 1983. A hysteresis effect in the generation of field-aligned irregularities by a high-power radio wave. Radio Sci. 18, 835–839.

Kagan, L.M., St.-Maurice, J.P., 2004. Impact of electron thermal effects on Farley-Buneman waves at arbitrary aspect angles. J. Geophys. Res. 109, A12302. https://doi.org/10.1029/2004JA010444.

Kelley, M.C., Riggin, D., Pfaff, R.F., Swartz, W.E., Providakes, J.F., Huang, C.S., 1995. Large amplitude quasi-periodic fluctuations associated with a midlatitude sporadic E layer. J. Atmos. Terr. Phys. 57, 1165.

Keskinen, M.J., Mitchell, H.G., Fedder, J.A., Satyanarayana, P., Zalesak, S.T., Huba, J.D., 1988. Nonlinear evolution of the Kelvin-Helmholtz instability in the high-latitude ionosphere. J. Geophys. Res. 93, 137.

Kosch, M.J., Pedersen, T., Hughes, J., Marshall, R., Gerken, E., Senior, A., Sentman, D., McCarrick, M., Djuth, F.T., 2005. Artificial optical emissions at HAARP for pump frequencies near the third and second electron gyro-harmonic. Ann. Geophys. 23, 1585–1592.

Kosch, M.J., Pedersen, T., Mishin, E., Oyama, S., Hughes, J., Senior, A., Watkins, B., Bristow, B., 2007. Coordinated optical and radar observations of ionospheric pumping for a frequency pass through the second electron gyroharmonic at HAARP. J. Geophys. Res. 112, A06325. https://doi.org/10.1029/2006JA012146.

Krall, J., Huba, J.D., Ossakow, S.L., Joyce, G., 2010. Why do equatorial bubbles stop rising? Geophys. Res. Lett. 37, L09105. https://doi.org/10.1029/2010GL043128.

Kuo, S.P., Lee, M.C., 1982. On the parametric excitation of plasma modes at upper hybrid resonance. Phys. Lett. A 91, 444–446.

Landau, L.D., Lifshitz, E.M., 1958. Statistical Physics. Pergamon Press, London.

Landau, L.D., Lifshitz, E.M., 1971. The Classical Theory of Fields, third ed. Pergamon Press, New York, NY.

Larsen, M.F., 2000. A shear instability seeding mechanism for quasi-periodic radar echoes. J. Geophys. Res. 105 (A11), 24931–24940.

Larsen, M.F., Hysell, D.L., Zhou, Q.H., Smith, S.M., Friedman, J., Bishop, R.L., 2007. Imaging coherent scatter radar, incoherent scatter radar, and optical observations of quasiperiodic structures associated with sporadic E layers. J. Geophys. Res. 112, A06321. https://doi.org/10.1029/2006JA012051.

Lee, M.C., Kuo, S.P., 1983. Excitation of upper hybrid waves by a thermal parametric instability. J. Plasma Phys. 30, 463–478.

Madsen, C.A., Dimant, Y.S., Oppenheim, M.M., Fontenla, J.M., 2013. The multi-species Farley-Buneman instability in the solar chromosphere. Astrophys. J. 783 (128). https://doi.org/10.1088/0004-637X/783/2/128.

Maruyama, T., Fukao, S., Yamamoto, M., 2000. A possible mechanism for echo striation generation of radar backscatter from midlatitude sporadic E. Radio Sci. 35 (5), 1155–1164.

Milikh, G.M., Dimant, Y.S., 2003. Model of anomalous electron heating in the E region: 2. Detailed numerical modeling. J. Geophys. Res. 108 (A9), 1351. https://doi.org/10.1029/2002JA009527.

Minkoff, J., Kreppel, R., 1976. Spectral analysis and step response to radio frequency scattering from a heated ionospheric volume. J. Geophys. Res. 81, 2844–2856.

Minkoff, J., Kugelman, P., Weissman, I., 1974. Radio frequency scattering from a heated ionospheric volume. 1.

VHF/UHF field-aligned and plasma line backscatter measurements. Radio Sci. 9, 941–955.

Mjølhus, E., 1990. On linear conversion in magnetized plasmas. Radio Sci. 6, 1321–1339.

Mjølhus, E., 1993. On the small scale striation effect in ionospheric modification experiments near harmonics of the electron gyro frequency. J. Atmos. Terr. Phys. 55 (6), 907–918.

Nielsen, E., Schlegel, K., 1983. A first comparison of STARE and EISCAT electron drift velocity measurements. J. Geophys. Res. 88, 5745.

Nielsen, E., Schlegel, K., 1985. Coherent radar Doppler measurements and their relationship to the ionospheric electron drift velocity. J. Geophys. Res. 90, 3498–3504.

Noble, S.T., Djuth, F.T., Jost, R.J., Gordon, W.E., Hedberg, A., Thide, B., Derblom, H., Bostrom, R., Nielsen, E., Stubbe, P., Kopka, H., 1987. Multiple-frequency radar observations of high-latitude E region irregularities in the HF modified ionosphere. J. Geophys. Res. 92, 13613–13627.

Oppenheim, M.M., Dimant, Y.S., 2013. Kinetic simulations of 3-D Farley-Buneman turbulence and anomalous electron heating. J. Geophys. Res. 118(1–13). https://doi.org/10.1002/jgra.50196.

Oppenheim, M.M., Dimant, Y.S., 2016. Photoelectron-induced waves: a likely source of 150 km radar echoes and enhanced electron modes. Geophys. Res. Lett. 43, 3637–3644. https://doi.org/10.1002/2016GL068179.

Ossakow, S.L., 1981. Spread F theories—a review. J. Atmos. Terr. Phys. 43, 437.

Pan, C.J., Liu, C.H., Röttger, J., Su, S.Y., 1994. A three dimensional study of E region irregularity patches in the equatorial aeronomy region using the Chung-Li VHF radar. Geophys. Res. Lett. 21, 1763.

Perkins, F., 1973. Spread F and ionospheric currents. J. Geophys. Res. 78, 218.

Perkins, F.W., Doles III, J.H., 1975. Velocity shear and the E×B instability. J. Geophys. Res. 80, 211.

Picone, J.M., Hedin, A.E., Drob, D.P., Aikin, A.C., 2002. NRLMSISE-00 empirical model of the atmosphere: statistical comparisons and scientific issues. J. Geophys. Res. 107 (A12). https://doi.org/10.1029/2002JA009430 SIA 15-1–SIA 15-16.

Rao, N.N., Kaup, D.J., 1990. Upper hybrid mode conversion and resonance excitation of Bernstein modes in ionospheric heating experiments. J. Geophys. Res. 95, 17245–17252.

Rappaport, H.L., 1998. Localized modes with zonal neutral wind, diffusion, and shear in equatorial spread F. J. Geophys. Res. 103, 29137.

Reddy, S.C., Henningson, D.S., 1993. Energy growth in viscous channel flows. J. Fluid Mech. 252, 209–238.

Retterer, J.M., 2010. Forecasting low-latitude radio scintillation with 3-D ionospheric plume models: 1. Plume model.

J. Geophys. Res. 115, A03306. https://doi.org/10.1029/2008JA013839.

Robinson, T.R., Yeoman, T.K., Dhillon, R.S., Lester, M., Thomas, E.C., Thornhill, J.D., Wright, D.M., van Eyken, A.P., McCrea, I., 2006. First observations of SPEAR induced artificial backscatter from CUTLASS and the EIS-CAT Svalbard radar. Ann. Geophys. 24, 291–309.

Rojas, E.L., Young, M.A., Hysell, D.L., 2016. Phase speed saturation of Farley-Buneman waves due to stochastic, self-induced fluctuations in the background flow. J. Geophys. Res. 121, 5785–5793. https://doi.org/10.1002/2016JA022710.

Ronchi, C., Similon, P.L., Sudan, R.N., 1989. A nonlocal linear theory of the gradient drift instability in the equatorial electrojet. J. Geophys. Res. 94, 1317.

Ronchi, C., Sudan, R.N., Farley, D.T., 1991. Numerical simulations of large-scale plasma turbulence in the daytime equatorial electrojet. J. Geophys. Res. 96, 21263.

Sahr, J.D., Fejer, B.G., 1996. Auroral electrojet plasma irregularity theory and experiment: a critical review of present understanding and future directions. J. Geophys. Res. 101, 26893–26909.

Saito, S., Yamamoto, M., Hashiguchi, H., Maegawa, A., 2006. Observation of three-dimensional signatures of quasi-periodic echoes associated with mid-latitude sporadic-E layers by MU radar ultra-multi-channel system. Geophys. Res. Lett. 33, L14109. https://doi.org/10.1029/2005GL025526.

Satyanarayana, P., Guzdar, P.N., Huba, J.D., Ossakow, S.L., 1984. Rayleigh-Taylor instability in the presence of a stratified shear layer. J. Geophys. Res. 89, 2945.

Satyanarayana, P., Lee, Y.C., Huba, J.D., 1987. The stability of a stratified shear layer. Phys. Fluids 30, 81.

Scherliess, L., Fejer, B.G., 1999. Radar and satellite global equatorial F region vertical drift model. J. Geophys. Res. 105, 6829–6842.

Schlegel, K., Haldoupis, C., 1994. Observation of the modified two-stream instability in the midlatitude E region. J. Geophys. Res. 99, 6219.

Sekar, R., Kelley, M.C., 1998. On the combined effects of vertical shear and zonal electric field patterns on nonlinear equatorial spread F evolution. J. Geophys. Res. 103, 20735.

Senior, A., Borisov, N.D., Kosch, M.J., Yeoman, T.K., Honary, F., Rietveld, M.T., 2004. Multi-frequency HF radar measurements of artificial F-region field-aligned irregularities. Ann. Geophys. 22, 3503–3511.

Shukla, P.K., Rahman, H.U., 1998. The Rayleigh-Taylor mode with sheared plasma flows. Phys. Scr. 57, 286.

Sinitsin, V.G., Kelley, M.C., Yampolsky, Y., Hysell, D.L., Zalizovski, A., Ponomarenko, P.V., 1999. Ionospheric conductivities according to Doppler radar observations of stimulated turbulence. J. Atmos. Sol. Terr. Phys.

61, 903–912. https://doi.org/10.1016/S1364-6826(99)00039-5.

St.-Maurice, J.P., Choudhary, R.K., Ecklund, W.L., Tsunoda, R.T., 2003. Fast type-1 waves in the equatorial electrojet: evidence for nonisothermal ion-acoustic speeds in the lower E region. J. Geophys. Res. 108 (A5), 1170. https://doi.org/10.1029/2002JA009648.

Stubbe, P., Kopka, H., Lauche, H., Rietveld, M.T., Brekke, A., Holt, O., Jones, T.B., Robinson, T., Hedberg, A., Thide, B., Crochet, M., Lotz, H.J., 1982. Ionospheric modification experiments in northern Scandinavia. J. Atmos. Terr. Phys. 44, 1025–1041.

Sudan, R.N., 1983. Nonlinear theory of type 1 irregularities in the equatorial electrojet. Geophys. Res. Lett. 10, 983.

Sultan, P.J., 1996. Linear theory and modeling of the Rayleigh-Taylor instability leading to the occurrence of equatorial spread F. J. Geophys. Res. 101, 26875.

Trefethen, L.N., Embree, M., 2005. Spectra and Pseudospectra: The Behavior of Nonnormal Matrices and Operators. Princeton University Press, Princeton, Oxford.

Trefethen, L.N., Trefethen, A.E., Reddy, S.C., Driscoll, T.A., 1993. Hydrodynamic stability without eigenvalues. Science 261, 578–584.

Vas'kov, V.V., Gurevich, A.V., 1977. Resonance instability of small-scale plasma perturbations. Sov. Phys. JETP 46, 487–494.

Woodman, R.F., La Hoz, C., 1976. Radar observations of F region equatorial irregularities. J. Geophys. Res. 81, 5447–5466.

Woodman, R.F., Yamamoto, M., Fukao, S., 1991. Gravity wave modulation of gradient drift instabilities in mid-latitude sporadic E irregularities. Geophys. Res. Lett. 18, 1197.

Wright, D.M., Davies, J.A., Yeoman, T.K., Robinson, T.R., Shergill, H., 2006. Saturation and hysteresis effects in ionospheric modification experiments observed by the CUTLASS and EISCAT radars. Ann. Geophys. 24, 543–553.

Yampolski, Y.M., Beley, V.S., Kascheev, S.B., Koloskov, A.V., Hysell, D.L., Isham, B., Kelley, M.C., 1997. Bistatic HF radar diagnostics of induced field-aligned irregularities. J. Geophys. Res. 102, 7461–7467.

Yokoyama, T., Shinagawa, H., Jin, H., 2014. Nonlinear growth, bifurcation and pinching of equatorial plasma bubble simulated by three-dimensional high-resolution bubble model. J. Geophys. Res. Space Phys. 119, 10474–10482.

Zargham, S., Seyler, C.E., 1987. Collisional interchange instability: 1. Numerical simulations of intermediate-scale irregularities. J. Geophys. Res. 92, 10073.

Zargham, S., Seyler, C.E., 1989. Collisional and inertial dynamics of the ionospheric interchange instability. J. Geophys. Res. 94, 9009.

Equatorial F region irregularities

Managlathayil Ali Abdu

National Institute for Space Research (Instituto Nacional de Pesquisas Espaciais-INPE), Sao Jose dos Campos, Brazil

1 Introduction

Plasma irregularities in the equatorial ionospheric F region develop under the unique condition of the low inclination geomagnetic field lines that confine the F region plasma to its low-latitude conjugate E layers. The vertical transport of plasma, which is basically responsible for the irregularity generation, takes place by the action of the zonal electric field generated by the dynamo action within the ionosphere or by electric field imposed from the magnetosphere. The irregularity generation in the F region is possible only under the condition that the driving electric field cannot be shorted out by the E layer conductivity, a condition that can exist after sunset. Under typical situations, the evening prereversal enhancement (PRE) in the plasma vertical drift is a necessary prerequisite for the irregularity development, which may be initiated from a seed perturbation in the form of a wave structure in the ambient plasma density. The post sunset F layer has a steep bottom-side density gradient shaped by the recombination and electrodynamical processes. During the PRE vertical drift, the layer rises to higher altitudes of decreasing ion-neutral collision frequency, whereby the instability growth rate initiated at the F layer bottom-side gradient region becomes enhanced.

1.1 Generation mechanism

The instability is believed to be driven by the gravitational Rayleigh-Taylor (GRT) mechanism characterized by a situation similar to a heavy fluid resting on a light fluid, as illustrated in Fig. 1.

In the presence of a perturbation at the upward gradient region of the ionization boundary in the bottom-side ionosphere (sketched in equatorial plane in Fig. 1), the action of gravity generates an eastward ion current, which could induce a polarization electric field, δE. The polarity of the δE is eastward (westward) in the reduced (enhanced) density regions so that $\delta E \times B$ vertical drift of the ions and electrons is upward (downward) in the lower (higher)-density regions of the perturbation. As the plasma in the reduced density region rises up, the amplitude of the reduction relative to the upward increasing background plasma density increases, which results in an increase of the eastward polarization electric field that causes accelerated upward motion of the rarified plasma leading to further increase in the

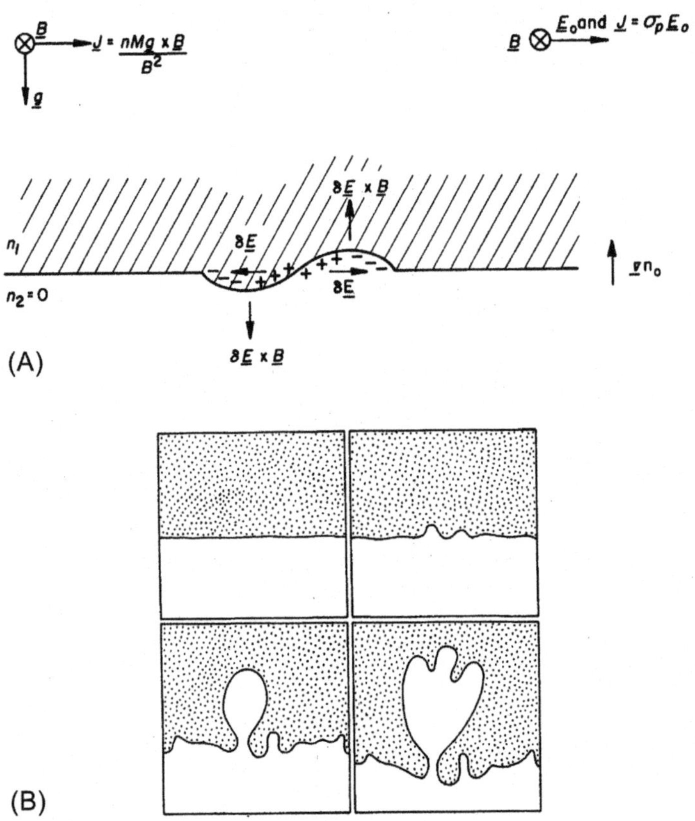

FIG. 1 An illustration of the instability growth at the F layer bottom-side gradient region. *Adapted from Kelley, M.C., 1989. The Earth's Ionosphere. Plasma Physics and Electrodynamics, Geophysics Series, vol. 43. Academic Press, San Diego, CA.*

polarization electric field and associated enhanced vertical drift, the process resulting in nonlinear growth of the instability. This process results in the formation of magnetic field aligned plasma-depleted regions, widely known as plasma bubbles. From the equations of continuity and current convergence, the local growth rate of the instability is given by:

$$\gamma'_L = \frac{\Delta n}{n}\left(\frac{g}{\nu_{in}}\right) - \beta_L \qquad (1)$$

where n is the electron density, β_L is the recombination rate, ν_{in} is the ion-neutral collision frequency.

The walls of these Equatorial Plasma Bubbles (EPB) are characterized by large density gradients that become unstable to perturbations in density and polarization electric field, as a result of which secondary irregularities may develop through cascading processes leading to formations of irregularities of scale sizes ranging from a few meters to several hundreds of kilometers. The composite characterization of these structures is widely known by the generic name, Equatorial Spread F irregularities, or, simply, ESF.

2 Observational techniques used in ESF investigation, and the irregularity types

Depending upon the scale size of the irregularities, different techniques are used to investigate the ESF irregularities. The earliest technique was based on vertical sounding of the ionosphere using ionosondes that measure the range of received echo as a function of the sounding frequency in the HF and VHF bands. Modern digital ionosonds/Digisondes provide detailed and rather complete data sets on the dynamical characteristics of the bottom-side ionosphere, such as the precise echo location, velocity components (east-west, north-south, and vertical) of the ionospheric motion, etc. Such data are obtained through angle-of-arrival measurements using Doppler triangulation technique employing spaced antenna system, etc. This method continues to be widely used today (Reinisch and Galkin, 2011; Abdu et al., 1981; Tsunoda and White, 1981). The scale sizes of the irregularities investigated by this technique vary from decameters to hundreds of kilometers.

Another widely used technique is based on measurement of the total electron content (TEC) of the ionosphere using signals transmitted by geostationary satellites, or Global Positioning System (GPS) constellation of satellites, that are received at ground stations. The height/path integrated modulations of the amplitude and phase of the received signals are dependent on the total electron content of the ionosphere that may suffer large changes due to forcing by geophysical drivers. The motion of the TEC irregularity structures across the satellites signal path may cause large fluctuations in amplitude and phase of the propagating signals, thereby producing scintillations of the received signal and fluctuations in the measured TEC. Measurements using the global network of GPS receivers have made it possible to produce global maps of the TEC and scintillation occurrence rates (see, for example, Li et al., 2010). Due to the nearly stationary locations of the GPS satellites, the signals received from these satellites are widely used to investigate the dynamics of the scintillation producing irregularities (of scale sizes of a few tens to hundreds of meters). Measurements of the horizontal motion of these irregularities may be used as tracers to monitor the dynamics of the ambient ionospheric plasma motion.

Radars in HF-VHF bands are powerful tools to study the dynamics of the irregularities at scale sizes matching the radar wavelength, usually in the range of a few to tens of meters. At VHF frequencies, the vertical distribution of the irregularities in the entire height region extending from the base of the ionosphere till about 2000 km can be investigated (see, for example, Woodman and La Hoz, 1976; Tsunoda and White, 1981). Radar investigations have made it possible to understand the vertical development and structuring of plasma bubbles starting from their initiation at the upward density gradient region of a rising bottom-side F layer.

3 Spatial and temporal distribution, and variabilities of the irregularities

The ESF irregularity development is strongly controlled by solar activity (represented by the F10.7 index) and season of the year. During the post sunset hours, the dependence of the ESF on solar EUV flux (F10.7 index) operates through the solar activity control of the evening prereversal vertical drift enhancement (Fejer et al., 1999; Abdu et al., 2010) that is known to control the ESF. It needs to be recognized that ESF observed in equatorial ionograms (especially close to dip equator) is indicative of the presence of bottom-side F region irregularities as well as that of the plasma bubbles extending to low latitudes that are mapped to higher equatorial heights, whereas the spread F observed in low-latitude ionograms is a direct indicator of the presence of plasma bubbles. It has been found that the occurrence rate of bottom-side spread F over the equatorial site Fortaleza is nearly 100% for all solar flux values (Abdu et al., 1985). It is well known that certain threshold levels of the F layer bottom-side height and the vertical drift need to be attained for the initiation of the ESF/bubble irregularity development, the threshold value being larger for the case of bubble development than for the bottom side irregularities. These threshold values have been found to be upward of 300 km for the height (Fejer et al., 1999) and 15 m/s for the vertical drift (Abdu et al., 1983).

The seasonal dependence of the ESF development varies with longitude. The main factor controlling the seasonal dependence in ESF is the alignment of the sunset terminator with magnetic meridian of the observing site. Close alignment conditions correspond to the sunset occurring nearly simultaneously at conjugate sites, which can cause largest degree of longitudinal gradient in the integrated conductivity, and hence large increase in the vertical drift that can results also in large degree of ESF development as was shown by Abdu et al. (1981). Because of the longitudinal variation in the magnetic declination angle, the season in which the terminator-magnetic meridian alignment can occur also depends on the longitude. For example, in the eastern Brazilian (Atlantic) sector where the magnetic declination angle is westward and large (22°W), close alignment occurs in December month, which therefore corresponds to the month of the annual maximum in the ESF occurrence. In contrast to this, two equinoctial maxima in ESF occur at most other longitude where the alignment between the terminator and magnetic meridian is close to being optimum in equinoctial months.

In addition to the vertical drift, a seed perturbation, in the form of gravity-wave-induced height/density oscillation is necessary for the initiation of the ESF. Thus, a season-dependent intensity of the seeding source is also known to control the seasonal variation in ESF occurrence. The intertropical convergence zone (ITCZ) is believed to be an important source of gravity wave generation (Waliser and Gautier, 1993) that may lead to the seeding of ESF development. Annual maximum in the ESF occurrence has been observed during solstice months (mainly) in Pacific longitude, which has been attributed to a corresponding seasonal maximum in the seeding source in the form of gravity waves, whose generation is believed to

be regulated by the proximity and alignment of the ITCZ with the magnetic equator.

While the nature of seasonal variation in the ESF occurrence is relatively better understood, the causes of the short-term and day-to-day variabilities in the ESF occurrence are far from being well clarified. This is because such short-term variabilities are driven by different mechanisms, such as: (1) upward propagating atmospheric waves (gravity waves, planetary waves, etc.) originating from lower atmospheric sources, or (2) equator-ward propagating disturbances in the form of penetration electric field or ionosphere-thermosphere (I-T) perturbations originating from magnetospheric disturbances. We will briefly explain the nature of such short-term variabilities.

3.1 Upward propagating wave disturbances

The winds and waves in upper layers of the atmosphere have sources in upward propagating tides, gravity waves, and planetary/Kelvin waves. They interact with the magnetized conducting ionosphere in the dynamo region generating the dynamo electric fields that drive the Sq and equatorial electrojet current systems, in the E region. In the F region, the zonal component of the electric field produces vertical E x B plasma drift whose well-known manifestation is the equatorial plasma fountain that drives the Equatorial Ionization Anomaly (EIA). The EIA is characterized by two global maxima of plasma densities (the ionization crests), formed at low latitudes, at ± 15–$18°$, on either sides of an ionization minimum over the equator (the ionization trough). The disturbances in the electric fields and winds control the ionospheric weather variations at low latitudes. Oscillations in the equatorial ionosphere at Planetary wave periods (2-day, 3–5-day, 10-day, and 16-day) have been well established, through observations of variations in the equatorial electrojet (EEJ) strength, mesospheric temperature, equatorial F-layer heights, vertical plasma drift, and ionization anomaly (EIA) (see for example, Forbes, 1996; Forbes and Leveroni, 1992; Chen, 1992; Pancheva et al., 2003; Takahashi et al., 2006, 2007; Abdu et al., 2006, 2015; Vineeth et al., 2007). Of particular interest for the equatorial region are the Kelvin waves, i.e., equatorially trapped eastward propagating planetary waves, also identified as ultra fast Kelvin (UFK) waves, of 3–4 day periodicity (Forbes, 1996). Due to their relatively longer vertical wavelengths, they may propagate up to the dynamo region where they modulate the tidal oscillations, thereby playing key roles in the electrodynamics of the vertical coupling.

An important consequence of such modulation is the modification of the evening prereversal electric field enhancement and F layer heights that play leading roles in the equatorial spread F/plasma bubble irregularity development and in their large degree of day-to-day variability. An example of the mesospheric wind oscillations of ~3–4 day periods, and their upward propagation to the dynamo region, is shown in Fig. 2.

Results of wavelet analysis of mesospheric winds over Cachoeira Paulista show oscillations of a few days (in the range of 3–4 days) periodicity (which is in the range of planetary wave periodicity) becoming enhanced with increasing height, leading to modification of the evening prereversal vertical drift (PRE) measured over Caximbo, a magnetic equatorial site, and over Campo Grande, a low-latitude site. The oscillation episodes are clearly observable during the interval of d305 – d320 (centered around d310). This is a clear example of the vertical coupling process through upward propagating planetary waves manifested in the form of mesospheric dynamics modulating the equatorial F region evening vertical drift. The modification in the vertical drift may in turn modulate the spread F/plasma bubble irregularity development process (although the latter aspect is not demonstrated here). The mechanism of such coupling process operates through ionospheric interaction with the upward propagating waves

(especially in the zonal wind component) in the presence of the sunset longitudinal gradient in the E layer conductivity. The wave-associated wind oscillations interact with the ion distribution by modifying its height and longitudinal gradients. The resulting change in the conductivity longitudinal gradient leads to modification of the vertical drift velocity, which in turn modifies the ESF development as explained by Abdu et al. (2006). In this way, the day-to-day and short-term variabilities in the ESF arising from planetary wave activity may be explained.

Upward propagating gravity waves are another important source of short-term variability in the ESF occurrence rate. The conditions for the upward propagation of such gravity waves to thermospheric heights have been widely discussed in the literature (e.g., Fritts et al., 2008; Vadas, 2007; Kherani and Abdu, 2011). On a statistical basis, an important contribution for shaping the seasonal-longitudinal patterns of ESF occurrence has been attributed to gravity wave generation from source associated to the Inter-Tropical Convergence Zone (McClure et al., 1998; Tsunoda, 2010). Some case studies have further supported the role of gravity waves as a necessary precursor for the development of the R-T instability leading to ESF irregularities. The precursor signature of the gravity waves in observational data can vary depending upon the techniques used in their diagnosis. Large-scale wave structure (LSWS) in the form of upwelling of the bottom-side plasma density prior to plume development was observed by incoherent scatter (IS) radar over Kwajelein (Tsunoda and White, 1981). The occurrence of satellite traces adjacent to the main F layer trace was found to be precursor to spread F development in ionogram (Abdu et al., 1981; Tsunoda, 2008; Li et al., 2012). In the Atmospheric Explorer-E satellite data, wave-like ion density structure was observed as precursor to plasma bubble development (Singh et al., 1997). Precursor wave structures, identified as LSWS, have been observed also in the form of longitudinal wave structure in ionospheric total

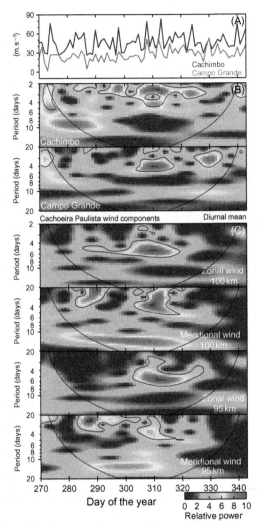

FIG. 2 (A) Vertical drift velocity (V_{zp}) variations over CX and CG for the period October–December 2002. (B) Morlet wavelet power spectral distribution of V_{zp} oscillations over the two stations. (C) Wavelet spectral distribution of the daily means of mesospheric zonal and meridional winds at 100 km over Cachoeira Paulista (the top two plots) and at 95 km (the bottom two plots).

electron content (TEC) measured on C/NOFS satellite passes that were observed well before the E layer sunset (e.g., Thampi et al., 2009; Tulasi Ram et al., 2014; and Tsunoda et al., 2011). Abdu et al. (2009) showed that height oscillations due to LSWS at F layer bottom-side, as observed by Digisonde, presented upward propagating gravity wave characteristics. They were present in the afternoon and continued into post sunset hours leading to the growth of R-T instability that resulted in plasma bubble irregularity development, as illustrated in Fig. 3. Oscillations in the frequency band of 15-min to 1.5h are present on all days with the amplitudes increasing toward sunset as a result of the afternoon E layer conductivity decaying toward sunset. The amplitude of the height oscillations (DhF) increasing toward sunset acts

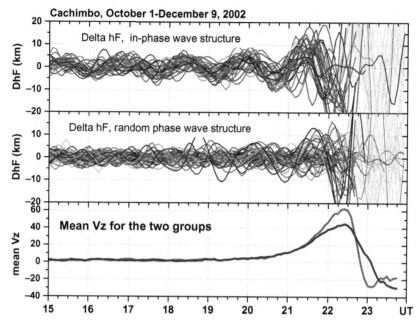

FIG. 3 Mass plots of the band-pass filtered F layer true height (hF) oscillations (DhF) at 5 and 8 MHz plasma frequencies for October–December 2002 over Cachimbo (top and middle plots). The gray segment of each DhF curve, starting on an average near 21:30 UT, represents the presence of range spread F (ESF). The cases of nonoccurrence of spread F are highlighted by the continuation of color-coded curves into the night.

like seed perturbations that could lead to the development of R-T instability. The resulting ESF development is indicated by the gray segment of the respective curves. Clearly, not all cases of the oscillation are followed by instability growth. The requirement of a sufficient (threshold) vertical drift, which is also a necessary condition for the initiation of ESF, is not identifiable in this plot, however.

3.2 Equator-ward propagating disturbances in the form of penetration electric field or ionosphere-thermosphere (I-T) perturbations, originating from magnetospheric disturbances

The solar-flare-associated coronal mass ejection impacting the earth's magnetosphere, under southward orientation of the interplanetary magnetic field (IMF) Bz, is the main driver for the development of disturbances in the form

of magnetic storms/substorms. These disturbances are marked by AE and Dst intensification due to motional electric field, that is, interplanetary electric field, IEF, $(-B_z.U_{sw})$, mapped to high latitudes as dawn-dusk electric field, which propagates to low latitudes as prompt penetration/undershielding electric field (PPEF) with eastward (westward) polarity on the day-side (nightside). With the B_z turning north, marking an AE recovery, the associated overshielding electric field, that has opposite polarity to that of the undershielding/convection electric field, penetrates to equatorial latitudes (e.g., Kikuchi et al., 2000; Abdu et al., 2003), with opposite polarities on the day and night sides. The penetration efficiency of the PPEF is variable so that the ionospheric electric field can be as much as 5 to10 percent of the IEF (see, for example, Kelley and Retterer, 2008; Huang et al., 2007, 2010). The intensity and polarity of the penetration electric fields will depend also on the large-scale conductivity gradients of the ionosphere that causes the daytime eastward polarity of the electric field to extend into post sunset hours. The intensity of this electric field maximizes at the time of the PRE vertical drift, prior to its westward reversal by ~21 LT. The night side electric field is westward, which attains peak value during presunrise hours prior to its eastward reversal to dayside by ~05–07 LT (Richmond et al., 2003; Fejer et al., 2008).

Another type of disturbance electric field in the equatorial region is the disturbance dynamo electric field (DDEF) produced by equator-ward propagating thermospheric disturbance winds that arises from the auroral heating process. It occurs with a time delay of several hours from the storm initiation (Richmond et al., 2003), and has the polarity local time dependence similar to that of the overshielding electric field. These disturbance electric fields are important drivers of the large variability observed in the low-latitude ionospheric irregularities, as briefly discussed later.

Among the variabilities impacting most the space-based communication and navigation systems is that of the plasma bubble/ESF irregularities arising from disturbance electric fields. Depending upon their local time-dependent polarity, such electric fields can cause anomalous development, suppression or disruption of the irregularity development simply through the vertical drift/F layer height variations brought about by these electric fields.

3.2.1 ESF/bubble development or disruption due to undershielding electric field

When a large-amplitude PPEF occurs almost in-phase with (around the time of) the quiet time PRE, the post sunset plasma bubble development can be greatly enhanced even during a season of small (or negligible) PRE vertical drift, such as June months in Brazil (Abdu et al., 2003). A case of post sunset bubble development due to undershielding electric field was presented by Li et al. (2009); the event was observed during the super storm of November 10, 2004, which occurred in a bubble nonoccurrence season in the Asian

longitude sector. The vertical growth of the bubble is faster with increase in the vertical drift. This is because the vertical drift directly contributes to the instability linear growth rate by raising the layer bottom side to reduced collision frequency domain where the gravity term further enhances the instability growth. Thus, the intensity of an ESF event can be greatly enhanced due to larger amplitudes of the PPEF, especially when it occurs at the time of the normal PRE vertical drift. However, there is an upper limit of the vertical drift beyond which bubble development may not occur, because of the possibly rapid enough change in the ambient conditions that could, with the increasing vertical drift, begin to adversely affect the instability growth conditions. For example, Abdu et al. (2008) found that bubble did not develop when the PPEF-induced vertical drift in the post sunset hours attained ~900 m/s over Brazil during the super storm of October 30, 2003. The abnormally large vertical drift, as explained earlier, was caused by polarization electric field arising from large enhancement in the ionospheric conductivity gradients due possibly to energetic particle precipitation in the SAMA region. It is not clear what can be the upper threshold limit of the vertical drift for such nondevelopment of bubble irregularities.

The PPEF has westward polarity after ~22 LT, and a substorm/storm development at these times can result in plasma downdraft and large depression in the F layer heights. Consequently, the bubble irregularities, in their development, or in developed phase, will be brought down to the height domain of higher collision frequency and recombination rates. As a result, an event in progress can be disrupted as was shown, for example, by Abdu et al. (2013) during, the October 29, 2003 super storm sequence.

4 Summary

In summary, we have presented a brief description of the nighttime equatorial ionosphere irregularities, widely known as spread F irregularities, highlighting the mechanism of their generation and development, techniques of their measurements, their spatial and temporal distributions, and short-term variability. In particular, it was noted that the plasma structuring and dynamics constitute an important component of the ionospheric weather, which may be defined as the short-term variabilities in the electric fields, currents, plasma drifts, and instabilities. Development of predictive capability on their occurrence is an important goal in the pursuit of investigation of these irregularities. Toward this end, a more detailed understanding of the variabilities in the different component parameters of the ionospheric structuring in terms of their phenomenology and cause-effect sequence at different time and space scales is a fundamental requirement.

References

Abdu, M.A., Batista, I.S., Bittencourt, J.A., 1981. Some characteristics of spread F at the magnetic equatorial station Fortaleza. J. Geophys. Res. 86, 6836. https://doi.org/10.1029/JA086iA08p06836.

Abdu, M.A., Medeiros, R.T., Bittencourt, J.A., Batista, I.S., 1983. Vertical ionization drift velocities and range types spread F in the evening equatorial ionosphere. J. Geophys. Res. 88 (A1), 399–402.

Abdu, M.A., Sobral, J.H.A., Nelson, O.R., Batista, I.S., 1985. Solar cycle related range type spread-F occurrence characteristics over equatorial and low latitude stations in Brazil. J. Atmos. Terr. Phys. 47 (8–10), 901–905.

Abdu, M.A., Batista, I.S., Takahashi, H., MacDougall, J., Sobral, J.H., Medeiros, A.F., Trivedi, N.B., 2003. Magnetospheric disturbance induced equatorial plasma bubble development and dynamics: a case study in Brazilian sector. J. Geophys. Res. 108 (A12), 1449. https://doi.org/10.1029/2002JA009721.

Abdu, M.A., Batista, P.P., Batista, I.S., Brum, C.G.M., Carrasco, A., Reinisch, B.W., 2006. Planetary wave oscillations in mesospheric winds, equatorial evening prereversal electric field and spread F. Geophys. Res. Lett. 33 (L07107), 1–4.

Abdu, M.A., et al., 2008. Abnormal evening vertical plasma drift and effects on ESF and EIA over Brazil-South Atlantic sector during the 30 October 2003 superstorm. J. Geophys. Res. 113.A07313. https://doi.org/10.1029/2007JA012844.

Abdu, M.A., Kherani, E.A., Batista, I.S., de Paula, E.R., Fritts, D.C., Sobral, J.H.A., 2009. Gravity wave initiation of equatorial spread F/plasma bubble irregularities based on observational data from the SpreadFEx campaign. Ann. Geophys. 27, 2607–2622.

Abdu, M.A., Batista, I.S., Brum, C.G.M., MacDougall, J.W., Santos, A.M., de Souza, J.R., Sobral, J.H.A., 2010. Solar flux effects on the equatorial evening vertical drift and meridional winds over Brazil: a comparison between observational data and the IRI model and the HWM representations. Adv. Space Res. 46, 1078–1085.

Abdu, M.A., Souza, J.R., Batista, I.S., Fejer, B.G., Sobral, J.H.A., 2013. Sporadic E layer development and disruption at low latitudes by prompt penetration electric fields during magnetic storms. J. Geophys. Res. 118, 2639–2647.

Abdu, M.A., de Souza, J.R., Kherani, E.A., Batista, I.S., MacDougall, J.W., Sobral, J.H.A., 2015. Wave structure and polarization electric field development in the bottomside F layer leading to postsunset equatorial spread F. J. Geophys. Res. Space Physics 120, 6930–6940. https://doi.org/10.1002/2015JA021235.

Chen, P.R., 1992. Two-day oscillation of the equatorial anomaly. J. Geophys. Res. 97, 6343–6357.

Fejer, B.G., Scherliess, L., de Paula, E.R., 1999. Effects of the vertical plasma drift velocity on the generation and evolution of equatorial spread F. J. Geophys. Res. 104 (A9), 19859–19869.

Fejer, B.G., Jensen, J.W., Su, S.-Y., 2008. Seasonal and longitudinal dependence of equatorial disturbance vertical plasma drifts. Geophys. Res. Lett. 35, L20106. https://doi.org/10.1029/2008GL035584.

Forbes, J.M., 1996. Planetary waves in the thermosphere-ionosphere system. J. Geomag. Geoelec. 48, 91–98.

Forbes, J.M., Leveroni, S., 1992. Quasi 16-day oscillation in the ionosphere. Geophys. Res. Lett. 19, 981–984.

Fritts, D.C., et al., 2008. Gravity wave and tidal influences on equatorial spread F based on observations during the spread F experiment (SpreadFEx). Ann. Geophys. 26, 3235–3252.

Huang, C.-S., Sazykin, S., Chau, J.L., Maruyama, N., Kelly, M.C., 2007. Penetration electric fields: efficiency and characteristic time scale. J. Atmos. Solar Terr. Phys. 69, 1135–1146.

Huang, C.-S., de La Beaujardiere, O., Pfaff, R.F., Retterer, J.M., Roddy, P.A., Hunton, D.E., Su, Y.-J., Su, S.-Y., Rich, F.J., 2010. Zonal drift of plasma particles inside equatorial plasma bubbles and its relation to the zonal drift of the bubble structure. J. Geophys. Res. 115. A07316. https://doi.org/10.1029/2010JA015324.

Kelley, M.C., Retterer, J., 2008. First successful prediction of a convective equatorial ionospheric storm using solar wind parameters. Space Weather 6, S08003. https://doi.org/10.1029/2007SW000381.

Kherani, E.A., Abdu, M.A., 2011. The acoustic gravity wave induced disturbances in the equatorial ionosphere. In: Aeronomy of the Earth's Atmosphere and Ionosphere, first ed. IAGA/IUGG, Springer. (Org.), Vol. 1. Springer Science + Business Media B.V, Dordrecht, pp. 141–162.

Kikuchi, T., Luhr, H., Schlegel, K., Tachihara, H., Shinohara, M., Kitamura, T.-I., 2000. Penetration of auroral electric fields to the equator during a substorm. J. Geophys. Res. 105, 23251–23261. https://doi.org/10.1029/2000JA900016.

Li, G., Ning, B., Zhao, B., Liu, L., Wan, W., Ding, F., Xu, J.S., Liu, J.Y., Yumoto, K., 2009. Characterizing the 10 November 2004 storm-time middle-latitude plasma bubble event in Southeast Asia using multi-instrument observations. J. Geophys. Res. 114, A07304. https://doi.org/10.1029/2009JA014057.

Li, G., Ning, B., Hu, L., Liu, L., Yue, X., Wan, W., et al., 2010. Longitudinal development of low-latitude ionospheric irregularities during the geomagnetic storms of July 2004. J. Geophys. Res. 115, A04304. https://doi.org/10.1029/2009JA014830.

Li, G., Ning, B., Abdu, M.A., Wan, W., Hu, L., 2012. Precursor signatures and evolution of post-sunset equatorial spread-F observed over Sanya. J. Geophys. Res. 117, A08321. https://doi.org/10.1029/2012JA017820.

McClure, J.P., Sing, S., Bamgboye, D.K., Johnson, F.S., Kil, H., 1998. Occurrence of equatorial F region irregularities: evidence for tropospheric seeding. J. Geophys. Res. 103, 29119–29135.

Pancheva, D., Haldoupis, C., Meek, C.E., Manson, A.H., Mitchell, N.J., 2003. Evidence of a role for modulated atmospheric tides in the dependence of sporadic E layers on planetary waves. J. Geophys. Res. 108 (A5), 1176. https://doi.org/10.1029/2002JA009788.

Reinisch, B.W., Galkin, I.A., 2011. Global ionospheric radio observatory (GIRO). Earth Planets Space 63 (4), 377–381. https://doi.org/10.5047/eps.2011.03.001.

Richmond, A.D., Peymirat, C., Roble, R.G., 2003. Long-lasting disturbances in the equatorial ionospheric electric field simulated with a coupled magnetosphere-ionosphere-thermosphere model. J. Geophys. Res. 108 (A3), 1118. https://doi.org/10.1029/2002JA009758.

Singh, S., Johnson, F.S., Power, R.A., 1997. Gravity wave seeding of equatorial plasma bubbles. J. Geophys. Res. 102 (A4), 7399–7410.

Takahashi, H., Wrasse, C.M., Pancheva, D., Abdu, M.A., Batista, I.S., Lima, L.M., Batista, P.P., Clemesha, B.R., Shiokawa, K., 2006. Signatures of 3– 6 day planetary waves in the equatorial mesosphere and ionosphere. Ann. Geophys. 24, 3343–3350.

Takahashi, H., Wrasse, C.M., Fechine, J., Pancheva, D., Abdu, M.A., Batista, I.S., Lima, L.M., Batista, P.P., Clemesha, B.R., Schuch, N.J., Shiokawa, K., Gobbi, D., Mlynczak, M.G., Russell, J.M., 2007. Signatures of ultra fast kelvin waves in the equatorial middle atmosphere and ionosphere. Geophys. Res. Lett. 34, L11108. https://doi.org/10.1029/2007GL029612.

Thampi, S.V., Yamamoto, M., Tsunoda, R.T., Otsuka, Y., Tsugawa, T., Uemoto, J., Ishii, M., 2009. First observations of large-scale wave structure and equatorial spread F using CERTO radio beacon on the C/NOFS satellite. Geophys. Res. Lett. 36, L18111. https://doi.org/10.1029/2009GL0398.

Tsunoda, R.T., 2008. Satellite traces: an ionogram signature for large-scale wave structure and a precursor for equatorial spread F. Geophys. Res. Lett. 35. https://doi.org/10.1029/2008GL035706.

Tsunoda, R.T., 2010. On seeding equatorial spread F during solstices. Geophys. Res. Lett. 37, L05102. https://doi.org/10.1029/2010GL042576.

Tsunoda, R.T., White, B.R., 1981. On the generation and growth of equatorial backscatter plumes: 1. Wave structure in the bottomside F layer. J. Geophys. Res. 86, 3610–3616. https://doi.org/10.1029/JA086iA05p03610.

Tsunoda, R.T., Yamamoto, M., Tsugawa, T., Hoang, T.L., Tulasi Ram, S., Thampi, S.V., Chau, H.D., Nagatsuma, T., 2011. On seeding, large scale wave structure, equatorial spread F, and scintillations over Vietnam. Geophys. Res. Lett. 38, L20102. https://doi.org/10.1029/2011GL049173.

Tulasi Ram, S., Yamamoto, M., Tsunoda, R.T., Chau, H.D., Hoang, T.L., Damtie, B., Wassaie, M., Yatini, C.Y., Manik, T., Tsugawa, T., 2014. Characteristics of large-scale wave structure observed from African and southeast Asian longitudinal sectors. J. Geophys. Res. Space Physics 119, 2288–2297. https://doi.org/10.1002/2013JA019712.

Vadas, S.L., 2007. Horizontal and vertical propagation and dissipation of gravity waves in the thermosphere from lower atmospheric and thermospheric sources. J. Geophys. Res. 112, A06305. https://doi.org/10.1029/2006JA011845.

Vineeth, C., Pant, T.K., Devasia, C.V., Sridharan, R., 2007. Atmosphere-ionosphere coupling observed over the dip equatorial MLTI region through the quasi 16-day wave. Geophys. Res. Lett. 34, L12102. https://doi.org/10.1029/2007GL030010.

Waliser, D.E., Gautier, C., 1993. A satellite-derived climatology of the ITCZ. J. Climate 6, 2162. https://doi.org/10.1175/1520-0442(1993)006<2162: ASDCOT>2.0.CO;2.

Woodman, R.F., La Hoz, C., 1976. Radar observations of F region equatorial irregularities. J. Geophys. Res. 81, 5447–5466.

Further reading

Kelley, M.C., 1989. The Earth's Ionosphere. Plasma Physics and Electrodynamics. Geophysics Series, vol. 43 Academic Press, San Diego, CA.

Scintillation theory

Dennis L. Knepp[a], Christopher J. Coleman[b]

aNorthWest Research Associates, Monterey, CA, United States
bThe University of Adelaide, Adelaide, SA, Australia

1 Introduction

Ionospheric turbulence can result in irregular structure that has a far-reaching impact on the propagation of electromagnetic waves. In particular, such irregularities can cause scintillation or rapid fading in amplitude and phase that disrupts satellite navigation systems and also degrades astronomical observations (the twinkle in stars). Consequently, the study of propagation through irregularities is important for understanding the performance of important sensing and communications systems. Unfortunately, such structure is far too complex to be usefully described in deterministic terms and a statistical description is the only practical alternative. The major problem is that of relating the statistics of the irregularities to the statistics of the propagating field. In the first part of this chapter we look at some traditional approaches to this problem, starting with the Born approximation. The Born approximation considers the irregularities that consist of a small perturbation to a background medium. The range of applicability of this approach can be extended through the Rytov approximation, but both techniques break down in strong turbulence.

The effect of irregularities is often described through the mutual coherence function (MCF), a function that describes the correlation between the field at different times, positions, and frequencies. In the case of strong turbulence, this function can be shown to satisfy a parabolic equation, providing we assume statistical independence in the direction of propagation (i.e., the Markov assumption). However, these equations are difficult to solve and to make any real progress, it is necessary to assume the irregularities are confined to a thin screen. This is known as the phase-screen approximation for which analytic solutions are available. Such solutions provide an invaluable insight into trans-ionospheric propagation (satellite communications and astronomy), but there are many circumstances where the irregularities are far too strong for this approach to be viable. Under these circumstances, however, the medium can be represented by a sequence of phase-changing screens. The latter approach has proven extremely effective.

The Dynamical Ionosphere
https://doi.org/10.1016/B978-0-12-814782-5.00013-3

The latter part of the chapter concentrates on the use of multiple phase screens (MPSs) to simulate propagation and, in particular, the simulation of propagation under turbulent conditions. Several examples are included to demonstrate the utility of the approach.

2 The Rytov approximation

We consider a propagation medium with a background permittivity ϵ that is perturbed by an amount $\delta\varepsilon$. The time harmonic components of the electric field will satisfy a Helmholtz equation of the form

$$\nabla^2 E + \omega^2\mu(\epsilon + \delta\varepsilon)E = 0 \qquad (1)$$

where E is the electric field with time dependence $\exp(i\omega t)$ suppressed. The inhomogeneous Helmholtz equation for a point source at $\mathbf{r_0}$ is

$$\nabla^2 K + \omega^2\mu\epsilon K = -\delta(\mathbf{r} - \mathbf{r_0}) \qquad (2)$$

and this can be integrated to yield

$$K(\mathbf{r}, \mathbf{r_0}) = \frac{\exp(-ik|\mathbf{r} - \mathbf{r_0}|)}{4\pi|\mathbf{r} - \mathbf{r_0}|} \qquad (3)$$

where $k^2 = \omega^2\mu\epsilon$. We can formally integrate Eq. (1) to yield the integral equation

$$E(\mathbf{r}) = \omega^2 \int_V \mu\,\delta\varepsilon(\mathbf{r_0})K(\mathbf{r},\mathbf{r_0})E(\mathbf{r_0})\,dV_0 \qquad (4)$$

where V is a volume that contains the permittivity perturbations. Now consider the case of a plane wave $\psi_0 = e^{-ikz}$ that is incident on the perturbations (we have implicitly assumed that propagation is in the positive z-direction). The integral equation represented by Eq. (4) can be solved in terms of a perturbation series

$$E = E_0 + E_1 + E_2 + E_3 + \cdots \qquad (5)$$

where

$$E_{i+1}(\mathbf{r}) = \omega^2 \int_V \mu\,\delta\varepsilon(\mathbf{r_0})K(\mathbf{r},\mathbf{r_0})E_i(\mathbf{r_0})\,dV_0 \qquad (6)$$

If $\delta\varepsilon$ is small, the first perturbation

$$E_1(\mathbf{r}) = \omega^2 \int_V \mu\,\delta\varepsilon(\mathbf{r_0})K(\mathbf{r},\mathbf{r_0})E_0(\mathbf{r_0})\,dV_0 \qquad (7)$$

where E_0 is the perturbation-free solution and is sometimes sufficient. This is known as the Born approximation (Ishimaru, 1999).

In theory we can gain an exact solution to Eq. (1) be summing the infinite series (5), but convergence can be quite slow. It is possible to accelerate the convergence by seeking an alternative expansion such as that introduced by Rytov. In this approach, we assume an ansatz of the form

$$E = \exp(\phi) \qquad (8)$$

and, from Eq. (1), this implies

$$\nabla^2\phi + \nabla\phi \cdot \nabla\phi + \omega^2\mu(\epsilon + \delta\varepsilon)\phi = 0 \qquad (9)$$

We now expand ϕ as a perturbation series in terms of the permittivity perturbation, that is,

$$\phi = \phi_0 + \phi_1 + \phi_2 + \phi_3 + \cdots \qquad (10)$$

where the equation at the zeroth order is

$$\nabla^2\phi_0 + \nabla\phi_0 \cdot \nabla\phi_0 + \omega^2\mu\epsilon\phi_0 = 0 \qquad (11)$$

and the equation at the first order is

$$\nabla^2\phi_1 + 2\nabla\phi_0 \cdot \nabla\phi_1 + \omega^2\mu\delta\varepsilon = 0 \qquad (12)$$

for the expansion terms ϕ_0 and ϕ_1, respectively. These equations will be satisfied if $\phi_0 = \ln(E_0)$ and $\phi_1 = E_1/E_0$ and so

$$E \cong E_0\exp(E_1/E_0) \qquad (13)$$

This is known as the Rytov approximation (Ishimaru, 1999) and will usually have a much greater range of validity than the two-term expansion $E = E_0 + E_1$.

Let E_0 be a plane wave that illuminates a patch of irregularities in a volume V. For simplicity we take the direction of propagation to be the z-direction and then $E_0 = \exp(-ikz)$. Consequently, the first perturbation to the electric field will be given by

$$E_1(\mathbf{r}) = \omega^2 \int_V \mu\delta\varepsilon(\mathbf{r_0}) \frac{\exp(-ik|\mathbf{r}-\mathbf{r_0}|)}{4\pi|\mathbf{r}-\mathbf{r_0}|} \exp(-ikz_0) \, dV_0$$

$$(14)$$

We will assume that the wavelength of the plane wave is very much smaller than the length scale of the permittivity perturbations. Then, due to the oscillatory nature of the integrand, the main contributions to the integral will come from around the z-axis. In this case, we can make use of the approximation

$$|\mathbf{r}-\mathbf{r_0}| \cong |z-z_0| + \frac{(x-x_0)^2 + (y-y_0)^2}{2|z-z_0|} \qquad (15)$$

and Eq. (14) will reduce to

$$E_1(\mathbf{r}) = \omega^2 \exp(-ikz)$$

$$\int_{V^+} \mu\delta\varepsilon(\mathbf{r_0}) \frac{\exp\left(-ik\frac{(x-x_0)^2 + (y-y_0)^2}{2|z-z_0|}\right)}{4\pi|z-z_0|} \, dV_0$$

$$(16)$$

where V^+ is that part of V for which $z > z_0$ (note that contributions for $z < z_0$ will be canceled out due to the oscillatory nature of the z_0 integral). Consequently, the Rytov approximation will take the form

$$E \cong \exp(-ikz)\exp(\phi_1) \qquad (17)$$

where

$$\phi_1 = \omega^2 \int_{V^+} \mu\delta\varepsilon(\mathbf{r_0}) \frac{\exp\left(-ik\frac{(x-x_0)^2 + (y-y_0)^2}{2|z-z_0|}\right)}{4\pi|z-z_0|} \, dV_0$$

$$(18)$$

If D is a typical length scale of the irregularity, and L is the propagation distance, the earlier integral will greatly simplify for $D^2 \gg \lambda L$, that is, the *geometric optics* limit. In this case, we can assume $\mu\delta\varepsilon(\mathbf{r_0}) \cong \mu\delta\varepsilon(x, y, z_0)$ and the integrals in the x- and y-directions can be evaluated analytically. Note that $\int_{-\infty}^{\infty} \exp(-iax^2)\, dx = \sqrt{\pi/ia}$, we obtain $\phi_1 = -ik\int_{-\infty}^{z}(\delta\varepsilon(x,y,z_0)/2\epsilon)\, dz_0$.

The term ϕ_2 in the Rytov expansion will satisfy the following equation:

$$\nabla^2\phi_2 + \nabla\phi_1 \cdot \nabla\phi_1 + 2\nabla\phi_0 \cdot \nabla\phi_2 = 0 \qquad (19)$$

and, by analogy with the equation for ϕ_1,

$$\phi_2 = \int_{V^+} \nabla\phi_1 \cdot \nabla\phi_1 \frac{\exp\left(-ik\frac{(x-x_0)^2 + (y-y_0)^2}{2|z-z_0|}\right)}{4\pi|z-z_0|} \, dV_0$$

$$(20)$$

3 Mutual coherence function

The irregularities are rarely known in any detail and we must often content ourselves with a statistical description. In this case, the major problem is that of relating the statistical properties of the irregularities to those of the resulting propagating wave. The effects on the propagating wave are usually described in terms of the statistics of the amplitude and phase perturbations. An important characterization is through the MCF, a function that measures the coherence between the wave that arrives at different locations, times, and frequencies. The MCF is defined by

$$\Gamma(\mathbf{r},t,\omega,\mathbf{r'},t',\omega') = \langle E(\mathbf{r},t,\omega)E^*(\mathbf{r'},t',\omega')\rangle = \langle EE'^*\rangle$$

$$(21)$$

where E' is the electric field evaluated at $(\mathbf{r'}, t', \omega')$. However, it is often more useful to consider the normalized MCF $\Gamma_a = \langle E_a E_a'^*\rangle$ where $E_a = E/E_0$ is the field when normalized by the unperturbed field. The significance of the normalized MCF can be seen by considering the sum of the field when sampled at two different locations ($\mathbf{r}$ and $\mathbf{r'}$). The intensity of the summed field $E^S = E + E'$ is given by

$$I^S = \langle E^S E^{S*}\rangle$$

$$= \langle EE^*\rangle + 2\mathcal{R}\{\langle EE'^*\rangle\} + \langle E'E'^*\rangle \qquad (22)$$

$$= I + 2\mathcal{R}\{\Gamma\} + I'$$

When $\Gamma_a = 0$, $I^S = I + I'$, and the powers are said to add incoherently. However, when $\Gamma_a = 1$, $I^S = E_0^S E_0^{S*}$, and the fields are said to add coherently. Spatial coherence is important in technologies that use multiple antenna systems (radio astronomy for example) and it is important to be able to assess the impact of irregularities on such systems.

In terms of the Rytov approximation

$$\Gamma_a = \langle \exp(\phi + \phi'^*) \rangle \qquad (23)$$

From Eq. (17), it will be noted that ϕ can be regarded as the sum of a large number of independent random variables (the values of $\delta\varepsilon$ at the various points of V) and so the central limit theorem suggests that ϕ will be a Gaussian random variable (it should be noted that this places some restrictions on the statistics of $\delta\varepsilon$ and these might not always apply). For a Gaussian random variable θ, it can be shown that

$$\langle \exp(\theta) \rangle = \exp(\langle\theta\rangle) \exp\left(\frac{1}{2}\langle(\theta - \langle\theta\rangle)^2\rangle\right) \qquad (24)$$

and it follows that

$$\Gamma_a = \exp\left(\langle\phi + \phi'^*\rangle - \frac{1}{2}\langle\phi + \phi'^*\rangle^2 + \frac{1}{2}\langle(\phi + \phi'^*)^2\rangle\right) \qquad (25)$$

It will be noted that $\langle\phi_0\rangle = 0$ and, since we assume that $\langle\delta\varepsilon\rangle = 0$, that $\langle\phi_1\rangle = 0$. As a consequence

$$\Gamma_a \simeq \exp\left(\langle\phi_2 + \phi_2'^*\rangle + \frac{1}{2}\langle(\phi_1 + \phi_1'^*)^2\rangle\right) \qquad (26)$$

From this it will be noted that the term ϕ_2 in the Rytov expansion is important in calculating the MCF. From the expression for ϕ_1 we have that the wavelength is much less than a typical irregularity scale.

$$\langle\phi_1^2\rangle = \omega^4\mu^2 \iint \langle\delta\varepsilon_1\,\delta\varepsilon_0\rangle$$
$$\frac{\exp\left(-ik\left(\frac{(x-x_1)^2+(y-y_1)^2}{2|z-z_1|} + \frac{(x-x_0)^2+(y-y_0)^2}{2|z-z_0|}\right)\right)}{16\pi^2|z-z_1||z-z_0|}\,dV_1\,dV_0 \qquad (27)$$

and

$$\langle\phi_1\phi_1'^*\rangle = \omega^4\mu^2 \iint \langle\delta\varepsilon_1\,\delta\varepsilon_0\rangle$$
$$\frac{\exp\left(-ik\left(\frac{(x-x_1)^2+(y-y_1)^2}{2|z-z_1|} - \frac{(x'-x_0)^2+(y'-y_0)^2}{2|z'-z_0|}\right)\right)}{16\pi^2|z-z_1||z-z_0|}\,dV_1\,dV_0 \qquad (28)$$

(note that ε_i denotes ε evaluated at point $\mathbf{r}_i$). In addition, we have

$$\langle\phi_2\rangle = \int_{V^+} \langle\nabla\phi_1 \cdot \nabla\phi_1\rangle \frac{\exp\left(-ik\frac{(x-x_0)^2+(y-y_0)^2}{2|z-z_0|}\right)}{4\pi|z-z_0|}\,dV_0 \qquad (29)$$

The previous expressions are difficult to evaluate but some headway can be made on noting that the major contributions to the x and y integrals come from around the z axes. This is due to the oscillatory nature of these integrals and the fact that we have assumed the length scale of variations in the permittivity is much greater than a wavelength. In the z-direction, the integrals are not oscillatory, but we will assume that the dimensions of V^+ are much greater than the correlation length and, in this case, $\langle\delta\varepsilon_1\,\delta\varepsilon_0\rangle$ will have the nature of a delta function in the z-direction. In this case, the autocorrelation of the permittivity perturbations will take the form

$$\langle\delta\varepsilon_1\,\delta\varepsilon_0\rangle = \langle\varepsilon^2\rangle\delta(z_1 - z_0)A(x_1 - x_0, y_1 - y_0) \qquad (30)$$

where A is known as the *structure function* and is given by

$$A(x_1 - x_0, y_1 - y_0, z_1)$$
$$= \int_{\infty}^{\infty} \langle\varepsilon^2\rangle^{-1}\langle\delta\varepsilon(x_1,y_1,z_1)\delta\varepsilon(x_0,y_0,z_0)\rangle\,dz_0 \qquad (31)$$

We now consider propagation through irregularities (Fig. 1) that are confined to a slab that is orthogonal to the z-axis (from $z = z_S$ to $z = z_F$). Then the previous integrals can be evaluated (Coleman, 2017; Ishimaru, 1999) and yield a normalized MCF of the form

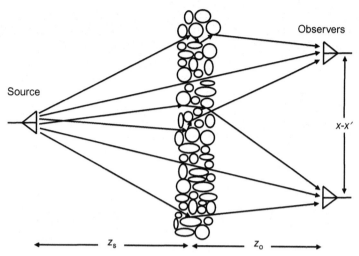

FIG. 1   Spatial decorrelation experienced by a wave propagating from a point source through a slab of ionospheric irregularities.

$$\Gamma_a \cong \exp\left(-(z_E - z_S)\frac{k^2}{4}(A(0,0) - A(x - x', y - y'))\right) \tag{32}$$

where $z_E = min(z, z_F)$. Although the earlier expression is strictly only valid for a plane wave that is incident upon the slab, it can be extended to a point source that is a finite distance z_S from the slab. Let z_0 be the distance of the observer from the slab, then

$$\Gamma_a = \exp\left(-\Delta z\frac{k^2}{4}(A(0,0) - A(\gamma(x - x'), \gamma(y - y')))\right) \tag{33}$$

where $\gamma = z_S/(z_S + z_0)$ and Δz is the thickness of the slab.

4   The parabolic wave equation

The Rytov approximation is useful for describing weak scintillation, but breaks down in the case of strong scintillation. However, when the paraxial approximation is valid, there is another approach that is also valid for strong scintillations. We once again consider a background field that is a plane wave traveling along the z-axis, that is, $E_0 = \exp(-ikz)$ and consider a normalized field such that $E = \exp(-ikz)E_a$. Substituting into Eq. (1), we obtain

$$-k^2 E_a - 2ik\frac{\partial E_a}{\partial z} + \frac{\partial^2 E_a}{\partial z^2} + \nabla_T^2 E_a + \omega^2\mu(\epsilon + \delta\epsilon)E_a = 0 \tag{34}$$

where ∇_T^2 is the Laplace operation transverse to the direction of propagation (i.e., $\partial^2/\partial x^2 + \partial^2/\partial y^2$). In the paraxial limit, the major z variation is contained in the E_0 term and so $\partial^2/\partial z^2$ can be ignored in comparison to other terms. As a consequence, the normalized field E_a will satisfy the parabolic equation

$$-2ik\frac{\partial E_a}{\partial z} + \nabla_T^2 E_a + \omega^2\mu\delta\epsilon E_a = 0 \tag{35}$$

In the case of the ionosphere and atmosphere $\mu = \mu_0$ and, for the ionosphere, we have that $\epsilon \cong \epsilon_0(1 - \omega_p^2/\omega^2)$ where ω_p is the plasma frequency of the ionosphere. It will be noted that $\omega_p^2 = N_e e^2/\epsilon_0 m$ where N_e is the electron density, e is the electron charge, ϵ_0 is the permittivity of free space, and m is the mass of an electron. As a consequence, in terms of perturbations δN_e in the electron density, $\delta\varepsilon = -\delta N_e e^2/m\omega^2$.

If we take the mean of Eq. (35), we obtain

$$-2ik\frac{\partial\langle E_a\rangle}{\partial z} + \nabla_T^2\langle E_a\rangle + \omega^2\mu\langle\delta\varepsilon E_a\rangle = 0 \qquad (36)$$

Assuming perturbations such that Eq. (30) holds, it can be shown that $\langle\delta\varepsilon E_a\rangle = -ik\epsilon A(0,0)\langle E_a\rangle/4$ and, as a consequence

$$-2ik\frac{\partial\langle E_a\rangle}{\partial z} + \nabla_T^2\langle E_a\rangle - \frac{ik^3}{4}A(0,0)\langle E_a\rangle = 0 \qquad (37)$$

For a medium that is uniformly irregular, this has the solution

$$\langle E_a\rangle = \exp\left(-\frac{k^2}{8}A(0,0)z\right) \qquad (38)$$

that is, a plane wave that travels through such a medium will suffer attenuation as it propagates.

The equation for the MCF is far more complex and we limit ourselves to transverse correlations (i.e., we assume $z = z'$). From Eq. (35), we find that

$$-2ik\frac{\partial\langle E_a E_a'\rangle}{\partial z} + \nabla_T^2\langle E_a E_a'\rangle - \nabla_T'^2\langle E_a E_a'\rangle$$

$$+ \omega^2\mu\langle(\delta\varepsilon - \delta\varepsilon')E_a E_a'\rangle = 0 \qquad (39)$$

Assuming perturbations such that Eq. (30) holds, it can be shown that $\langle(\delta\varepsilon - \delta\varepsilon')E_a E_a'\rangle = -ik\epsilon(A(0,0) - A(x-x',y-y'))\langle E_a E_a'\rangle/2$ and, as a consequence

$$-2ik\frac{\partial\Gamma_a}{\partial z} + \nabla_T^2\Gamma_a - \nabla_T'^2\Gamma_a - \frac{ik^3}{2}(A(0,0)$$

$$-A(x-x',y-y'))\Gamma_a = 0 \qquad (40)$$

For a medium that is uniformly irregular, Eq. (40) has the solution

$$\Gamma_a(x-x',y-y',z) = \exp\left(-\frac{zk^2}{4}(A(0,0)\right.$$

$$\left. -A(x-x',y-y'))\right) \qquad (41)$$

In order to consider coherence in terms of time and frequency, Eq. (40) needs to be generalized. We assume that

$$\langle\delta\varepsilon,\delta\varepsilon'\rangle = \langle\epsilon^2\rangle\delta(z-z')A(x-x',y-y',t-t')B(\omega)B(\omega') \qquad (42)$$

where $B(\omega) = -\omega_p^2/(\omega^2 - \omega_p^2)$ in the case of an ionospheric medium. Eq. (40) then generalizes (Knepp, 1983a; Dana, 1986) to

$$0 = \frac{\partial\Gamma_a}{\partial z} + \frac{i}{2k}\nabla_T^2\Gamma_a - \frac{i}{2k}\nabla_T'^2\Gamma_a + \left(\frac{k^2 B^2}{8} + \frac{k^2 B'^2}{8}\right)A(0,0,0)\Gamma_a$$

$$-\frac{kk'BB'}{4}A(x-x',y-y',t-t')\Gamma_a \qquad (43)$$

The solution of Eq. (43) is complex but, in the case where the irregularities are confined to a thin slab, an analytic result can be derived (Knepp, 1983a). Such an approach is known as the phase-screen approximation and can be used in extended regions by means of the MPS approximation. While the MCF provides us with useful information about the statistics of propagation, we need to find a way of simulating the actual propagating wave and it is here that the MPS technique (Knepp, 1983b) proves to be of great utility.

5 MPS calculation

Here a numerical/analytical propagation technique is presented to describe the effects of ionized structure on a propagating electromagnetic signal. The examples here show the propagation of wide bandwidth signals through a hypothetical ionospheric environment. The MPS technique is quite general and may be easily applied to problems involving numerous, separated, layers of ionization characterized by spatially varying electron density. The electron density fluctuations can be described statistically, by their power spectral density (PSD), or by realizations obtained from separate calculations or measurements. The corresponding phase of a layer is obtained by integrating through the structure in the direction of propagation. This process may give correlated phase screens. Although we consider only statistical descriptions of the in situ structure here, we discuss the more general case in the following. MPS techniques can handle all levels of ionospheric disturbances from the least severe, where only minor phase variations occur, to the most severe case of frequency selective scintillation where the spectral components can decorrelate over the bandwidth of the propagating signal.

Since a direct solution to the parabolic wave equation is obtained, the results are exact given the description of the propagation environment. Thus the applicability of the MPS simulation depends on the accuracy of the description of the scattering medium. In the example results presented here, the electron density environment is described by a one-dimensional q^{-3} power-law phase power spectrum with an inner scale cutoff. This spectrum corresponds to a q^{-4} three-dimensional (3D) power-law PSD for in situ electron density fluctuations, which is representative of many in situ measurements. In our notation, $q/2\pi$ is the wave number of the ionization irregularities.

Consider the propagation of an EM signal through electron density structure produced by equatorial plasma instabilities, which cause the ionization to breakup into long filaments, or striations, aligned with the Earth's magnetic field lines. Let us model this as a thick medium composed of random fluctuations in the index of refraction. Consider a plane, unmodulated carrier wave at a constant transmission frequency that traverses the striated region. The wave first suffers random perturbations in the phase due to variations in the phase velocity within the medium. These phase variations in the propagating wave front introduce small changes in the direction of propagation of the wave. Portions of the once plane wave front now propagate in different directions relative to other portions. As the wave propagates farther, diffraction or angular scattering causes constructive and destructive interference which introduces fluctuations in amplitude as well as phase. These spatially varying amplitude and phase fluctuations represent an undesired complex modulation of the wave. If the medium or the line of sight from transmitter to receiver moves with time, the spatial variation is converted into temporal variation of the propagating wave.

This chapter allows for full 3D propagation of a spherical wave, where the phase screens are defined in the $x-y$ plane and propagation is in the z-direction and the parabolic approximation is used to describe the spherical wave. The present work is more general than that of a similar paper by Grimault (1998) where he derives and solves the parabolic wave equation in cylindrical coordinates using MPS techniques. The application here to the case of extended source regions allows much more direct consideration of different transmitter configurations than in *Grimault's* extension of his multiple phase circles technique.

The related paper of Carrano et al. (2012) also applies to 3D plane waves propagating at a fixed angle with respect to the phase screen. They use the spherical wave correction to account for the curvature of the propagating wave front. In a paper very similar to that of Carrano et al. (2012), Deshpande et al. (2014) apply MPS techniques to model propagation along the magnetic field direction, where the magnetic field line is close to or in the same direction as the propagating wave. However, the latter paper confuses the MPS technique with its application to media where the correlation distance of the electron density fluctuations is large in the direction of propagation and the resulting phase screens are correlated. There is no inherent limitation in MPS propagation calculations that require statistically independent phase screens. In fact, given a 3D description of electron density with large correlated structures, phase screens can be generated by numerical integration along the direction of propagation. The phase screens must be correctly spaced to properly represent the ionized medium. If they are closer than the electron-density correlation distance, the screens will be correlated.

A number of authors have discussed the requirement for the statistical independence along the direction of wave propagation (referred to as the Markov assumption). Brown (1971) states that "Beran (1970, p. 1058) has derived a partial differential equation for the second moment (mutual coherence function). A *crucial* (Our emphasis) step in the derivation of this equation is the assumption of local independence. More precisely, Beran (1970) assumes that the field on a plane $z = z_0$ is uncorrelated with the index inhomogeneities between z_0 and $z_0 + \Delta z$, where Δz is much larger than the scale of the inhomogeneities." On the same approximation, Fante (1975, p. 12) states "We can solve the parabolic wave equation for the mean field and the second moment (mutual coherence function) by a method known as the 'Markov approximation.' In this method it is first assumed that the index of refraction fluctuation is delta correlated in the direction of propagation, so that the turbulent eddies look like flat discs oriented normally to the propagation path." Ishimaru (1999, p. 409) invokes the same assumption in solving for the average field and the MCF. Thus the literature clearly states that the Markov approximation is a requirement for the *analytic* solution for the moments of the field governed by the parabolic wave equation.

In related works (Gherm et al., 2005; Gherm and Zernov, 2015; Zernov and Gherm, 2015), compute realizations of signals after 3D spherical-wave propagation using Markov's assumption and the Rytov approximation to analytically calculate the second and fourth moments of the field and the correlation of the real and imaginary parts of the propagating signal. The Markov assumption is a requirement for the *analytic* solution for the moments of the field governed by the parabolic wave equation

(Fante, 1975), but not for the numerical MPS solution. Thus their realizations are generated more efficiently but are less physics based than that of the MPS solution.

6 Formulation

In this section, we describe the theory to allow for spherical wave propagation using MPS techniques. In the remainder of this chapter, we follow the notation of Yeh and Liu (1977) wherein the scalar Helmholtz equation for a propagating wave can be written as follows:

$$\nabla^2 E + k^2(1 + \beta\epsilon)E = 0 \tag{44}$$

where

$$\beta = \frac{-\omega_p^2}{\omega^2 - \omega_p^2} \tag{45}$$

$$\epsilon = \frac{\Delta N_e}{\langle N_e \rangle} \tag{46}$$

$$k = \frac{2\pi}{\lambda} \tag{47}$$

In Eq. (45), ω is the transmission frequency, $\omega_p^2 = 4\pi c^2 r_e \langle N_e \rangle$, where ω_p is the plasma frequency, r_e is the classical electron radius, $\langle N_e \rangle$ is the mean electron density, and ΔN_e is the deviation in electron density. The symbol k is the wave number and c is the speed of light in a vacuum. The product $\beta\epsilon$ of this current section is equivalent to $\delta\varepsilon/\epsilon$ of the previous sections. In the following, we will assume that $\omega \gg \omega_p$. Let the propagating wave front have the form

$$E(x,y,z) = \frac{U(x,y,z)}{(z - z_t)} \exp\{-ik(z - z_t)\}$$
$$\exp\left\{\frac{-ik((x - x_t)^2 + (y - y_t)^2)}{2(z - z_t)}\right\} \tag{48}$$

which is the Fresnel-region or parabolic approximation for a spherical wave that originates at $(x_t,\ y_t,\ z_t)$ and propagates in the positive z-direction. Here the time dependence $\exp(i\omega t)$ is suppressed. In the parabolic or narrow-angle approximation, the wavelength is small with respect to the irregularity scale size so that there is no backscatter from the irregularities. Assume the propagation distance is large and substitute the expression for E into Eq. (44) to obtain the equation for $U(x, y, z)$

$$0 = \frac{\partial^2 U}{\partial x^2} + \frac{\partial^2 U}{\partial y^2} - \frac{i2k}{(z - z_t)} \times \left((x - x_t)\frac{\partial U}{\partial x} + (y - y_t)\frac{\partial U}{\partial y} \right)$$
$$- i2k\frac{\partial U}{\partial z} + k^2\beta\epsilon U \tag{49}$$

Now make the substitutions

$$\theta = \frac{(x - x_t)}{z'} \tag{50}$$

$$\phi = \frac{(y - y_t)}{z'} \tag{51}$$

$$z' = z - z_t \tag{52}$$

and ignore higher-order terms in $1/(z - z_t)$ to obtain a simplified expression for $U(\theta, \phi, z')$ given by

$$\left(\frac{1}{z'^2}\right)\left(\frac{\partial^2 U}{\partial \theta^2} + \frac{\partial^2 U}{\partial \phi^2}\right) - i2k\frac{\partial U}{\partial z'} + k^2\beta\epsilon U = 0 \tag{53}$$

Neglecting the higher-order terms in $1/(z - z_t)$ gives results for the propagating field that is consistent with the Fresnel approximation (Silver, 1965).

Consider for convenience only the two-dimensional (2D) problem, corresponding to infinitely elongated irregularities in the y-direction, with only θ and z' variation. It is easy to extend the solution here to include the omitted terms. This geometry is appropriate for propagation in the Earth's equatorial region if the field lines are perpendicular to the direction of propagation. For this equatorial geometry, there is no possibility for propagation in the direction of the magnetic field lines, where the correlation distance of electron density fluctuations is large.

However, propagation along the magnetic field is easily handled in the 3D MPS case through the use of correlated phase screens that are generated by direct integration of electron density along the direction of propagation. Now separate the thick ionized layer into a number of thin layers, each of which is modeled as a central phase changing screen (phase screen) surrounded by free space.

6.1 Propagation through a phase screen

Consider a layer of thickness $\Delta z'$ centered at zero z'. For small $\Delta z'$ the equation for propagation through this layer is obtained from Eq. (53) with the first (diffraction) term neglected.

The resulting equation is written as

$$\frac{\partial U}{U} = -i\frac{k\beta\epsilon}{2}\partial z' \tag{54}$$

Now integrate over $\Delta z'$ to obtain the solution

$$U\left(\theta, \frac{\Delta z'}{2}\right) = U\left(\theta, \frac{-\Delta z'}{2}\right) \times \exp\left\{-i\frac{k}{2}\int_{-\frac{\Delta z'}{2}}^{\frac{\Delta z'}{2}}\beta\epsilon(\theta, \zeta)d\zeta\right\} \tag{55}$$

Recognizing that the deviation in the index of refraction is $\Delta n = 2\pi r_e \Delta N_e/k^2$ and using Eqs. (45), (46), one can describe the effect of the phase screen (ϕ_{sc}) as

$$U\left(\theta, \frac{\Delta z'}{2}\right) = U\left(\theta, \frac{-\Delta z'}{2}\right)\exp\left\{-ik\int_{-\frac{\Delta z'}{2}}^{\frac{\Delta z'}{2}}\Delta n(\theta, \zeta)\,d\zeta\right\}$$

$$= U\left(\theta, \frac{-\Delta z'}{2}\right)\exp\{-i\phi_{sc}(\theta)\} \tag{56}$$

This implies that

$$\phi_{sc}(\theta) = k\int_{-\frac{\Delta z'}{2}}^{\frac{\Delta z'}{2}}\Delta n(\theta, \zeta)\,d\zeta \tag{57}$$

for 2D problems.

6.2 Free-space propagation

Eq. (53) is valid for free-space propagation between the phase screens if the last term is neglected. To solve the resulting equation, introduce the Fourier transform pair

$$U(\theta,z') = \int_{-\infty}^{\infty} \hat{U}(q_\theta,z')\exp{(i2\pi q_\theta\theta)}dq_\theta \quad (58)$$

$$\hat{U}(q_\theta,z') = \int_{-\infty}^{\infty} U(\theta,z')\exp{(-i2\pi q_\theta\theta)}d\theta \quad (59)$$

Substitute Eq. (58) into Eq. (53) with the last term (which is zero in free space) and the term dependent on ϕ (which was omitted for the convenience of the reader) omitted to obtain the equation for $\hat{U}(q_\theta,z')$

$$\frac{\partial \hat{U}}{\partial z'} = \left(\frac{i2\pi^2 q_\theta^2}{kz'^2}\right)\hat{U} \quad (60)$$

or

$$\frac{\partial \hat{U}}{\hat{U}} = \left(\frac{i2\pi^2 q_\theta^2}{k}\right)\frac{\partial z'}{z'^2} \quad (61)$$

Now integrate from z_1' to z_2' to obtain the solution for free-space propagation

$$\hat{U}(q_\theta,z_2') = \hat{U}(q_\theta,z_1')\exp\left\{\frac{i2\pi^2 q_\theta^2}{k}\left(\frac{1}{z_1'} - \frac{1}{z_2'}\right)\right\} \quad (62)$$

The solution for the electric field is then obtained using Eq. (62) to propagate in free space from screen to screen and Eq. (56) to propagate through each phase screen. This simply involves a series of Fourier transforms that are implemented numerically using fast Fourier transforms (FFTs).

6.3 Phase-screen generation

Given a description of the electron density as a function of the variables z, θ, and ϕ in the 3D case, Eq. (56) describes the method to directly obtain phase screens (ϕ_{sc}) to describe the propagation environment for numerical calculation.

The phase screens need not be homogeneous and can be elongated in the direction of propagation and do not need to satisfy the Markov approximation.

In the examples here, each phase screen consists of a random portion described by its PSD, and a multiplicative tapering window described in the following. The random portion is defined over the entire extent of the MPS grid and is calculated by the discrete equivalent of the Fourier transform

$$\phi_r(x,z) = \sigma_\phi \int_{-\infty}^{\infty} \sqrt{S_\phi(q,z)}g(q) \times \exp{(i2\pi qx)}\,dq \quad (63)$$

where $S_\phi(q, z)$ is the PSD of the phase and $g(q)$ is a complex uncorrelated Gaussian variate with zero-mean and unity variance. The consequences of using another probability density function for g are not known, but could easily be determined numerically on a case-by-case basis. Generally the central limit theorem predicts that the spectral components (g) obtained from a large distribution of irregularities would be Gaussian distributed. The actual phase screen used here is the product of the random portion and a tapering window, that is, $\phi_{sc}(x,z) = \phi_r(x, z) \times w_t(x)$. The purpose of the tapering window is to allow an undisturbed spatial region at the edge of the calculational grid. Thus one can visually monitor possible spatial overlapping in the FFT at the wave propagates. The quantity σ_ϕ^2 is the variance of the phase comprising the entire phase screen, although the total variance is reduced because of the tapering window.

From Eq. (57) it is straightforward to express the 2D PSD of phase fluctuations in terms of the PSD of 3D fluctuations in in situ electron density.

$$S_\phi(q_x,q_y) = \lambda^2 r_e^2 \Delta z' S_{\Delta N_e}(q_x,q_y,q_z = 0) \quad (64)$$

In the examples that follow, we use 2D MPS grids that are situated in the $x-y$ plane with the Shkarofsky (1968) form of the power-law PSD where the PSD has a q^{-n} form for scales between the inner scale l_i and outer scale L_x, L_y with an inner scale that is not dependent on direction.

$$S(q_x, q_y) = \frac{2\pi\sqrt{L_x L_y}\, l_i}{K_{\frac{n-2}{2}}\left(\sqrt{\frac{4L_x L_y\, l_i^2}{4L_x^2 L_y^2 - L_{xy}^2}}\right)}$$

$$\left(1 + 4\pi^2(L_x^2 q_x^2 + L_{xy} q_x q_y + L_y^2 q_y^2)\right)^{-n/4}$$

$$= K_{n/2}\left(\left[\sqrt{\frac{4L_x L_y\, l_i^2}{4L_x^2 L_y^2 - L_{xy}^2}}\right]\right.$$

$$\left.\sqrt{\left(1 + 4\pi^2(L_x^2 q_x^2 + L_{xy} q_x q_y + L_y^2 q_y^2)\right)}\right). \quad (65)$$

For values of wave number q_x or q_y larger than $1/l_i$, the PSD falls off exponentially. In Eq. (65), the infinite integrals over q_x and q_y are unity, so that the standard deviation of the phase is controlled by the quantity σ_ϕ in Eq. (63).

The autocorrelation function corresponding to the PSD in Eq. (65) is given by

$$\rho(x,y) = \frac{1}{\left(\frac{4L_x L_y\, l_i^2}{4L_x^2 L_y^2 - L_{xy}^2}\right)^{\frac{n-2}{4}} K_{\frac{n-2}{2}}\left(\sqrt{\frac{4L_x L_y\, l_i^2}{4L_x^2 L_y^2 - L_{xy}^2}}\right)}$$

$$\times \left(\frac{4L_x L_y l_i^2 + 4L_y^2 x^2 - 4L_{xy} xy + 4L_x^2 y^2}{4L_x^2 L_y^2 - L_{xy}^2}\right)^{\frac{n-2}{4}}$$

$$K_{\frac{n-2}{2}}\left(\sqrt{\frac{4L_x L_y l_i^2 + 4L_y^2 x^2 - 4L_{xy} xy + 4L_x^2 y^2}{4L_x^2 L_y^2 - L_{xy}^2}}\right)$$

$$(66)$$

This combination of PSD and autocorrelation function has several advantages. First, the PSD has the power-law form between outer scale and inner scale that matches most of the in situ observations. Furthermore, the relationship between PSD and autocorrelation function is analytically tractable and the inner-scale cuts off integrals over the PSD.

6.4 Example of wide bandwidth propagation

To obtain the effect of the ionosphere on wide bandwidth signals, one can apply the MPS code at a number of discrete frequencies across a signaling bandwidth. In the following examples, we consider spherical wave propagation through two phase screens to a receiver plane. The first example uses phase screens composed of deterministic Gaussians. The second combines the deterministic screens with random small-scale structure. In both examples, the center frequency is 100 MHz with a bandwidth of 40 MHz. A spherical wave propagates from a transmitter located at $z = -300$ km through two phase screens, located at $z = 100$ km, and $z = 150$ km, and then to the receiver plane located at $z = 200$ km. For this calculation, the size of the first phase screen in the $x–y$ plane, at $z = 100$ km, is set to be 50×50 km. Because the wave front expands as it propagates, the receiver plane expands to 62.5×62.5 km in extent. There are 4000 sample points in the x-direction and 4096 sample points in the y-direction. The MPS code calculates 64 spectral components spanning the range from 80 to 120 MHz.

Two phase screens are given by a Gaussian function as

$$\phi(x,y) = \phi_0 \exp\left\{-\left(\frac{x-x_0}{L_{0x}}\right)^2\right.$$

$$\left. -B_{xy}\left(\frac{(x-x_0)(y-y_0)}{L_{0x}L_{0y}}\right) - \left(\frac{y-y_0}{L_{0y}}\right)^2\right\}$$

$$(67)$$

The phase screen at $z = 100$ km has $\phi_0 = 100$ rad at the center frequency of 100 MHz, $L_{0x} = 5$ km, $L_{0y} = 7$ km, $B_{xy} = 1$, $x_0 = 500$ m, and $y_0 = 300$ m. The phase screen at $z = 150$ km has $\phi_0 = 100$ rad at the center frequency, $L_{0x} = 5$ km, $L_{0y} = 7$ km, $B_{xy} = 1$, $x_0 = 300$ m, and $y_0 = 200$ m.

In the MPS calculation, the peak phase of the deterministic Gaussian lens (ϕ_0) is specified at the carrier frequency. The phase of each of the

64 phase screens across the entire bandwidth is then obtained by scaling inversely with frequency. Fig. 2 (top) shows the amplitude of the propagating field in decibels in the receive plane. Fig. 2 (bottom) shows the phase of the Gaussian phase screen over a portion of the entire MPS $x-y$ grid at $z = 150$ km at the transmission frequency of 100 MHz. The large Gaussian blob of electron density is acting as a divergent lens with a reduction in power of almost 4 dB near the center of the receive plane. The power that was originally near the center of the MPS grid is now spread outward from the center in an oval pattern.

Fig. 3 shows the amplitude and phase of the electric field along horizontal and vertical cuts through the center of the receiver plane. Only the single spectral component at 100 MHz is shown. Note that the field is not symmetric about the center of the grid because of the non-zero values of the offsets x_0 and y_0.

The impulse response is calculated as the integral over the transmission bandwidth

$$h(x,y,t) = \int_{-\frac{B}{2}}^{-\frac{B}{2}} w(f)U(x,y,z_r,f)e^{i2\pi ft}\, df \quad (68)$$

where $w(f)$ are frequency-domain weights and $U(x, y, z_r, f)$ is the transfer function as obtained from a multiple-frequency MPS calculation. If the ionosphere acts as a phase-changing screen with $\phi(x, y) = \lambda r_e N_t(x, y)$, where r_e is the classical electron radius (2.82×10^{-15} m) and N_t is the total electron content, assuming straight line propagation through a thin phase screen, then the transfer function is given by

$$U(x,y,z_r,f) = \exp\left(-i\phi(x,y)\right) = \exp\left(-i\lambda r_e N_t(x,y)\right) \quad (69)$$

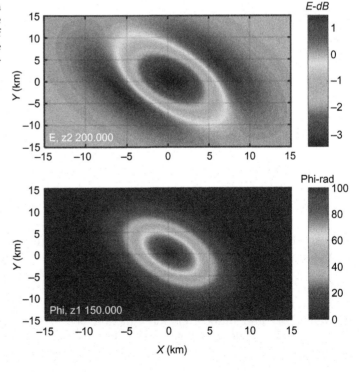

FIG. 2   (*Top*) Electric field at the transmission frequency of 100 MHz in the $x-y$ plane at the receiver location, $z = 200$ km. (*Bottom*) Phase of the large Gaussian lens in the $x-y$ plane at the phase screen located at $z = 150$ km. Only the central portion of the MPS grid is shown.

III. Local irregularities

Then, if the weights are unity, the impulse response function to first order is

$$h(x,y,t) = \frac{\sin\left[\left(\pi B\left(t - \frac{\lambda_0^2 r_e N_t}{2\pi c}\right)\right)\right]}{\pi\left(t - \frac{\lambda_0^2 r_e N_t}{2\pi c}\right)} \quad (70)$$

where λ_0 is the wavelength at the carrier frequency. From Eq. (70), the ionization causes a delay in the received signal. Eq. (70) is valid for delay only; however, additional propagation effects are included in the MPS calculation.

Fig. 4 shows examples of the impulse response function calculated using Eq. (68) with Hanning weights. For this figure we sample 100 equally spaced MPS sample points in the receive plane along the horizontal line ($y = 0$) through the center of the MPS grid. This line has values of x ranging from about -31.25 to 31.25 km, with the sample points evenly spaced by about 500 m. Fig. 4 shows the amplitude on a linear scale of the impulse response function for these 100 sample points plotted as a waterfall plot with each successive function plotted behind the previous, removing the hidden lines where the succeeding function is less than the previous.

Note the time delay experienced near the center of the MPS grid in comparison to the time at the peak of the impulse response near the edges of the MPS grid where there is no phase perturbation. The two edges of the MPS grid appear at the top and bottom of Fig. 4. At the center of the MPS grid, near the center of the figure, the most time delay occurs. For the case that the phase is 200 rad, the delay multiplied by the 40-MHz

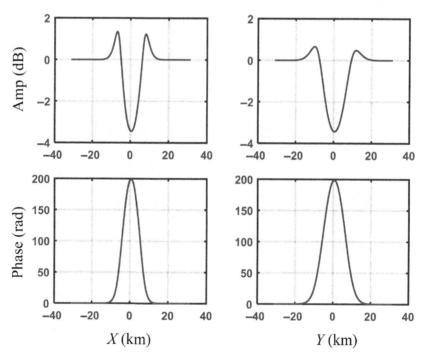

FIG. 3 Phase screens consisting of large Gaussians: Horizontal ($y = 0$) and vertical ($x = 0$) cuts through the MPS grid in the receiver plane at a transmission frequency of 100 MHz. The left-hand frames show the amplitude (*top*) and phase (*bottom*) of the received field as a function of distance along the MPS grid in the x-direction. The right-hand frames show the amplitude and phase of the received field as a function of distance in the y-direction.

III. Local irregularities

bandwidth gives 12.7. This agrees with the maximum delay shown in Fig. 4. Also note the pulse spreading due to dispersion in which different spectral components of the propagating wave propagate at different velocities.

We also performed a similar MPS calculation to obtain the impulse response function for the case of a propagation environment consisting of large divergent lenses and superimposed small-scale structure. The large Gaussian lenses in this example are the same as those used in the previous example. The structure is specified by a Wittwer PSD (65) at each of the two phase screens. For each screen, $\sigma_\phi = 5$ rad. The spectral index for the phase screen at $z = 150$ km is 3. The spectral index for the phase screen at $z = 150$ km is 3.5. Note for the natural ionosphere with an in situ spectral index of electron density fluctuations of 4, the spectral index of phase fluctuations is 3. For both phase screens, the outer scales are 3 km in the x-direction and 5 km in the y-direction. The inner scale is 10 m

in each direction, and the value of the correlation coefficient L_{xy} is zero.

To ensure that the random portion of the composite phase screens is limited in extent to roughly that of the large smooth Gaussian lens, we applied a mask to the phase screen. The mask consists of a Gaussian multiplicative factor, consisting of a central flat region of value unity comprising 20% of the MPS grid. The region outside the flat central region consists of another 20% of the MPS grid characterized by a Gaussian fall-off with a one-sided $1/e$ point equal to 10% of the MPS grid.

Fig. 5 (top) shows the amplitude in decibels of the field in the receiver plane. Fig. 5 (bottom) shows the phase of the composite Gaussian lens plus the masked random phase over a portion of the MPS $x-y$ grid at the transmission frequency of 100 MHz. The large composite structures act as perturbed divergent lenses with random fluctuations.

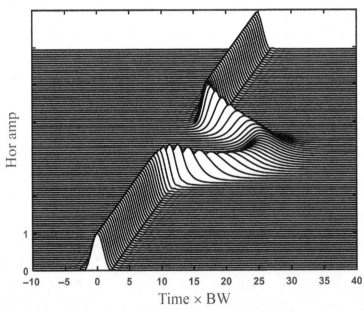

FIG. 4 Phase screens consisting of large Gaussians: Amplitude of the impulse response function versus distance along the horizontal line at $y = 0$ in the 2D MPS grid as a function of the product of time and the bandwidth of the signal. The length of the entire 62.5-km MPS grid is shown.

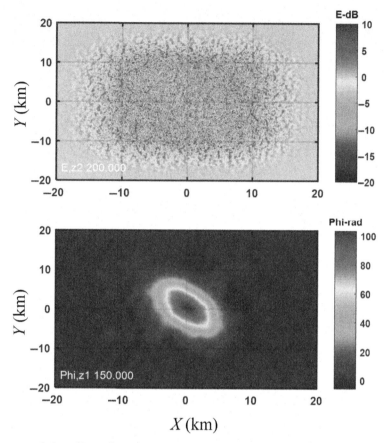

FIG. 5 Phase screens consisting of large Gaussians and random small-scale structure: (*Top*) Electric field at the carrier frequency of 100 MHz in the *x−y* plane at the receiver location $z = 200$ km. (*Bottom*) Phase of the phase screen at $z = 150$ km.

Fig. 6 shows the amplitude and phase of the electric field in the receiver plane. Only the single spectral component of the field at 100 MHz is shown. Fig. 6 shows the field along the vertical and horizontal lines that pass through the center of the 2D MPS grid. The composite lenses cause the field in the receiver plane to experience severe scintillation near the center with some fades of 30 dB and greater. There is a reduction in power in the center and a small amount of focusing visible near the edges of the central region. The field on the far edges of the MPS grid is not affected.

Fig. 7 shows examples of the impulse response function for this example in the same format as that of Fig. 4. Fig. 7 shows the amplitude of the impulse response function for 100 equally spaced points along a horizontal line ($y = 0$) through the center of the MPS grid at the receiver plane. Note the increase in delay caused by the large amount of ionization near the center of the MPS grid. Scintillation is also severe throughout the central region of the MPS grid with additional time-delay jitter caused by the small-scale structure.

III. Local irregularities

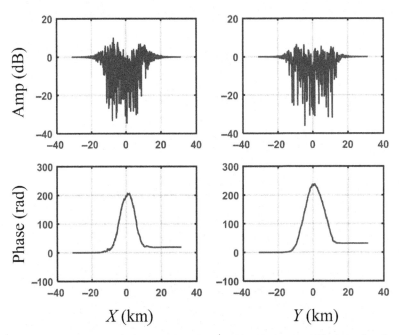

FIG. 6 Phase screens consisting of large Gaussians and random small-scale structure: Horizontal and vertical cuts through the MPS grid in the receiver plane at a transmission frequency of 100 MHz. The left-hand frames show the amplitude (*top*) and phase (*bottom*) of the received field as a function of distance along the MPS grid in the *x*-direction. The right-hand frames show the amplitude and phase of the received field as a function of distance in the *y*-direction.

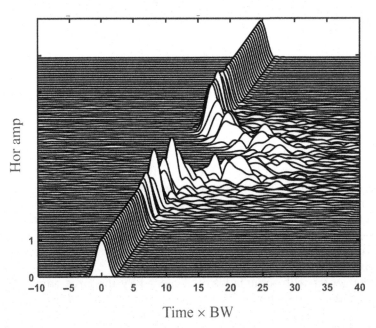

FIG. 7 Phase screens consisting of large Gaussians and random small-scale structure: Amplitude of the impulse response function versus distance along the line $y = 0$ in the 2D MPS grid as a function of time multiplied by the bandwidth of the signal. The length of the entire 62.5-km MPS grid is shown.

III. Local irregularities

7 Conclusion

In this chapter, we discuss the effects of ionospheric structure on the propagation of plane and spherical waves. The beginning of this chapter considers backscatter from ionospheric structure. Here the Rytov approximation is used in the place of the usual first Born approximation to describe scattering from small irregularities. The next several segments utilize the Markov approximation to analytically solve for the mean field and the MCF. The Markov approximation requires the ionospheric structure to be delta correlated along the direction of propagation and is necessary to obtain an analytic solution to the equations for the moments of the propagating field. These solutions give the moments of the field and not the field itself.

In contrast, the MPS propagation technique gives a numerical solution for the propagating field in situations where the Markov approximation is not necessarily valid. The MPS solution is quite general and may be easily applied to problems involving numerous, separated, layers of ionization, characterized either by deterministic-like specifications or by spatially varying PSD. This technique can also handle geometries that violate the Markov approximation where the ionization is highly elongated along the direction of propagation. MPS techniques can handle all levels of ionospheric disturbances from the least severe, where only minor phase fluctuations occur, to the most severe cases of frequency-selective scintillation.

The solution techniques discussed in this chapter are intended to provide different methodologies to handle various aspects of the problems relating to ionospheric scintillation. Understanding the capabilities and limitations of the propagation theory is important for the development of communications and sensing systems that are affected by ionospheric disturbances. For example, a space-based radar system may be impacted because the signals reflected by closely spaced objects in the scene

being imaged are decorrelated by ionospheric scattering. It may be possible to determine whether or not this decorrelation is important by calculating spatial correlation function (i.e., the MCF). On the other hand, a communications signal from a satellite may experience scintillation that disrupts various functions of the receiver system. It may be necessary to obtain an explicit solution for the received field via MPS techniques in order to evaluate the performance of the receiver. Mitigation technique development is only possible through the knowledgeable application of propagation techniques that fully determine the impact of the ionosphere on the propagating signal.

References

Beran, M., 1970. Propagation of a finite beam in a random medium. J. Opt. Soc. Am. 60 (4), 518–521.

Brown, W.P., 1971. Second moment of a wave propagating in a random medium. J. Opt. Soc. Am. 61 (8), 1051–1059.

Carrano, C.S., Groves, K.M., Caton, R.G., 2012. Simulating the impacts of ionospheric scintillation on L-band SAR image formation. Radio Sci. 47, RS0L20. https://doi.org/10.1029/2011RS004956.

Coleman, C.J., 2017. Analysis and Modeling of Radio Wave Propagation. Cambridge University Press, Cambridge.

Dana, R.A., 1986. Propagation of RF signals through ionization. Rep. DNA-TR-86-156, Defense Nuclear Agency, Washington DC, May.

Deshpande, K.B., Bust, G.S., Clauer, C.R., Rino, C.L., Carrano, C.S., 2014. Satellite-beacon ionospheric-scintillation global model of the upper atmosphere (SIGMA) I: high-latitude sensitivity study of the model parameters. J. Geophys. Res. Space Phys. 119, 4026–4043. https://doi.org/10.1002/2013JA019699.

Fante, R.L., 1975. Optical bean propagation in turbulent media. Air Force Cambridge Res Lab, AD/A-018 061, Hanscom AFB, MA.

Gherm, V.E., Zernov, N.N., 2015. Strong scintillation of GNSS signals in the inhomogeneous ionosphere: 2, simulator of transionospheric channel. Radio Sci. 50, 168–176. https://doi.org/10.1002/2014RS005604.

Gherm, V.E., Zernov, N.N., Strangeways, H.J., 2005. Propagation model for transionospheric fluctuating paths of propagation: simulator of the transionospheric channel. Radio Sci. 40, RS1003. https://doi.org/10.1029/2004RS003097.

Grimault, C., 1998. A multiple phase screen technique for electromagnetic wave propagation through random ionospheric irregularities. Radio Sci. 33 (3), 595–605. https://doi.org/10.1029/97RS03552.

Ishimaru, A., 1999. Wave Propagation and Scattering in Random Media, vol. 2, Multiple Scattering, Turbulence, Rough Surfaces, and Remote Sensing. IEEE/OUP Series in Electromagnetic Wave Theory, IEEE Press and Oxford University Press, Oxford and New York.

Knepp, D.L., 1983a. Analytic solution for the two-frequency mutual coherence function for spherical wave propagation. Radio Sci. 18 (4), 535–549. https://doi.org/10.1029/RS018i004p00535.

Knepp, D.L., 1983b. Multiple phase-screen calculation of the temporal behavior of stochastic waves. Proc. IEEE 71, 722–737.

Shkarofsky, I.P., 1968. Generalized turbulence space-correlation and wave-number spectrum-function pairs. Can. J. Phys. 46, 2133–2153. https://doi.org/10.1139/p68-562.

Silver, S., 1965. Microwave Antenna Theory and Design. Dover, New York, NY.

Yeh, K.C., Liu, C.H., 1977. An investigation of temporal moments of stochastic waves. Radio Sci. 12 (5), 671–680. https://doi.org/10.1029/RS012i005p00671.

Zernov, N.N., Gherm, V.E., 2015. Strong scintillation of GNSS signals in the inhomogeneous ionosphere: 1, theoretical background. Radio Sci. 50, 153–167. https://doi.org/10.1002/2014RS005603.

PART IV

The future era of ionospheric science

The complex ionosphere

Massimo Materassi

Institute for Complex Systems of the National Research Council (ISC-CNR), Florence, Italy

1 Introduction

Probably, the best way I have to explain why I think the Earth's ionosphere (EI) should be described as a *complex system* is to quote the words of Professor Kenneth Davies, the one who taught me the physics of ionospheric radio propagation through his book (Davies, 1990), leaving us in 2015 (Professor Kenneth Davies' Obituary, n.d.) after a life proficient as a scientist and a teacher, and generous as a man.

While Professor Davies was visiting the Institute for Applied Physics "Nello Carrara" (IFAC-CNR) in Florence, I asked him what the "big unsolved question" was about the physics of the EI, apparently so well explained in terms of concepts dating back more than 100 years (Classical Mechanics, Maxwell Equations, and some statistics). Kenneth Davies replied that what he had always been wondering was *why the total electron content (TEC) on the top of the same location in the same helio-geophysical conditions at the same local time can be so different every day*, despite everything determining the TEC being supposed to be "classical," hence deterministic. Davies' question is not ill posed at all, neither was it a rhetorical question to introduce his clever answer: *he really had no answer.*

I want to observe that Davies' question is most of all a statement *ruling out simple behaviors* of the EI: even a very limited portion of the EI is composed by such a huge amount of particles that a reasonable control on it in terms of initial conditions and forces must be excluded. The point is that *a purely deterministic mind is not adequate* to face the challenge of theoretical modeling of the EI, and ultimately answer Davies' question. The near-Earth plasma system, and the EI itself, shows the emergence of *coherent collective behaviors* (Heelis et al., 1982; Watkins et al., 1999), *multiscale covariance (multifractality)* (Parkinson, 2008; Chang, 1999), *self-organized criticality* (Chang, 1999), *short-time stochasticity* (Materassi et al., 2011), *percolation properties* (Zelenyj and Milovanov, 2004), and so on. Moreover, the ionospheric medium (IM) sounded via radio signals turns out to show apparently random features needing subtle propagation modeling to be mimicked (Alfonsi et al., 2003; Wernik et al., 2007), and presenting multiscale relationships (Materassi et al., 2005; Materassi and Mitchell, 2007) between fluctuations of different duration. All these features reported for the EI are among those expected in *complex systems*: these are (physical, but not only) *composite systems, the many components of which interact with each other* (Wikipedia, n.d.). The EI is a giant system of many subsystems interacting with each other (single particles, mesoscopic structures, currents, vortices, convective cells, different chemical species), and is not isolated at all (it interacts with all the other sectors of the Earth's atmosphere, and is forced by the solar radiation and plasma, and by cosmic rays).

Turning back to Davies' question: it is of use to underline that *irregular time and space variability* and *apparent randomness* are among the behaviors that complex systems can show. The "tiny" detail unpredictability that Davies' question was mentioning is substantially of the same origin of the unexpected day-to-day weather variability in the troposphere. The question *why* weather has elements of unpredictability should be answered: *because what determines the weather is a complex system*. Meteorologists have accepted to take seriously this intrinsic condition of their object of study (Lorenz, 1963; Liang, 2013), and ionospheric physicists should equally do so: any even "small" part of the EI is characterized by *complexity*, making it hopeless to expect to have totally predictable behavior without an unphysical perfect knowledge of the initial conditions of the system, and of the forces acting on it.

If complexity is recognized as a paramount feature of any aspect of the EI dynamics, it is desirable to take profit of the whole machinery imported from the world of Complexity Sciences (CS) to study it. Complex systems have their own science, developing since the early mid 20th century: as shown in "Castellani's map of complexity" (Castellani, 2009), reported in Fig. 1, complex systems span many sectors of sciences, but the many similarities among them made science produce both theoretical and data analysis tools to study their common features. For instance, we have the theory of renormalization group (Ma, 1973; Chang et al., 1978, 1992) discussing scaling issues; nonlinearities giving rise to highly structured phase spaces are studied by the theory of chaos (Lichtenberg and Liebermann, 1992); information theory tools

have been employed to follow information flow and causality (Liang, 2013; Bossomaier et al., 2016); but the condition of out of equilibrium, instead, does not yet have a unifying framework (Prigogine, 1980; Beretta, 1987; Beretta et al., 1984; Livi and Politi, 2017; Öttinger, 2005).

In this chapter we will focus on how a *dynamical theory*(DYT) of the IM may be formulated, borrowing some mathematical tools originally conceived in CS. The aspect of ionospheric complexity on which we focus mainly here is *plasma turbulence*, which is strictly related to all the phenomenology of *ionospheric irregularities* (see Chapters 11 and 18).

Turbulence is a real paradigm for complexity: its phenomenology still represents an important challenge to the traditional theoretical approach (TTA), because a self-consistent theory of it has not been formulated yet, and probably it cannot be in terms of TTAs. Moreover, the difficulties rendering TTAs not sufficient to describe turbulence precisely lie in the nonnegligible interactions among modes of the system characterized by several different space and time scales, which is the essence of complexity itself. In a sentence, we may say the interest for (plasma) turbulence lies in the fact that it goes beyond the TTAs *because it is complex*.

The two TTAs we mention are *fluid dynamics* (FD) and *kinetic theory* (KT), each of which is not fully able yet to describe the complete phenomenology of turbulence. The claim proposed in this chapter is what is needed to describe turbulence better than FD and KT can do, which can probably be constructed from the theoretical tools presented as *subfluid representation* (SFR) in Materassi et al. (2012), generally inspired from CS.

The aim of SFRs is to represent the "midland" between regarding the IM as a continuum (as done in FD), and a set of many point particles interacting rather weakly (which is what KT does). Indeed, particles do not only interact via two-body collisions perturbing the single particle KT equations via collision integrals. Instead, they undergo multiparticle collisions, form clusters

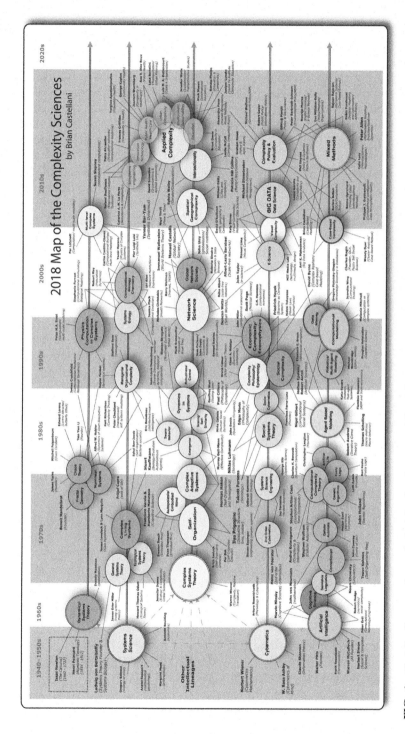

FIG. 1 Castellani's map of complexity: the development of the different lines of research of "science of complexity," and of the network of their "interactions," from the 1940s to the present day (Castellani, 2009).

IV. The future era of ionospheric science

that interact on a wider level, and produce many-body coherent structures like vortices, filaments, and currents. Subfluid models keep the simplicity of treating the medium "as a whole," but alter the point of view of FD, enriching it with details that encode some aspects of the "mesoscopic world" most affecting plasma turbulence.

In this chapter, an SFR for the IM, believed to be relevant for a (future) plasma turbulence theory, is presented, with an application.

First of all, a critical review of the FD approach to the IM is presented in Section 2, where the "philosophy" of fluid mechanics is discussed in terms of what fundamental assumptions must hold in order for the FD picture to make sense; these assumptions are violated in turbulence conditions. The appearance of complexity within the FD description of the IM, and beyond it, is sketched.

In Section 3, the IM is represented as a fluid system undergoing a strongly time- and space-irregular forcing, essentially *noise*, encoding the effects of short living, small-scale fluctuations on the dynamics of fluid variables. This representation is referred to as *stochastic fluid mechanics* (Materassi and Consolini, 2015), and the formalism adopted is that of *stochastic field theory* (SFT). The practical example to which SFT is applied is the problem of Equatorial Spread F "seeding" (Kelley, 1989): in this case, the fluctuations in the local time variability of the electric charge density and in the nighttime recombination rate justify the initiation of irregularities evolving into plasma bubbles and plumes (see Kelley, 1989 and references therein, and Chapter 11).

Finally, what this SFR teaches about a future plasma turbulence theory is resumed in Section 4. A little note about Davies' question closes the chapter.

2 Smooth and deterministic: FD of the EI

An extended material system represented as a single fluid is a *continuous system*: at every time t there exists a given (finite, but not necessarily) portion of the physical space, namely $\mathcal{C}(t) \subseteq \mathbb{R}^3$,

at every point of which matter of the system exists. The continuity of the system consists in the fact that $\mathcal{C}$ *is indeed a manifold* (Burke, 1985); the manifold $\mathcal{C}$ can stretch and roll, it can be deformed by tensions and torsions, its center of mass moves under the action of external forces, but *it cannot be ripped up or punctured* (Fasano and Marmi, 2006). All in all, the evolution of the continuum should be represented as the curve $\mathcal{C}(t)$, from the time interval of interest onto the set of parts of $\mathbb{R}^3$. Despite this general essence of a continuous medium configuration, physics is more interested in *the local description* of its configuration as time passes, and appeals to much simpler, and less abstract, mathematical tools.

2.1 The "parcel philosophy"

In order to describe locally the continuum $\mathcal{C}$, the central concept of *fluid parcel* is defined: a certain "small" portion $\delta\mathcal{C}$ of the whole manifold is considered as the system, and the evolution of the matter occupying the initial configuration of this portion, $\delta\mathcal{C}(t_0)$, is followed, after the initial time t_0. In Fig. 2, the fluid parcel is pictorially represented, as encircled by a solid line within the larger shape of the whole continuum.

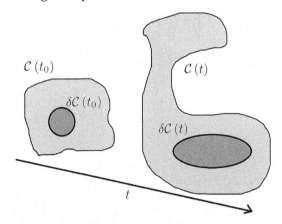

FIG. 2 A pictorial representation of fluid parcel evolution: at the initial time, the whole fluid system is the manifold $\mathcal{C}(t_0)$, and its portion $\delta\mathcal{C}(t_0)$ is chosen as "a parcel" (encircled by a solid line within the larger shape). Then, as time t flows, the continuum evolves into $\mathcal{C}(t)$, while the parcel evolves into $\delta\mathcal{C}(t)$.

In classical mechanics, the degrees of freedom of a fluid parcel are the δN positions and δN momenta of the δN particles in δC: in what we will refer to as the *basic fluid approach* (BFA), those $6 \times \delta N$ dynamical variables are then replaced by the six variables $\vec{r}(\delta C)$ and $\vec{p}(\delta C)$, the *position and momentum of the center of mass* of δC, respectively, and a very small number of variables representing *the equilibrium thermodynamics* of the δN particles (Materassi, 2015). In order for such a scheme to make sense, one must choose δC compatibly with the assumption of thermodynamic equilibrium for the δN particles. This means that, on the one hand, δN must be *large enough* to be able to speak about thermodynamics properly; on the other hand, as no phenomenon of macroscopic transport and relative motion should show up *within* the parcel, the difference between the velocity $\vec{v}_{i=1,...,\delta N}$ of any of the δN particles and $\frac{d}{dt}\vec{r}(\delta C)$ should vanish faster than the parcel size $\ell(\delta C)$. As these two following conditions hold

$$\delta N = \mathbb{O}(N_A), \quad \left| \vec{v}_{i=1,...,\delta N} - \frac{d}{dt}\vec{r}(\delta C) \right| = o(\ell(\delta C))$$

(1)

(being here N_A Avogadro's number), the BFA may be used. Within such a framework, the degrees of freedom of the single δC will be the set of $6 + m$ variables

$$\vec{r}(\delta C), \quad \vec{p}(\delta C), \quad \mathbf{\Theta}(\delta C),$$

being $\mathbf{\Theta}(\delta C)$ the collection of the m thermodynamic coordinates describing the equilibrium thermodynamics of the matter in δC. As $\mathbf{\Theta}(\delta C)$ depends on the parcel chosen, one speaks about *local thermal equilibrium*; these $\mathbf{\Theta}(\delta C)$ will encode the whole world of "relative-to-the-center-of-mass" variables for δC (see Materassi, 1996 and references therein). Note that "proper thermodynamics" here are referred to "large system thermodynamics," as those treated in Zemansky (1957). Recently, "small system thermodynamics" are being developed (Hill, 1964; Gross, 2001)

for systems where the first assumption of Eq. (1) is not satisfied, in which surprising effects due to fluctuations may appear.

The second condition in Eq. (1) makes the whole parcel δC move (almost) as a uniform translation, at least at the space resolution $\ell(\delta C)$: this will not hold when the fluid has a certain level of vorticity, in particular in turbulent regimes (Frisch, 2010), when one has rather $\left| \vec{v}_{i=1,...,\delta N} - \frac{d}{dt}\vec{r}(\delta C) \right| \propto \ell^\alpha(\delta C)$, with $\alpha < 1$, and when the space variability of the medium velocity is very similar to a noisy signal (Materassi and Consolini, 2015).

The equations of motion for the fluid are those obtained by applying mass conservation, $\frac{d\vec{p}}{dt} = \vec{F}$ and energy balance to the matter in δC, plus some more constraints related to the equilibrium thermodynamics of the δN particles in the parcel, for each parcel of the continuum. Mass conservation, energy balance, and momentum balance are general conditions for systems undergoing the Galileo symmetry group, while thermodynamic equilibrium constraints define the chemical species at hand more specifically.

At any time t at each point $\vec{\zeta}_t \in C(t)$ the center of mass of a parcel $\delta C\left(\vec{\zeta}_t\right)$ may be located, and the size and the shape of $\delta C\left(\vec{\zeta}_t\right)$ will be chosen so that conditions (1) are satisfied; this may be repeated for every point of $C(t)$, and it is clear that this will describe the whole material configuration of the continuum at time t. In this way, it is rather obvious that the fluid variables are a set of as many groups $\left(\vec{r}(\delta C), \vec{p}(\delta C), \mathbf{\Theta}(\delta C)\right)$ as the $\vec{\zeta}_t \in C(t)$ are, that is, a *continuous set* $\left(\vec{r}\left(\vec{\zeta}_t\right) = \vec{\zeta}_t, \vec{p}\left(\vec{\zeta}_t\right), \mathbf{\Theta}\left(\vec{\zeta}_t\right)\right)$ as

$$C(t) = \bigcup_{\vec{\zeta}_t \in C(t)} \delta C\left(\vec{\zeta}_t\right).$$

(2)

As we need to attach a material identity to parcels, so that each δC will be a properly identified physical system, we will assign a subdivision of the continuum at the initial time t_0, defining $\left\{ \vec{a} \right\}$ the set of all the initial positions of the centers of mass of the parcels:

$$\vec{\zeta}_{t_0} = \vec{a}, \quad C(t_0) = \bigcup_{\vec{a} \in C(t_0)} \delta C\left(\vec{a}\right).$$

Then, the decomposition in parcels of the continuum at time t is simply described by Eq. (2), but considering $\vec{\zeta}_t$ as the evolution of one precise initial position $\vec{a}$:

$$C(t) = \bigcup_{\vec{a} \in C(t_0)} \delta C\left(M_t\left(\vec{a}\right)\right),$$

being $M_t : \mathbb{R}^3 \mapsto \mathbb{R}^3$ the *diffeomorfism* mapping any initial position $\vec{a}$ of some material point of the continuum into its *evolved version* at time t (i.e., the parcel that is in $\vec{a}$ at the time t_0 will be in $\vec{\zeta}_t = M_t\left(\vec{a}\right)$ at time $t > t_0$).[a] In this way, the dynamical variables describing $\delta C\left(\vec{a},t\right) = \delta C\left(M_t\left(\vec{a}\right)\right)$ are simply the $6 + m$ variables $\left(\vec{r}\left(\vec{a},t\right), \vec{p}\left(\vec{a},t\right), \Theta\left(\vec{a},t\right)\right)$. Each set of dynamical variables of the fluid is labeled by the 3D continuous space variable $\vec{a}$. The pictorial sketch of this reasoning is given in Fig. 3. This way of representing the fluid is referred to as *material* or *Lagrangian description* (Antoniou and Pronko, 2002).

From this construction, it is clear that FD is a *field theory*: it describes the state of spatially extended material systems via a *field variable* $X\left(\vec{a},t\right) = \left(\vec{r}\left(\vec{a},t\right), \vec{p}\left(\vec{a},t\right), \Theta\left(\vec{a},t\right)\right)$, depending on time t but also on a continuous index

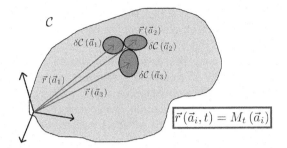

FIG. 3 Three nearby parcels considered out of a continuum C, indicated by the indices $\vec{a}_{i=1,2,3}$. The formula in the *inset* makes it clear that $\vec{a}_i$ plays the role of the initial position of $\vec{r}\ (\rightarrow a_i, t)$.

$\vec{a} \in C(t_0)$. The equations of motion for X will then read

$$\partial_t X = f\left(X, \frac{\partial X}{\partial \vec{a}}\right), \quad \vec{a} \in C(t_0), \qquad (3)$$

where ∂_t is the partial derivative with respect to t. The solution of Eq. (3) is some field $X\left(\vec{a},t\right)$.

Eq. (3) should be commented on stating that, in the case of *classical fluids*, one generally looks for *smooth solutions*

$$X \in C^{\infty}\left(\mathbb{R}^3 \times \mathbb{R}, \Gamma\right) \qquad (4)$$

mapping from the space-time domain $\mathbb{R}^3 \times \mathbb{R}$ to the space of physical states of the system Γ.

Eq. (3) is still written in the Lagrangian representation, that is, the motion of the $\vec{a}$th parcel is described; however, fluids are mostly described in the *Eulerian* or *local* representation, that is, via quantities thought of as properties of the spatial point (Fasano and Marmi, 2006; Choudhuri, 1998; Leitinger et al., 1996). When some quantity $\mathcal{G}$ is attributed to a fluid, the relationship

---

[a] In order to be completely rigorous, the diffeomorphism should be indicated as M_{t,t_0}, as the map from $C(t_0)$ onto $C(t)$ will depend both on the initial *and* on the final time. However, we omit the initial time assuming to fix it once and forever, in order to have a lighter notation.

between the Lagrangian version $\mathcal{G}_L$ and the Eulerian version $\mathcal{G}_E$ of it is given in terms of time derivatives, and reads

$$\frac{d}{dt}\mathcal{G}_L = \partial_t \mathcal{G}_E + \vec{V} \cdot \vec{\partial}\, \mathcal{G}_E, \tag{5}$$

with $\vec{V}$ being the *local speed* of the fluid to which the quantity $\mathcal{G}$ is attributed, and $\vec{\partial} \overset{\text{def}}{=} \frac{\partial}{\partial \vec{r}}$ its space gradient (not to be confused with the derivative $\frac{\partial}{\partial \vec{a}}$). Rigorously speaking, this velocity field $\vec{V}\left(\vec{r},t\right)$ is the speed of the center of mass of the parcel $\delta\mathcal{C}\left(\vec{a},t\right)$ such that, at time t, the evolution of $\vec{a}$ brings that initial position at $\vec{r}$:

$$M_t\left(\vec{a}\right) = \vec{r} \implies \vec{V}\left(\vec{r},t\right) = \frac{\vec{p}\left(\vec{a},t\right)}{m\left(\delta\mathcal{C}\left(\vec{a},t\right)\right)},$$

being $m\left(\delta\mathcal{C}\left(\vec{a},t\right)\right)$ the mass of the parcel under examination. In the Lagrangian and in the Eulerian representations, the state X of the continuum is not the same, so in Eq. (3) one should use the more proper symbol X_L; when the Eulerian representation is adopted, one will formulate some PDE as

$$\partial_t X_E = f_E\left(X_E, \vec{\partial}\, X_E\right), \quad \vec{r} \in \mathbb{R}^3. \tag{6}$$

In Eq. (6), the state is some $X_E = X_E\left(\vec{r},t\right)$, depending on time and on a physical position in the space, $\vec{r}$. Moreover, the two states X_L and X_E defining the parcel dynamics in the Lagrangian and Eulerian representations, respectively, contain different physical quantities. Indeed, while one has $X_L\left(\vec{a},t\right) = \left(\vec{r}\left(\vec{a},t\right), \vec{p}\left(\vec{a},t\right), \Theta\left(\vec{a},t\right)\right)$ as seen before, once the Eulerian representation is adopted, the fields of use are $X_E\left(\vec{r},t\right) = \left(\vec{V}\left(\vec{r},t\right), \Theta\left(\vec{r},t\right)\right)$, with $\Theta\left(\vec{r},t\right)$ being the local field corresponding to the

equilibrium thermodynamic quantities $\Theta\left(\vec{a},t\right)$ in the Eulerian representation. It is useful to note that when the different atmospheric components are supposed to be treatable as fluids at local thermal equilibrium, they have different temperatures and are not in thermal equilibrium *with each other* (Schlegel and St.-Maurice, 1981; Duhau and Azpiazu, 1981).

2.2 The ionospheric fluid equations

The fluid description of the EI is done considering a distinct continuum $\mathcal{C}_I$ for each chemical species present, with $I = 1, 2, \ldots, W$ running over all the chemically different species. The species composing the EI are generally as follows:

- *Neutral particles.* Various molecules or atoms (e.g., O_2, N_2, CO_2, NH_4, O, N) with zero electric charge: this is the major part, both numerically as well as massively, of all the upper atmosphere, up to the level of about 1000 km height, where the *plasmasphere* begins.
- *Negative ions.* Neutrals which have captured one or more electrons, relatively very rare, their presence is practically negligible in general. Nevertheless, one should consider them when dealing with the lower part of the EI, as the region D and the lower E (Leitinger et al., 1996; Jursa, 1985). Their abundance is determinant only under 95 km.
- *Positive ions.* Neutrals that have lost one or more electrons: the 2- or 3-valent positive ions are very rare, so we can restrict to 1-valent positive ion, for example, NO^+, O^+.
- *Electrons.* Electrons are the most important element for *ionospheric radio phenomenology*, because of their very small mass which gives them great mobility under the effect of superimposed electromagnetic fields (Davies, 1990; Ciraolo, 1993), at least for GNSS signals.

A reasonable collection of continua of the EI is

$$C_{O_2}, \ C_{N_2}, \ C_{CO_2}, \ C_{NH_4}, \ C_O, \ C_N, \ C_{NO^+}, \ C_{O^+}, \ C_{e^-},$$
(7)

with $W = 9$. Each C_I needs to be treated as we have discussed for the fluid C in Section 2.1: a proper Eulerian representation will be constructed, with some local $X_E^{(I)}(\vec{r},t)$ such that

$$\partial_t X_E^{(I)}(\vec{r},t) = f_E^{(I)}\Big(\underline{X}_E(\vec{r},t), \vec{\partial} \ \underline{X}_E(\vec{r},t), \vec{r},t\Big),$$
$$\vec{r} \in \mathbb{R}^3, \ t \in \mathcal{I},$$
(8)

adapting Eq. (6) to the Ith continuum (underlined quantities as $\underline{X}_E$ are collections of the W fields, one for each component C_I). Each of these fluids C_I occupies, in principle, the whole available space, so they are copresent in the same region of $\mathbb{R}^3$; the Eulerian representation in this case is particularly suitable in order to represent simply and transparently chemical reactions: two parcels δC_I and δC_J of the species in Eq. (7) will react when they are "at the same space point $\vec{r}$." The reference frame used is in general the Earth-corotating one.

Every fluid C_I is described as made of particles of mass m_I and electrical charge q_I: we will use ρ_I to indicate the mass volume density, N_I for numerical density and δ_I for charge density:

$$\rho_I = m_I N_I, \quad \delta_I = q_I N_I.$$
(9)

A given parcel δC_I centered in $\vec{r}$ at the time t has mass $m(\delta C_I) = \rho_I(\vec{r},t) dv_I(\vec{r},t)$ and charge $q(\delta C_I) = \delta_I(\vec{r},t) dv_I(\vec{r},t)$, with $dv_I(\vec{r},t)$ being the volume of the parcel at hand. In most cases, on these δ_Is a constraint of *local neutrality* holds[b]

$$\sum_I \delta_I = 0,$$
(10)

besides the global one:

$$\sum_I \int_{\mathbb{R}^3} \delta_I(\vec{r},t) d^3 r = 0.$$
(11)

As the configuration of $\delta C_I(\vec{a},t)$ is represented through the dynamical variables of its center of mass and the thermodynamic coordinates of the particles it contains, the equations of motion of the FD of the EI will undergo the *laws of point particle mechanics* (i.e., classical Newtonian physics) plus *classical equilibrium thermodynamics*. In the case of the EI, all the chemicals (7) are treated as *gases*, so the thermodynamic variables may be the number density and temperature:

$$X_E^{(I)}(\vec{r},t) = \Big(\vec{V}_I(\vec{r},t), N_I(\vec{r},t), T_I(\vec{r},t)\Big).$$
(12)

The goal is to obtain Eq. (8) for this collection of variables (one per chemical species): we need to end up with something as $\partial_t N_I(\vec{r},t) = \cdots$, $\partial_t \vec{V}_I(\vec{r},t) = \cdots$, and $\partial_t T_I(\vec{r},t) = \cdots$. As already observed, Newton's dynamics plus equilibrium thermodynamics will be used. The expression for $\partial_t N_I(\vec{r},t)$ is given through a stoichiometric reasoning, appealing to the existence of *generally additive conserved scalars*, as the mass in nonrelativistic physics, or the electric charge. The expression for $\partial_t \vec{V}_I(\vec{r},t)$ is obtained from the application of $\frac{d\vec{p}}{dt} = \vec{F}$ to the parcel center of mass, while $\partial_t T_I(\vec{r},t)$ will be formulated according to internal energy balance of the parcel.

For what is needed to our reasoning here, we "simply" report the full DYT for the fluid ionosphere as

---

[b] This is locally violated in the model presented in Section 3.

$$
\begin{cases}
\partial_t N_I + \vec{\partial} \cdot \left(N_I \vec{V}_I\right) = \Sigma_I\left(\underline{N}, \underline{T}; \vec{x}, t\right), \\[4pt]
\partial_t \vec{V}_I + \left(\vec{V}_I \cdot \vec{\partial}\right)\vec{V}_I = \dfrac{q_I}{m_I}\left(\vec{E} + \vec{V}_I \times \vec{B}\right) \\[6pt]
\quad + \vec{g}_\oplus + \delta \vec{g}_{\text{tide}} + \vec{g}_\Omega - 2\vec{\Omega} \times \vec{V}_I - \dfrac{k_B}{m_I N_I}\vec{\partial}\,(N_I T_I) - \dfrac{\vec{\mathcal{L}}_I\left(\partial \vec{V}_I\right)}{m_I N_I} \\[10pt]
\quad + \displaystyle\sum_{J \neq I} \nu_{IJ}^{\text{micro}}\left(\vec{V}_I - \vec{V}_J\right) - \dfrac{\vec{\partial} \cdot \boldsymbol{\tau}_I^{\text{wav}}}{m_I N_I} - \dfrac{\Sigma_I\left(\underline{N}, \underline{T}; \vec{x}, t\right)}{N_I}\vec{V}_I, \\[12pt]
\partial_t T_I + \vec{V}_I \cdot \vec{\partial}\, T_I = -\dfrac{\Sigma_I\left(\underline{N}, \underline{T}; \vec{x}, t\right)}{N_I c_{VI}^{\text{micro}}(T_I)}\displaystyle\int_{T_0}^{T_I} c_{VI}^{\text{micro}}(T_I')\,dT_I' \\[10pt]
\quad + \dfrac{\vec{\mathcal{L}}_I\left(\partial \vec{V}_I\right)\cdot \vec{V}_I}{m_I N_I c_{VI}^{\text{micro}}(T_I)} + \dfrac{\Psi_I^{\text{micro}}}{c_{VI}^{\text{micro}}(T_I)} - \dfrac{\kappa_I^{\text{micro}}\partial^2 T_I}{m_I N_I c_{VI}^{\text{micro}}(T_I)} - \dfrac{k_B T_I \vec{\partial}\cdot \vec{V}_I}{m_I c_{VI}^{\text{micro}}(T_I)},
\end{cases}
\tag{13}
$$

where $\vec{E}$ and $\vec{B}$ are the electric and magnetic fields in the EI, considered to be rigid external terms for the needs of our reasoning.

In Eq. (13), the dynamical variables are just N_I, $\vec{V}_I$, and T_I, and all the other terms appearing must either be expressed in terms of these $X_E^{(I)}$, or assigned in some other way.

The Σ_I is the source term in the balance equation of the Ith chemical, depending on the photochemistry of the continua forming the EI.

The accelerations $\vec{g}_\oplus$, $\delta\vec{g}_{\text{tide}}$, and $\vec{g}_\Omega$ are, respectively, the Earth's gravity, the fluctuations of it due to the tidal phenomena, and the correction due to the inertial forces in the Earth-corotating frame; $-2\vec{\Omega} \times \vec{V}_I$ is Coriolis acceleration.

$\vec{\mathcal{L}}_I\left(\partial \vec{V}_I\right)$ is the linear function of the velocity gradient $\vec{\partial} \otimes \vec{V}_I$ representing the stress tensor force (Landau and Lifshiz, 1971; Materassi, n.d.), ν_{IJ}^{micro} is the collision frequency between particles of species I and particles of species J, while $\vec{\partial} \cdot \boldsymbol{\tau}_I^{\text{wav}}$ is the force applied on $\mathcal{C}_I$ by gravity waves.[c]

The constant volume-specific heat for the Ith species is $c_{VI}^{\text{micro}}(T_I)$, κ_I^{micro} is the thermal conductivity for the Ith species, while Ψ_I^{micro} is a heat production rate due to inner processes of $\mathcal{C}_I$.

All the quantities indicated with the superscript "micro," that is, ν_{IJ}^{micro}, $c_{VI}^{\text{micro}}(T_I)$, Ψ_I^{micro}, and κ_I^{micro}, should be calculated from a proper *microscopic theory* of the Ith component of the EI, which is outside the scope of this chapter.

2.3 Complexity within and beyond the basic fluid approximation

"Complexity" is expected to arise in two ways for the dynamical theory sketched up to now: *within the fluid approximation* (1) (the BFA) *and beyond it.*

---

[c] This is expressed as the divergence of a suitable stress tensor τ_I^{wav} (Kelley, 1989).

While the BFA holds, one has the system (13): this is a system of five *coupled* field equations for each continuum $\mathcal{C}_I$, and different I continua are coupled among themselves via $\Sigma_I\left(\underline{N},\underline{T};\vec{x},t\right)$, $-\sum_{J\neq I}\nu_{IJ}^{\text{micro}}\left(\vec{V}_I-\vec{V}_J\right)$, $\Psi_I^{\text{micro}}...$; each of those equations is nonlinear, for example, in the Σ_I terms, in the advection terms $\left(\vec{V}_I\cdot\vec{\partial}\right)\vec{V}_I$, in the factors N_I^{-1}, in the dependence $c_{VI}^{\text{micro}}(T_I)...$; last but not least, the physical $\mathbb{R}^3$ space in which all those variables live is *nonhomogeneous*, because the coefficients of all these PDEs depend on $\vec{r}$ explicitly and on time too. So, *well within the fluid approximation*, one ends up with a theory of variables $X_E^{(I)}$ forming a network of nonlinearly interacting components, *expected to show complexity*. Fluid models of the EI are well able to describe a very rich environment in which diversity of local conditions, as those experimentally described in Mahajan (1967), are met; moreover, the different fluids interwoven with each other are not in mutual thermal equilibrium, as experimentally proved in Duhau and Azpiazu (1981). This all supports the *global complexity* described in Part II of this book, in which a rich hierarchy of the parts of the geospace is discussed (Kelley, 1989; Schunk and Nagy, 2009; Kallenrode, 2000).

What if we cannot rely on the conditions (1) and the BFA breaks down?

This happens in turbulence: in this case, *another degree of complexity arises*, essentially the one discussed in Part III of the book. What we lose studying irregular media with respect to fluids is *smoothness*, as happens in the ionospheric regions, giving rise to radio scintillation (see the discussions in Chapters 13 and 18). This means that the "fields" $X_E^{(I)}\left(\vec{x},t\right)$ are not any more smooth in space or in time. This is because the fluctuations of velocities *within the parcel* are so great that the mechanical degrees of freedom

relative to the center of mass cannot be neglected or simply treated as at thermal equilibrium.

If smoothness is released, there may be time intervals in which the evolution of the local proxies of the EI resemble an erratic time sequence, because their fluctuations are intense but short lived: this may give rise to models as discussed in Newman Coffey (2009), in which stochastic acceleration mechanisms are found important in ion heating in the EI.

The necessity of going beyond the BFA appears to be due to the granular nature of matter. Granularity is simply swept under the carpet of the smooth fluid approximation, but its effects cannot always be forgotten. First of all, the within-the-parcel dynamics of a system $\mathcal{C}$ does make the difference, determining the form of fluid theory itself. Ohde et al. (1997) show that the macroscopic ionization/recombination proxies (in our Σ_Is) are deeply influenced by the presence of many-body effects in the microscopic theory of the medium, as well as in diffusion (i.e., the form of $\vec{\mathcal{L}}_I$) and thermal conductivity (our κ_I^{micro}). When the role of *fluctuations in turbulent plasma* is recognized (Consolini et al., 2006), the macroscopic level requires *singular* diffusion and transport equations, which can characterize a properly nonequilibrium thermodynamics point of view. Such mathematical tools become necessary when the emergence of non-Gaussian statistics is recognized, as in composite thermodynamic systems. The supreme effect of matter granularity is that the point particles group themselves hierarchically in coherent structures due to their interactions, forming structures of great diversity (de Anna et al., 2014), which are highlighted in turbulent conditions (Chang et al., 2004). The importance of the mesoscopic level of interactions renders both FD and KT unsuitable to describe turbulence.

It always bears repeating that *smooth fields in Eq. (4) come into the play as an extremization of these quasistatic-equilibrium assumptions*: condition (4) not only renders the fields $\vec{V}$ and Θ constant within the single parcel, but make the differences $d\vec{V}$ and $d\Theta$ from the parcel $\delta\mathcal{C}\left(\vec{r},t\right)$ to the $\vec{r}$- or t-nearby parcel $\delta\mathcal{C}\left(\vec{r}+d\vec{r},t+dt\right)$ as small as $\left|d\vec{r}\right|$ or $\left|dt\right|$, because *the fluid fields vary smoothly* throughout the space and a long time. In turbulence (Materassi et al., 2012; Frisch, 2010; Kallenrode, 2000), the particles move together in mesoscopic groups that are largely independent from each other, to determine extremely severe space gradient and time variability conditions in the fields of the medium both "within the parcel" and "from a parcel to the nearby ones." Turbulence is characterized by *fractal behavior* of the fields; one will have $\left|\vec{v}_{i=1,\dots,\delta N}-\frac{d}{dt}\vec{r}\left(\delta\mathcal{C}\right)\right|\propto\ell^{\alpha}(\delta\mathcal{C})$ for $\alpha\in(0,1)$ *within* the single parcel or $\left|X\left(\vec{r}+d\vec{r}\right)-X\left(\vec{r}\right)\right|\propto\left|d\vec{r}\right|^{\alpha}$ for *nearby* parcels: this clearly violates the second hypothesis in Eq. (1). Mathematically speaking, this renders nonsmooth the proxies $X\left(\vec{r},t\right)$ in both space and time, as one cannot make the derivative of a real function $f(x)$ in x_0 if $f(x_0+\ell)-f(x_0)=\mathbb{O}(\ell^{\alpha})$, with $\alpha\in(0,1)$, as

$$\lim_{\ell\to0}\left|\frac{f(x_0+\ell)-f(x_0)}{\ell}\right|=\lim_{\ell\to0}\mathbb{O}\left(\frac{1}{\ell^{1-\alpha}}\right)\to+\infty$$

diverges as $\ell^{\alpha-1}$, and f has *a cusp* in x_0. Including fractality in plasma physics may give intriguing results (Materassi and Consolini, 2007).

As announced in Section 1, here we present *a possible way of going beyond the BFA* and taking account of this turbulent roughness effect, without renouncing the use of the fluid dynamical variables (12), because of their intuitive nature. This is done by *including noise terms* in the fluid equations, attributing such mathematically irregular characters to quantities for which it is physically sensible to do so: this is done in Section 3.

3 Noisy turbulence in the equatorial spread F

As the value of a *noise* is not given deterministically, but rather distributed according to a certain probability law, *noisy signals may be very irregular*.

Let us consider a noise ζ depending on time t, so that the value $\zeta(t)$ is extracted from the sample space Ω_t of given probability distribution function (PDF) $P_t(\zeta(t))$: at a given time t_1, the value $\zeta(t_1)$ is extracted as some $\zeta(t_1)=\zeta_1$ out of the support $\Omega_{t_1}\equiv\{\zeta\in\mathbb{R}/P_{t_1}(\zeta)\neq0\}$; at a second nearby time $t_2=t_1+dt$, the value $\zeta(t_2)$ is not given by some law inferring it from what $\zeta(t_1)$ is, but instead extracted out of Ω_{t_2} in the way dictated by the proper $P_{t_2}(\zeta)$. Clearly, the difference

$$d\zeta(t_1)\stackrel{\text{def}}{=}\zeta(t_1+dt)-\zeta(t_1)$$

has no need of becoming smaller and smaller as dt is reduced, because the "extraction" of any $\zeta(t)$ out of Ω_t has its own independence, according to what correlations $\langle\zeta(t_1)\zeta(t_2)\rangle$ are given by the PDFs $P_t(\zeta(t))$. For instance, uncorrelated noises

$$\langle\zeta(t_1)\zeta(t_2)\rangle=0\ \ \forall t_1\neq t_2 \tag{14}$$

in general have no hope to be time-continuous, while correlated noises could be, but very likely are nondifferentiable. This irregularity condition of noises may be in both time and space coordinates, as stochastic "fields" $\zeta\left(\vec{r},t\right)$ (Frisch and Matsumoto, 2002).

Equations containing noises have very irregular coefficients, and this character is passed to their solutions: this is why a school of thought has developed, according to which *the irregularity of fluid quantities in a turbulent regime is due to the appearance of noise terms in the PDEs of turbulent fluids* (Carley, n.d.). The point is *what physical meaning* one should give to noise terms in FD equations.

Let us go back to the system of PDEs (13): in addition to the physical proxies of the system state N_I, $\vec{V}_I$, and T_I, there appear those quantities $\Phi^{\text{micro}} = \left(\nu_{IJ}^{\text{micro}}, c_{VI}^{\text{micro}}(T_I), \Psi_I^{\text{micro}}, \kappa_I^{\text{micro}} \right)$, which need to be predicted, in principle, from modeling the way particles interact. The "parcel philosophy" sketched in Section 2.1, on which FD is based, describes the particles forming the parcel just as a gas at thermal equilibrium; as we mentioned in Section 2.3, this approximation is too crude to describe the phenomenology of turbulence, which is caused by the effects of those fluctuations neglected in the assumption of local thermal equilibrium. In general, "microscopic theories" predicting the quantities Φ^{micro} regard them as fluctuating variables, and replace them in the PDEs (13) with their average $\left\langle \Phi^{\text{micro}} \right\rangle$ over the ensemble of microscopic states of the medium macroscopically equivalent to the $X_E^{(I)} \left(\vec{r}, t \right)$ at hand:

$$\text{FD}: \Phi^{\text{micro}} \left(\vec{r}, t \right) \stackrel{\text{def}}{=} \left\langle \Phi^{\text{micro}} \right\rangle_{X_E^{(I)}(\vec{r},t)}.$$

A less crude approximation includes fluctuations: this is like using in Eq. (13) statements of the form

$$\Phi^{\text{micro}} \left(\vec{r}, t \right) = \left\langle \Phi^{\text{micro}} \right\rangle_{X_E^{(I)}(\vec{r},t)} + \delta\Phi^{\text{micro}} \left(\vec{r}, t \right), \tag{15}$$

where $\delta\Phi^{\text{micro}}$ is a *stochastic variable* representing the fluctuations of Φ^{micro} around $\left\langle \Phi^{\text{micro}} \right\rangle$, which are distributed according to some $P_{\text{micro}}(\delta\Phi^{\text{micro}})$. In a single word, $\delta\Phi^{\text{micro}}$ are *noise fields*: hence, if terms (15) are included in the equations of the medium (13), the latter become *stochastic PDEs* (SPDEs), and allow for highly irregular field configurations, like those one may observe in turbulence.

The $\delta\Phi^{\text{micro}}$ statistics should be in principle obtained from the knowledge of a reasonable microscopic theory of the IM; in the literature of (plasma) turbulence treated via SPDEs,

$P_{\text{micro}}(\delta\Phi^{\text{micro}})$ is rather *assumed of a reasonable form*, matching what we grossly know about the medium, and what we are able to treat in the equations (Carley, n.d.).

As $\delta\Phi^{\text{micro}}$ are stochastic terms, N_I, $\vec{V}$, and T_I are necessarily *stochastic fields*: given the noise statistics, the problem becomes being able to predict the consequent statistics of the fluid variables N_I, $\vec{V}$, and T_I in Eq. (13). In some cases, as in the example we are going to describe here, the passage

$$P_{\text{micro}}(\delta\Phi^{\text{micro}}) \mapsto P(\underline{X}_E)$$

from noise to variable statistics may be performed through the *functional formalism* we are going to present in a few pages. About the noises $\delta\Phi^{\text{micro}}$, a last note must be given: at each point $\vec{r}$ and time t the quantity $\delta\Phi^{\text{micro}}\left(\vec{r}, t \right)$ is a different random variable with its own PDF, say $Q_{\vec{r},t}(\delta\Phi^{\text{micro}})$. As time flows, and point by point, an infinite number of possible time-varying configurations $\delta\Phi^{\text{micro}}\left(\vec{r}, t \right)$ may be realized by the stochastic process $\delta\Phi^{\text{micro}}$: each of them has its own probability to take place, hence a realization probability distribution *functional* (RPDF) will be defined, as the statistical weight of the particular "history" $\delta\Phi^{\text{micro}}\left(\vec{r}, t \right)$ in the sample space of realizations. The relationship between this RPDF and the PDFs $Q_{\vec{r},t}(\delta\Phi^{\text{micro}})$ is all but trivial, depending on the nonlocal correlations $\left\langle \delta\Phi^{\text{micro}}\left(\vec{r}, t \right) \delta\Phi^{\text{micro}}\left(\vec{r}', t' \right) \right\rangle$ (see Section 3.3 for the practical example of interest).

In the analytical study reported in Materassi (2019), and presented here as an example of how stochastic equations may be used in ionospheric turbulence, this formalism is applied to the so-called *equatorial spread F* (ESF, see Kelley, 1989 and references therein). ESF is the development of wide and deep ionization depletion in the F layer of the equatorial ionosphere at postsunset local time, basically due to a

Rayleigh-Taylor plasma instability. Not only do regions of smaller ionization density develop on the top of an otherwise smooth background, but moreover they expand and tend to move upward across the nighttime F layer toward the topside ionosphere. As these depletions (plasma bubbles) expand and go up, their local electron density $N_e(\vec{r}, t)$ becomes rougher and rougher, giving rise to plasma irregularities (Chapter 11). The development of irregularities out of instabilities leads the solution of the PDE for $N_e(\vec{r}, t)$ to break the smoothness condition (4): this may be a good case to apply the "noisy turbulence" point of view.

Noise terms appearing in the PDE for N_e have a precise physical meaning: they represent what Kelley refers to as *perturbations seeding the ESF*, in Kelley (1989). These perturbations may have various origins, for example, gravity waves of the underlying thermosphere; they may have different degree of time or space irregularity, but in general their mathematical nature is a *prior assumption* placed "rigidly" in the ESF evolution models. Instead, in this stochastic approach to the ESF, the triggers of irregular perturbations on the smooth background are encoded in noise terms in the electron density equation (Materassi, 2019).

3.1 The system

The physical system is formed by the two continua C_e (electrons) and C_i (single charge ions). The presence of neutrals is fully encoded in the neutral-ion collision term that appears in the equation for C_i, via the neutral-ion collision frequency ν_{in}. The equations for $X_E^{(e)}$ and $X_E^{(i)}$ are developed in the reference frame moving with the neutrals (Kelley, 1989); moreover, neutrals are considered so dense that $\nu_{in}\vec{V}_{e,i}$ is much larger than $\partial_t\vec{V}_{e,i}$, and acceleration terms will be neglected in the PDEs of velocities.

Ionospheric fluids move along the equatorial vertical plane (EVP) xOz, with x being an eastward horizontal axis and z an ascending vertical axis, located around zero magnetic latitude. The coordinate system is the Earth-corotating one, completed by the northward horizontal axis y (see Fig. 4). The relevant quantities in the EVP are the magnetic field $\vec{B} = B\hat{y}$ and gravity $\vec{g} = -g\hat{z}$ (with B and g positive). B and g are taken as constant; moreover, $\vec{g}$ is expected to include the pure positional inertial accelerations, while the Coriolis acceleration is neglected.

The free electron density $N_e(x,z,t)$ along the EVP is the only independent dynamical variable in $X_E^{(e)}$, while one works to get rid of $\vec{V}_e$, T_e, and $X_E^{(i)}$. The equation for N_e is its local balance equation

$$\partial_t N_e + \vec{\partial} \cdot \left(N_e \vec{V}_e\right) = \Sigma_e, \qquad (16)$$

with Σ_e being the ionization source term. The density N_e is the sum of a smooth background $N_0(z)$ plus fluctuations $n(x,t)$:

$$N_e(x,z,t) = N_0(z) + n(x,t). \qquad (17)$$

The drastic simplification $\partial_z n = 0$ is the assumption done in some basic ESF theory reported in Kelley (1989). The smooth (faint) background N_0 is assumed to be the nighttime stable

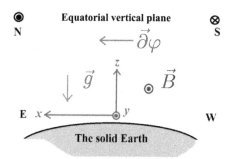

FIG. 4 The EVP along which the whole plasma dynamics takes place, with the fields $\vec{g}$, $\vec{B}$, and $\vec{\partial}\varphi$ drawn explicitly. The Cartesian system of reference and the Cardinal points are indicated.

equilibrium configuration due to recombination: if $\Sigma_e(n) = -Rn$ with $R \geq 0$, one has

$$\partial_t n + \vec{\partial} \cdot \left[(N_0 + n) \vec{V}_e \right] = -Rn. \qquad (18)$$

The velocity $\vec{V}_e$ is determined through the null local electron acceleration hypothesis, plus the high-electron-mobility approximation (Kelley, 1989): in practice, one assumes $\vec{V}_e$ to be given just by the E-cross-B plasma motion $\vec{V}_e = \frac{\vec{E} \times \vec{B}}{B^2}$, so that if $\vec{B}$ is constant and $\vec{E}$ is conservative $\vec{E} = -\vec{\partial} \phi$ the field $\vec{V}_e$ has null divergence, and the equation for n reads

$$\partial_t n + \frac{1}{B^2} \left(\vec{B} \times \vec{\partial} \phi \right) \cdot \vec{\partial} (N_0 + n) = -Rn. \qquad (19)$$

The electric potential ϕ may be expressed in favor of n and the other nondynamical quantities by taking into account the *electric charge continuity*. We assume *quasineutrality* in the formulation of the local electric current density expression

$$\vec{J} = e N_e \left(\vec{V}_i - \vec{V}_e \right), \qquad (20)$$

then the ion velocity is calculated for *intermediate ion-conductivity* (Kelley, 1989):

$$\vec{V}_i = \frac{1}{B^2} \left(\frac{M}{e} \vec{g} - \vec{\partial} \phi \right) \times \vec{B} + \frac{\nu_{in} M}{eB^2} \left(\frac{M}{e} \vec{g} - \vec{\partial} \phi \right)$$

(Here ν_{in} is the ion-neutral collision frequency, while M is the ion mass.) Placing this $\vec{V}_i$ and $\vec{V}_e = \frac{\vec{E} \times \vec{B}}{B^2}$ into Eq. (20), one obtains

$$\vec{J} = \frac{M}{B^2} N_e \vec{g} \times \vec{B} + \frac{N_e \nu_{in} M}{B^2} \left(\frac{M}{e} \vec{g} - \vec{\partial} \phi \right).$$

The electrostatic potential is the sum of a deterministic value ϕ_0 equilibrating gravity for ions

$$M \vec{g} - e \vec{\partial} \phi_0 = 0, \qquad (21)$$

plus fluctuations φ; the electric current density reads

$$\vec{J} = \frac{M}{B^2} (N_0 + n) \left(\vec{g} \times \vec{B} - \nu_{in} \vec{\partial} \varphi \right), \qquad (22)$$

while the PDE for n reads

$$\partial_t n + \frac{1}{B^2} \left[\vec{B} \times \left(\vec{\partial} \phi_0 + \vec{\partial} \varphi \right) \right] \cdot \vec{\partial} n$$
$$+ \frac{1}{B^2} \left[\vec{B} \times \left(\vec{\partial} \phi_0 + \vec{\partial} \varphi \right) \right] \cdot \vec{\partial} N_0 + Rn = 0. \qquad (23)$$

$\vec{\partial} \varphi$ is expressed in terms of n via the electric charge conservation $\vec{\partial} \cdot \vec{J} = -\partial_t \delta$: this is where a first noise term enters the play, considering the local time derivative of the charge density δ so irregular to be represented as a space-time noise $\xi(\vec{r}, t) \overset{\text{def}}{=} \partial_t \delta(\vec{r}, t)$. One then has

$$\vec{\partial} \cdot \vec{J} = -\xi, \; \xi = \partial_t \delta. \qquad (24)$$

As $\langle \delta \rangle = 0$ and $\langle \xi \rangle = 0$ are assumed, the neutrality of the turbulent plasma is realized on the (ensemble) average at all points and times. The charge continuity equation for the current (22) and the conditions (24) reads

$$(N_0 + n) \partial^2 \varphi + \left(\vec{\partial} N_0 + \vec{\partial} n \right) \cdot \vec{\partial} \varphi$$
$$= \frac{B^2 \xi}{M \nu_{in}} + \frac{\left(\vec{\partial} N_0 + \vec{\partial} n \right) \cdot \left(\vec{g} \times \vec{B} \right)}{\nu_{in}}. \qquad (25)$$

In order to invert Eq. (25) in favor of $\vec{\partial} \varphi$, one assumes $\varphi = \varphi(x, t)$ as done in many nonstochastic theories of the ESF, plus a plane modulation

$$\vec{k} = k\hat{x}, \quad \vec{\partial} \varphi = \partial_x \varphi \hat{x}, \quad \partial^2 \varphi = ik \partial_x \varphi,$$

so that one writes

$$\partial_x \varphi = \frac{B^2 \xi + gBM \partial_x n}{M \nu_{in} [ik(N_0 + n) + \partial_x n]}. \qquad (26)$$

The whole PDE for n reads

$$\partial_t n = \frac{g}{B} \left\{ \frac{M}{e} + \frac{\partial_z N_0}{\nu_{in} [ik(N_0 + n) + \partial_x n]} \right\} \partial_x n$$
$$+ \frac{\partial_z N_0}{M \nu_{in} [ik(N_0 + n) + \partial_x n]} \xi - Rn. \qquad (27)$$

In the SPDE (27), the terms R and ξ are both *multiplicative noises*, while the theory in Phythian (1977), which we want to make use of to construct the SFT for n, requires the existence of as many *additive* noises as the dynamical variables are. In order for the formalism defined in Phythian (1977) to be of use here, the choice is to define a new variable

$$\psi = \ln\left(\frac{n}{N_0}\right), \quad n = N_0 e^{\psi}. \tag{28}$$

As N_0 does not depend on t and x, one has

$$\frac{\partial_t n}{n} = \partial_t \psi, \quad \frac{\partial_x n}{n} = \partial_x \psi.$$

The PDE for ψ is written as

$$\partial_t \psi = \frac{g}{B}\left\{\frac{M}{e} + \frac{\partial_z N_0}{\nu_{in} N_0[ik(1+e^{\psi}) + e^{\psi}\partial_x \psi]}\right\}\partial_x \psi \tag{29}$$
$$+ \frac{\partial_z N_0}{M\nu_{in} N_0^2 e^{\psi}[ik(1+e^{\psi}) + e^{\psi}\partial_x \psi]}\xi - R.$$

For future convenience, we define a function

$$G(\psi, \partial_x \psi, k) = \frac{\partial_z N_0}{\nu_{in} N_0[ik(1+e^{\psi}) + e^{\psi}\partial_x \psi]}, \tag{30}$$

so that one may rewrite the SPDE as

$$\partial_t \psi = \frac{g}{B}\left[\frac{M}{e} + G(\psi, \partial_x \psi, k)\right]\partial_x \psi \tag{31}$$
$$+ G(\psi, \partial_x \psi, k)\frac{e^{-\psi}}{N_0 M}\xi - R.$$

Eq. (31) is the SPDE, for the field ψ, with noises ξ and R, to be treated as described in Section 3.2.

3.2 Functional formalism

In the absence of noise, with $\xi = 0$ and $R = 0$, Eq. (27) would simply be a nonlinear model for the development of the ESF. When the equation takes the form (31), the model with the variable ψ defined in Eq. (28) may be applied to the framework developed in Phythian (1977), and specialized to field theories in Materassi and Consolini (2008).

Let us consider a field $\psi = \psi(x,t)$ undergoing the stochastic equation

$$\partial_t \psi(x,t) = \Lambda[\psi;x,t] + \int_{\mathbb{R}} g(y,t)\Gamma[\psi;x,y,t]dy + f(x,t). \tag{32}$$

The expressions Λ and Γ are deterministic terms, while f and g are noises. In Λ and Γ, the hybrid $[\alpha;\beta]$ bracket indicates a nonlocal dependence on the arguments α and a local dependence on the βs; in particular, $\Lambda[\psi;x,t]$ means that Λ may depend on the value of ψ at all the coordinates x', while it depends explicitly only on the point x and time t. Equally, $\Gamma[\psi;x,y,t]$ may depend on all the values of ψ at all the points of its domain, but just on the two space points x and y and on the time t. Eq. (32) must be compared to Eq. (31): the terms Λ, Γ, f, and g of Eq. (32) are hence identified as

$$\Lambda[\psi;x,t] \overset{\text{def}}{=} \frac{g}{B}\left[\frac{M}{e} + G(\psi(x,t), \partial_x \psi(x,t), k)\right]\partial_x \psi(x,t),$$

$$\Gamma[\psi;x,y,t] = \gamma(\psi(y,t), \partial_y \psi(y,t))\delta(y-x),$$

$$\gamma(\psi(y,t), \partial_y \psi(y,t))$$
$$\overset{\text{def}}{=} \frac{g}{B}\left[\frac{M}{e} + G(\psi(y,t), \partial_y \psi(y,t), k)\right]\frac{e^{-\psi(y,t)}}{N_0 M},$$

$$g(y,t) \overset{\text{def}}{=} \xi(y,t), \quad f(x,t) \overset{\text{def}}{=} -R(x,t). \tag{33}$$

We are now in the position to apply the functional formalism to the SPDE (31), that is, to form an SFT for Eq. (31).

An SFT predicts the RPDF of the process $\psi(x,t)$ between an initial time t_i and a final time t_f, from the knowledge of the RPDF of noises, and the dynamics $\partial_t \psi$ in terms of an SPDE: in our case this means going from the RPDF $P[R,\xi;t_i, t_f]$ to the RPDF $\mathcal{A}[\psi;t_i, t_f]$ taking into account Eq. (31). The construction of the SFT begins with the introduction of an *auxiliary process* $\chi(x,t)$ that is conjugated to the additive noise $-R(x,t)$, in the sense that the RPDF of χ is the functional Fourier transform of the one of $-R$. As an intermediate step, one obtains a theory for the fields $(\psi(x,t),\chi(x,t))$, in the sense that the RPDF of the process (ψ,χ) between t_i and t_f is given as the kernel $A[\psi,\chi;t_i, t_f]$:

$$
\left\{
\begin{aligned}
&A[\psi,\chi;t_i,t_f) \\
&= A_0(t_i,t_f)C[\chi,\Gamma;t_i,t_f)e^{-i\int_{t_i}^{t_f} dt \int dx\left[\partial_t\psi(x,t)\chi(x,t)-\Lambda[\psi;x,t)\chi(x,t)-\frac{i}{2}\frac{\delta\Lambda[\psi;x,t)}{\delta\psi(x,t)}\right]}, \\[2em]
&C[\chi,\Gamma;t_i,t_f) \\
&\stackrel{\text{def}}{=} \int [dR]\int [d\xi]P[R,\xi;t_i,t_f)e^{i\int_{t_i}^{t_f} dt \int dx\left[-R(x,t)\chi(x,t)+\int dy\xi(x,t)\chi(y,t)\Gamma[\psi;x,y,t)+\xi(x,t)\int dy\frac{\delta\Gamma[\psi;x,y,t)}{\delta\psi(y)}\right]}.
\end{aligned}
\right.
$$

$$(34)$$

Of course, one would like to get rid of the χ-dependence in the SFT, calculating the physical kernel

$$
\mathcal{A}[\psi;t_i,t_f) = \int [d\chi]A[\psi,\chi;t_i,t_f), \tag{35}
$$

which is the statistical weight of the realization $\psi(x,t)$; the operation (35) turns out to be feasible only when the way of $A[\psi,\chi;t_i,t_f)$ to depend on χ is "easy" enough, for example, if the kernel $A[\psi,\chi;t_i,t_f)$ is Gaussian in χ. In turn, this proves to be the case when ξ is Gaussian and R undergoes an exponential distribution: that is the only case reported here, but many more may be of interest.

If the formulation (34) is applied to the problem of the ESF described by the SPDE (31), the kernel $A[\psi,\chi;t_i,t_f)$ is calculated as

$$
A[\psi,\chi;t_i,t_f)
$$
$$
= A_0(t_i,t_f)C[\chi,\Gamma;t_i,t_f)e^{-i\int_{t_i}^{t_f} dt \int dx\left\{\chi\partial_t\psi-\chi\Lambda[\psi;x,t)+\frac{i}{2}\frac{g\partial_z N_0(ike^\psi+e^\psi\partial_x\psi)\partial_x\psi}{B\nu_{in}N_0[ik(1+e^\psi)+e^\psi\partial_x\psi]^2}\right\}}.
$$

Moreover, considering the expression of $\Lambda[\psi;x,t)$ in Eq. (33), with the expression of G in Eq. (30), one has

$$
A[\psi,\chi;t_i,t_f) = A_0(t_i,t_f)C[\chi,\Gamma;t_i,t_f)
$$
$$
\times e^{-i\int_{t_i}^{t_f} dt \int dx\left\{\chi\partial_t\psi-\chi\left[\frac{Mg}{eB}+\frac{g\partial_z N_0}{B\nu_{in}N_0[ik(1+e^\psi)+e^\psi\partial_x\psi]}\right]\partial_x\psi+\frac{i}{2}\frac{g\partial_z N_0(ike^\psi+e^\psi\partial_x\psi)\partial_x\psi}{B\nu_{in}N_0[ik(1+e^\psi)+e^\psi\partial_x\psi]^2}\right\}},
$$

$$(36)$$

where the factor $C[\chi,\Gamma;t_i,t_f)$ in which the noise RPDF enters directly must be calculated with the following ingredients:

$$
C[\chi,\Gamma;t_i,t_f] = \int [dR] \int [d\xi] P[\xi,R;t_i,t_f] e^{\,i \int_{t_i}^{t_f} dt \int dx[-\chi R + W(\chi,\psi,\partial_x\psi)\xi]}\,,
$$

$$
W(\chi,\psi,\partial_x\psi) = \chi\gamma(\psi,\partial_x\psi) + \varsigma(\psi,\partial_x\psi),
$$

$$
\gamma(\psi,\partial_x\psi) = \frac{g}{B}\left[\frac{M}{e} + G(\psi,\partial_x\psi,k)\right]\frac{e^{-\psi}}{N_0 M},
$$

$$
\varsigma(\psi,\partial_x\psi) \overset{def}{=} -\frac{g\partial_z N_0(ik + \partial_x\psi)}{BM\nu_{in}N_0^2[ik(1+e^\psi) + e^\psi\partial_x\psi]^2} - \frac{ge^{-\psi}}{eBN_0} - \frac{g}{BMN_0}G(\psi,\partial_x\psi,k)e^{-\psi},
$$

$$
G(\psi,\partial_x\psi,k) = \frac{\partial_z N_0}{\nu_{in}N_0[ik(1+e^\psi) + e^\psi\partial_x\psi]}.
$$

(37)

This $C[\chi,\Gamma;t_i,t_f]$ in general is difficult, or impossible, to calculate: the point is to be able to make the integrations in $[d\xi]$ and $[dR]$, where in general the RPDF $P[\xi,R;t_i,t_f]$ multiplied by $e^{\,i \int_{t_i}^{t_f} dt \int dx[-\chi R + W(\chi,\psi,\partial_x\psi)\xi]}$ may have a form so as to give rise to "tough" integrands, the primitive of which remains unknown. In Materassi (2019),

some feasible examples are given, one of which is reported in Section 3.3.

3.3 Kernel $\mathcal{A}[\psi;t_i,t_f]$ for a calculable C case

The calculation of $C[\chi,\Gamma;t_i,t_f]$ as defined in Eq. (34) is all but straightforward; it can be done only if the whole expression

$$
P[R,\xi;t_i,t_f] e^{\,i \int_{t_i}^{t_f} dt \int dx\left[-R(x,t)\chi(x,t) + \int dy\,\xi(x,t)\chi(y,t)\Gamma[\psi;x,y,t] + \xi(x,t)\int dy\frac{\delta\Gamma[\psi;x,y,t]}{\delta\psi(y)}\right]}
$$

can be integrated, that is, if a primitive of its part in $\xi(x,t)$ and $R(x,t)$ may be retrieved explicitly. With some special, but physically reasonable, assumptions on $P[R,\xi;t_i,t_f]$, this graceful condition is met.

First of all, let us suppose the noises to be uncorrelated with each other:

$$
\langle R(x,t)\xi(x',t')\rangle = 0 \quad \forall\, x,x',t,t'.
$$

Then the RPDF $P[R,\xi;t_i,t_f]$ is the product of two R-dependent and ξ-dependent terms,

$$
P[R,\xi;t_i,t_f] = P_R[R;t_i,t_f]P_\xi[\xi;t_i,t_f]. \tag{38}
$$

Let us also assume both noises to be each δ-correlated in time and space:

$$
\begin{cases}
\langle R(x,t)R(x',t')\rangle \propto \delta(x-x')\delta(t-t'), \\
\langle \xi(x,t)\xi(x',t')\rangle \propto \delta(x-x')\delta(t-t').
\end{cases}
$$

This means that each factor in Eq. (38) is the infinite product of x- and t-fixed PDFs, as

$$
\begin{cases}
P_R[R;t_i,t_f] = \prod_{\substack{x\in\mathbb{R} \\ t\in[t_i,t_f]}} Q_R(R(x,t))^{dxdt}, \\[4mm]
P_\xi[\xi;t_i,t_f] = \prod_{\substack{x\in\mathbb{R} \\ t\in[t_i,t_f]}} Q_\xi(\xi(x,t))^{dxdt}.
\end{cases}
$$

(39)

The mathematical consistency of Eq. (39) is a bona fide assumption, for the moment being, because we are using infinitesimal powers $dxdt$ for finite factors and then making continuously infinite products $\prod_{\substack{x \in \mathbb{R} \\ t \in [t_i, t_f]}}$ ... of them: this is basically what is done in quantum mechanics (QM) when the path integral approach is adopted (Feynman and Hibbs, 1965).

Let us now consider some reasonable (but not precisely physics-based) assumptions on the RPDFs of R and ξ. As R is a positive quantity (see the reasonings at page 42), the noise R must be a positive defined stochastic variable; moreover, as in general almost no recombination appears in the scenario described in Section 3.1, the most likely value of R has to be $R = 0$, that is, the PDF $Q_R(R(x,t))$ must peak at null R. As a further requirement, $Q_R(R)$ has to be an integrable function over $\mathbb{R}^+$, so it has to tend to zero for great R. Last but not least, we would like to introduce a form of $P[R,\xi;t_i, t_f]$ in Eq. (37) that is reasonably easy to integrate in each $dR(x,t)$. The solution proposed is *a decreasing exponential distribution* for each $R(x,t)$: the RPDF for the noise R hence reads

$$
\begin{cases}
P_R[R;t_i, t_f] = H_R e^{\displaystyle -\int_{t_i}^{t_f} dt \int_{\mathbb{R}} dx \mu_R(x,t)R(x,t)}, \\[2mm]
H_R = \prod_{\substack{t \in [t_i, t_f] \\ x \in \mathbb{R}}} \mu_R(x,t)\theta(R(x,t))dxdt, \\[2mm]
\int [dR]P_R[R] = 1,
\end{cases}
$$

(40)

where the factor $\theta(R(x,t))$ in Eq. (40) is the Heaviside step function, so that $\theta(R(x,t)) = 0$ for $R(x,t) < 0$, while $\theta(R(x,t)) = 1$ for $R(x,t) \geq 0$.

As far as the other noise is concerned, $\xi = \partial_t \delta$, its physical meaning suggests it cannot be positive definite, but rather fluctuates around zero in a presumably symmetric way. So, why not suppose it to be Gaussian-distributed? As a

sidenote, the Gaussian functional will be easily integrable in each $d\xi(x,t)$, rendering the use of the RPDF of ξ in Eq. (37) suitable. Due to these reasons, we assume

$$
\begin{cases}
P_\xi[\xi;t_i, t_f] = K_\xi e^{\displaystyle -\int_{t_i}^{t_f} dt \int_{\mathbb{R}} dx \lambda_\xi(x,t)\xi^2(x,t)}, \\[2mm]
K_\xi = \prod_{\substack{t \in [t_i, t_f] \\ x \in \mathbb{R}}} \sqrt{\frac{\lambda_\xi(x,t)dxdt}{\pi}}, \\[2mm]
\int [d\xi]P_\xi[\xi] = 1.
\end{cases}
$$

(41)

With the RPDFs (40), (41), one may calculate the local average and variance of the noises as follows:

$$
\begin{cases}
\langle R(x,t)\rangle = \dfrac{1}{\mu_R(x,t)}, \quad \langle \xi(x,t)\rangle = 0, \\[3mm]
\left\langle \left(R(x,t) - \dfrac{1}{\mu_R(x,t)}\right)^2 \right\rangle = \dfrac{3}{\mu_R^2(x,t)}, \\[3mm]
\langle \xi^2(x,t)\rangle = \dfrac{1}{2\lambda_\xi(x,t)}.
\end{cases}
$$

In Materassi (2019), the calculations of $C[\chi,\Gamma;t_i, t_f]$ in Eq. (34) are performed step by step, with the final result as follows:

$$
C[\chi,\Gamma;t_i, t_f] = e^{\displaystyle -\int_{\mathbb{R}} dx \int_{t_0}^{t_f} dt \frac{W^2(\chi,\psi,\partial_x\psi)}{4\lambda_\xi(x,t)}} \prod_{\substack{t \in [t_i, t_f] \\ x \in \mathbb{R}}} \frac{\mu_R(x,t)}{\mu_R(x,t) + i\chi(x,t)}.
$$

(42)

As this mathematical quantity

$$
\exists \, \mathcal{Y}[\chi;t_i, t_f] \,/\, \mathcal{Y}[\chi;t_i, t_f] \overset{\text{def}}{=} \prod_{\substack{t \in [t_i, t_f] \\ x \in \mathbb{R}}} \frac{\mu_R(x,t)}{\mu_R(x,t) + i\chi(x,t)}
$$

is defined, and assumed to be reasonable finite (Materassi, 2019), it depends on the value of χ

at every point x and time t, that is, it will be a functional of χ. The expression of the stochastic factor $C[\chi,\Gamma;t_i,t_f]$ for the exponential distribution of R and Gaussian distribution of ξ is

$$C[\chi,\Gamma;t_i,t_f] = \mathcal{Y}[\chi;t_i,t_f]e^{-\int_{\mathbb{R}}dx\int_{t_i}^{t_f}d\tau\frac{W^2(\chi,\psi,\partial_x\psi)}{4\lambda_\xi(x,t)}} \quad (43)$$

Assuming that the quantity $\mathcal{Y}[\chi;t_i,t_f]$ converges, one may try to go ahead and integrate the kernel

$$A[\psi,\chi;t_i,t_f] = A_0(t_i,t_f)\mathcal{Y}[\chi;t_i,t_f]e^{-\int_{\mathbb{R}}dx\int_{t_i}^{t_f}dt\frac{W^2(\chi,\psi,\partial_x\psi)}{4\lambda_\xi(x,t)}}$$

$$\times e^{-i\int_{\mathbb{R}}dx\int_{t_i}^{t_f}dt\left\{\chi\partial_t\psi - \chi\left[\frac{Mg}{eB} + \frac{g\partial_z N_0}{B\nu_{in}N_0[ik(1+e^\psi)+e^\psi\partial_x\psi]}\right]\partial_x\psi + \frac{i}{2}\frac{g\partial_z N_0(ike^\psi+e^\psi\partial_x\psi)\partial_x\psi}{B\nu_{in}N_0[ik(1+e^\psi)+e^\psi\partial_x\psi]^2}\right\}} \quad (44)$$

with respect to the auxiliary variable χ: this may be done in a reasonable way as one adopts the convention

$$\int [d\chi]\ldots = \prod_{t,x}\int \frac{d\chi(x,t)}{\mathcal{H}(\chi;x,t)}\ldots\Big/\mathcal{H}(\chi;x,t)$$

$$\overset{\text{def}}{=}\frac{\mu_R(x,t)+i\chi(x,t)}{\mu_R(x,t)}, \quad (45)$$

yielding the final result

$$\mathcal{A}[\psi;t_i,t_f] = \tilde{A}_0(t_i,t_f)e^{\frac{1}{2}\int_{\mathbb{R}}dx\int_{t_i}^{t_f}dt\left\{\frac{g\partial_z N_0(ike^\psi+e^\psi\partial_x\psi)\partial_x\psi}{B\nu_{in}N_0[ik(1+e^\psi)+e^\psi\partial_x\psi]^2}+\frac{\varsigma^2(\psi,\partial_x\psi)}{2\lambda_\xi}\right\}}$$

$$\times e^{i\int_{\mathbb{R}}dx\int_{t_i}^{t_f}dt\frac{\lambda_\xi}{\gamma^2(\psi,\partial_x\psi)}\left[\partial_t\psi + \frac{\gamma(\psi,\partial_x\psi)\varsigma(\psi,\partial_x\psi)}{2\lambda_\xi} - \left(\frac{Mg}{eB}+\frac{g\partial_z N_0}{B\nu_{in}N_0[ik(1+e^\psi)+e^\psi\partial_x\psi]}\right)\partial_x\psi\right]^2} \quad (46)$$

with functional measure:

$$\int [d\psi]\ldots = \prod_{t,x}\int \frac{d\psi(x,t)}{\mathcal{J}(\psi;x,t)}\ldots\Big/\mathcal{J}(\psi;x,t)\overset{\text{def}}{=}\frac{\gamma(\psi;x,t)}{2}\sqrt{\frac{idtdx}{\pi\lambda_\xi(x,t)}}. \quad (47)$$

The result of applying the functional formalism to the SPDE of the electron density of the ESF is that the logarithmic fluctuation ψ evolves probabilistically, so that the probability density of the realization $\psi(x,t)$ between t_i and t_f is the $\mathcal{A}[\psi;t_i,t_f]$ in Eq. (46); when integrations of this complicated functional are performed, the functional measure $[d\psi]$ defined in Eq. (47) must be used. In Materassi (2019), the expression for the RPDF $\mathcal{A}[n;t_i,t_f]$ is reported, as obtained from this $\mathcal{A}[\psi;t_i,t_f]$.

3.4 Problems and perspectives

The functional $\mathcal{A}[\psi;t_i,t_f]$ in Eq. (46) represents the density of probability that the stochastic process ψ follows the realization $\psi(x,t)$ between t_i and t_f, as it is forced by the stochastic fields R and ξ according to Eq. (31): this is the result of applying the functional formalism à la Phythian (Phythian, 1977) to the problem of seeding the ESF (Kelley, 1989). Once one has got this $\mathcal{A}[\psi;t_i,t_f]$, the proper SFT should be developed, with a certain number of results that can be expressed in terms of the kernel.

First of all, it is possible to calculate the probability to have a final configuration $\psi_f(x)$ at time t_f from a given initial $\psi_i(x)$ at time $t_i < t_f$

$$\begin{cases} \psi(x, t_i) = \psi_i(x), \\ \psi(x, t_f) = \psi_f(x), \end{cases}$$

just integrating $\mathcal{A}[\psi; t_i, t_f]$ over all the intermediate time configuration $\psi(x, t)$ with $t \in (t_i, t_f)$, with the functional measure $[d\psi]$ in Eq. (47):

$$\mathcal{P}_{\psi_i \to \psi_f}(t_i, t_f) = \prod_{\substack{t \in (t_i, t_f) \\ x \in \mathbb{R}}} \int \frac{d\psi(x, t)}{\mathcal{J}(\psi; x, t)} \mathcal{A}[\psi; t_i, t_f] \quad (48)$$

(In this expression, one integrates over all the configurations $\psi(x, t)$ with t from $t_i + \varepsilon$ to $t_f - \varepsilon$, for arbitrarily small ε, while the final and initial configurations are kept fixed. Indeed, the normalization condition

$$\prod_{x \in \mathbb{R}} \int \frac{d\psi(x, t_i)}{\mathcal{J}(\psi; x, t_i)} \int \frac{d\psi(x, t_f)}{\mathcal{J}(\psi; x, t_f)} \mathcal{P}_{\psi_i \to \psi_f}(t_i, t_f) = 1$$

will hold; see the discretized version of such expressions in Materassi and Consolini (2008); note that the quantity $\mathcal{P}_{\psi_i \to \psi_f}(t_i, t_f)$ is correctly normalized, if $\mathcal{A}[\psi; t_i, t_f]$ is.)

Any quantity $F[\psi; t_i, t_f]$ depending on the whole realization of the stochastic process is a random variable, the average of which is calculated as

$$\langle F \rangle = \int [d\psi] F[\psi; t_i, t_f] \mathcal{A}[\psi; t_i, t_f]. \quad (49)$$

Equally, the whole statistical dynamics of the system described by ψ, in terms of *fluctuations* δF, *response functions*, or *correlation functions*, may be calculated using $\mathcal{A}[\psi; t_i, t_f]$, as indicated in Phythian (1977), so that one is in the position of constructing the whole SFT of the ESF in Eq. (31).

Is it really true? This is true only as far as the requested manipulations of $\mathcal{A}[\psi; t_i, t_f]$ are

analytically feasible, which cannot be the case for such a complicated functional as the one reported in Eq. (46). QM and quantum field theory (QFT) could be of good inspiration to deal with such a problem. However, looking at the integrand defining the exponent for the exponential in $\mathcal{A}[\psi; t_i, t_f]$,

$$\mathcal{A}[\psi; t_i, t_f] = \tilde{A}_0(t_i, t_f) e^{i \int_\mathbb{R} dx \int_{t_i}^{t_f} dt \mathcal{L}(\psi, \partial \psi)} \quad (50)$$

namely the *stochastic Lagrangian*,

$$\mathcal{L}(\psi, \partial \psi) = -\frac{i}{2} \left\{ \frac{g \partial_z N_0 (ike^\psi + e^\psi \partial_x \psi) \partial_x \psi}{B\nu_{in} N_0 [ik(1 + e^\psi) + e^\psi \partial_x \psi]^2} + \frac{\varsigma^2(\psi, \partial_x \psi)}{2\lambda_\xi} \right\} + \frac{\lambda_\xi}{\gamma^2(\psi, \partial_x \psi)} \left[\partial_\tau \psi + \frac{\gamma(\psi, \partial_x \psi) \varsigma(\psi, \partial_x \psi)}{2\lambda_\xi} - \left(\frac{Mg}{eB} + \frac{g \partial_z N_0}{B\nu_{in} N_0 [ik(1 + e^\psi) + e^\psi \partial_x \psi]} \right) \partial_x \psi \right]^2, \quad (51)$$

one recognizes it as the ratio between an expression that is quadratic in $\partial_t \psi$ and $\partial_x \psi$, and exponential in ψ, and an expression that is quadratic in $\partial_x \psi$, and exponential in ψ. The only thing that this Lagrangian density (51) has in common with the Lagrangian densities of QM (Feynman and Hibbs, 1965) or QFT (Weinberg, 2005) is being an integer expression in $(\partial_t \psi)^2$ and $\partial_t \psi$, while the way $\mathcal{L}$ depends on $\partial_x \psi$ and ψ is far more involuted than in those quantum Lagrangians. This means that once we are left with $\mathcal{A}[\psi; t_i, t_f]$, we are a little bit in an unexplored land, and the whole traditional machinery of perturbative techniques, Feynman diagrams, and so on is not immediately applicable.

This does not mean that the SFT for ionospheric turbulence does not take to any result, or that the reasonings proposed here are just "nice but useless," because many things can be taken from these developments. Indeed:

1. A sort of Langevin-like field equation is presented for the electron density fluctuations $n(x, t)$, that is, the SPDE (27), or its

equivalent logarithmic form (29). This equation may be the starting point for analytical studies and numerical simulations.

2. From the Langevin field equation, one can try to obtain a Fokker-Planck-like equation that should describe how the PDF of the value $n(x)$ or $\psi(x)$ evolves with time.

3. Holding the SFT kernel $\mathcal{A}$ in Eq. (46), one may explore some perturbative limit, working on λ_ξ, μ_R, or other parameters, to see whether the stochastic Lagrangian density $\mathcal{L}$ may show simpler expressions: doing so, one may hope to be able to apply perturbative techniques borrowed from quantum physics.

4 A few take-home messages

Before closing the chapter, I would like to highlight a few take-home messages I intend to convey about ionospheric complexity, and the limit of smooth and deterministic fluid representation due to *turbulence*.

First of all, the hypotheses under which the IM may be represented as a multiple fluid should be considered with great care, in order to represent the EI phenomena properly. Throughout the introduction, and in Section 2, we have tried to explain how the fluid representation corresponds to stating that a parcel of any ionospheric chemical is like *a point-like particle*, endowed with "internal degrees of freedom" that have *the same force and energy balance* of a gas at *thermal equilibrium*. This allows for the multifluid formulation, leading to the system of coupled PDEs (13), which are Navier-Stokes-like equations. The configurations for the fields $X_E^{(I)}\left(\vec{r},t\right) = \left(\vec{V}_I\left(\vec{r},t\right), N_I\left(\vec{r},t\right), T_I\left(\vec{r},t\right)\right)$ solving Eq. (13) for smooth initial and boundary conditions, and for smooth coefficients Φ^{micro}, are presumably smooth, in both time and space, at least within finite time. This does not rule out "bizarre" behaviors in terms of high sensitivity to initial conditions, rendering partially unpredictable the evolution of $X_E^{(I)}\left(\vec{r},t\right)$ in the still smooth and deterministic (Eq. 13); rather, one is just warned about the noncorrectness of expecting Eq. (13) to hold when the assumptions (1) do not hold.

The second take-home message to convey is that in those situations usually referred to as "ionospheric turbulence," the fields $X_E^{(I)}\left(\vec{r},t\right)$ may be very rough in both time and space: clearly, conditions (1) are violated in those cases. In principle, hence, it should not be taken for granted that ionospheric turbulence is correctly represented by Eq. (13) (even if nonsmooth boundary conditions, which necessarily "seed" irregularities as developed by Navier-Stokes-like PDEs, can be considered). On the other hand, computer simulations of different kinds of fluid equations may lead to hierarchies of structures, nested one into the other from large to virtually infinitely small scales, in principle all arbitrarily smooth (Leonard, 1975; Materassi et al., 2014): one must be conscious of the fact that it is not possible to go to "zero scale" believing fluid equations to hold.[d] The border between solidly believable fluid representation and "something else needed" has been discussed in Section 2.3: when the matter in the parcel δC_I is not at thermal equilibrium, within this δC_I relative nonmicroscopic motion will take place (Materassi and Consolini, 2015), and inhomogeneities in density may appear so that subparcel structures and fluctuations show up (Materassi and Consolini, 2007).

Out-of-equilibrium conditions invoke SFRs (Materassi et al., 2012): the third and last

---

[d] Still, interscale coupling given by nonlinearities in Eq. (13) render it possible to treat a part of turbulence phenomenology *within* the fluid approach.

take-home message is that an extension of the fluid representation can still be tried, saving the use of fluid quantities as $\vec{V}_I(\vec{r},t)$, $N_I(\vec{r},t)$, and $T_I(\vec{r},t)$, and yet introducing *explicitly irregular fluctuations*. This can be done by considering the coefficients Φ^{micro} as *stochastic fields* in the fluid equations, namely $\vec{r}$- and *t*-dependent *noises*, which is retaining the fluctuations that a purely fluid dynamical theory would neglect.

The stochastic fluid representation is illustrated here for the case of *electron density fluctuations n in the Equatorial Spread F*, in Section 3. One starts from the local balance equation for the electron density along the EVP, and assumes that the local recombination rate R and the electric current flux $\xi \overset{\text{def}}{=} -\vec{\partial} \cdot \vec{J}$ are noise terms: the probability that a certain specified realization $n(\vec{r},t)$ of n as a stochastic process, between two times t_i and t_f, is expressed as the functional integrand in Eqs. (50), (51), thanks to the functional formalism described in Phythian (1977). As the discussion in Section 3.4 has illustrated, even if its practical use is still a very long-term objective, the SFT for ionospheric turbulence contains in principle the ingredients to render feasible probabilistic calculations that cannot be done within FD and KT.

I would like to close this chapter with an attempt to give a concise answer to Kenneth Davies' question, which opened my reasonings.

The TEC day-to-day irregular variability is an effect of the interactions among structures of particles taking place at mesoscopic level, namely of ionospheric complexity. This means that the local out-of-equilibrium state of the IM renders the fluctuations of the medium properties relevant for the macroscopic dynamics of the EI, and ionospheric observables Q evolve as

$$\dot{Q} = \Lambda(Q) + \eta,$$

with a deterministic part Λ and a stochastic part η, similar to the general form (32). The solutions will result in $Q(t) = Q_0(t) + \delta Q(t)$, that is, some "deterministic background" Q_0 encoding the dynamics Λ, plus probabilistic fluctuations δQ determined by the interplay between Λ and noises η (see, for instance, Materassi et al. (2011) for an attempt to include this point of view in data analysis).

It is the intrinsic complexity of the EI, which we have tried to motivate in this chapter, that ultimately determines such a dynamics for any ionospheric physical quantity, including the TEC of Professor Kenneth Davies' question.

References

Alfonsi, L., Materassi, M., Wernik, A.W., 2003. Distribution of scintillation parameters calculated from in-situ data: preliminary results. In: Invited paper at the Atmospheric Remote Sensing using Satellite Navigation Systems, Special Symposium of the URSI Joint Working Group FG, October 13–15, 2003, ASI Centro di Geodesia Spaziale "Giuseppe Colombo", Matera, Italy.

Antoniou, I., Pronko, G.P., 2002. On the Hamiltonian description of fluid mechanics. ArXiv: hep-th/0106119 v2, and references therein.

Beretta, G.P., 1987. Quantum thermodynamics of nonequilibrium. Onsager reciprocity and dispersion-dissipation relations. Found. Phys. 17, 365.

Beretta, G.P., Gyftopoulos, E.P., Park, J.L., Hatsopoulos, G.N., 1984. Quantum thermodynamics. A new equation of motion for a single constituent of matter. Nuovo Cimento B 82, 169.

Bossomaier, T., Barnett, L., Harré, M., Lizier, J.T., 2016. An Introduction to Transfer Entropy. Springer International Publishing, Cham.

Burke, W.L., 1985. Applied Differential Geometry. Cambridge University Press, Cambridge.

Carley, M., n.d. Turbulence and Noise. Available from: http://people.bath.ac.uk/ensmjc/Notes/tnoise.pdf (notes on a course at the University of Bath. The reader will find interesting also the references within this booklet).

Castellani, B., 2009. Map of Complexity Science. Art & Science Factory, Cleveland, OH.

Chang, T., 1999. Sporadic localized reconnections and multiscale intermittent turbulence in the magnetotail. Phys. Plasmas 6, 4137. https://doi.org/10.1063/1.873678.

Chang, T.S., Nicoll, J.F., Young, J.E., 1978. A closed-form differential renormalization-group generator for critical dynamics. Phys. Lett. 67A (4), 287–290.

Chang, T.S., Vvedensky, D.D., Nicoll, J.F., 1992. Differential renormalization-group generators for static and dynamic critical phenomena. Phys. Rep. 217 (6), 279–360.

Chang, T., Tam, S.W.Y., Wu, C.-C., 2004. Complexity induced anisotropic bimodal intermittent turbulence in space plasmas. Phys. Plasmas 6, 1287–1299.

Choudhuri, A.R., 1998. The Physics of Fluids and Plasmas. Cambridge University Press, Cambridge.

Ciraolo, L., 1993. Elevation of GPS L2-L1 biases and related daily TEC profiles. Presented in Workshop on Morkling of the Ionosphere for GPS Application, Neusterlitz, September 20–30.

Consolini, G., De Michelis, P., Kretzschmar, M., 2006. Thermodynamic approach to the magnetospheric complexity: the role of fluctuations. Space Sci. Rev. 122, 293–299.

Davies, K., 1990. Ionospheric Radio. Peter Peregrinus Ltd, London.

de Anna, P., Dentz, M., Tartakovsky, A., Le Borgne, T., 2014. The filamentary structure of mixing fronts and its control on reaction kinetics in porous media flows. Geophys. Res. Lett. 41, 4586–4593. https://doi.org/10.1002/2014GL060068.

Duhau, S., Azpiazu, M.C., 1981. Non-thermal equilibrium between electrons and neutrals at ionospheric E-region heights. Geophys. Res. Lett. 8 (7), 819–822.

Fasano, A., Marmi, S., 2006. Analytical Mechanics. Oxford Graduate Texts, Oxford University Press, Oxford.

Feynman, R.P., Hibbs, A.R., 1965. Quantum Mechanics and Path Integrals. In: Earth and Planetary Sciences, first ed. McGraw-Hill College.

Frisch, U., 2010. Turbulence: The Legacy of A.N. Kolmogorov. Cambridge University Press. ISBN-10: 0521457130, ISBN-13: 978-0521457132.

Frisch, U., Matsumoto, T., 2002. On multifractality and fractional derivatives. J. Stat. Phys. 108, 1181.

Gross, D.H.E., 2001. Microcanonical Thermodynamics-Phase Transitions in "Small" Systems. World Scientific Publishing Co. Pte Ltd, Singapore.

Heelis, R.A., Lowell, J.K., Spiro, R.W., 1982. A model of the high-latitude ionospheric convection pattern. J. Geophys. Res. 87 (A8), 6339–6345.

Hill, T.L., 1964. Thermodynamics of Small Systems, I and II. Dover Publications Inc., New York, NY.

Jursa, A.S., 1985. Handbook of Geophysics and the Space Environment. Air Force Geophysics Laboratory, Air Force Systems Command, USA Air Force, Defense Technical Information Center, Fort Belvoir, VA.

Kallenrode, M.-B., 2000. Space Physics. Springer-Verlag, Berlin, Heidelberg.

Kelley, M.C., 1989. The Earth's Ionosphere. Academic Press Inc., New York, NY.

Landau, L.D., Lifshiz, E.M., 1971. Mecanique des Fluides´. Mir, Moscou.

Leitinger, R., et al., 1996. The Upper Atmosphere. Springer, New York, NY.

Leonard, A., 1975. Energy cascade in large-eddy simulations of turbulent fluid flows. Adv. Geophys. 18 (Pt. A), 237–248.

Liang, X.S., 2013. The Liang-Kleeman information flow: theory and applications. Entropy 15, 327–360. https://doi.org/10.3390/e15010327.

Lichtenberg, A.J., Liebermann, M.A., 1992. Regular and Chaotic Dynamics. Applied Mathematical Sciences, vol. 38. Springer, New York, NY.

Livi, R., Politi, P., 2017. Nonequilibrium Statistical Physics—A Modern Perspective. Cambridge University Press. ISBN 978-1-107-04954-3.

Lorenz, E.N., 1963. Deterministic nonperiodic flow. J. Atmos. Sci. 20, 130–141.

Ma, S.-K., 1973. Introduction to the renormalization group. Rev. Mod. Phys. 45 (4), 589.

Mahajan, K.K., 1967. Extent of thermal non-equilibrium in the ionosphere. J. Atmos. Terr. Phys. 29, 1137–1151.

Materassi, M., n.d. Matter dynamics' equations in ionospheric physics. CNR Scientific Report. Available from: www.materassiphysics.com.

Materassi, M., 1996. Variabili canoniche collettive e relative per il campo di Klein-Gordon classico (Degree thesis). University of Firenze, Italy.

Materassi, M., 2015. Metriplectic algebra for dissipative fluids in Lagrangian formulation. Entropy 17 (3), 1329–1346.

Materassi, M., 2019. Stochastic field theory for the ionospheric fluctuations in equatorial spread F. Chaos Solitons Fractals 121, 186–210

Materassi, M., Consolini, G., 2007. Magnetic reconnection rate in space plasmas: a fractal approach. Phys. Rev. Lett. 99 (17), 175002.

Materassi, M., Consolini, G., 2008. Turning the resistive MHD into a stochastic field theory. Nonlinear Process. Geophys. 15 (4), 701–709.

Materassi, M., Consolini, G., 2015. The stochastic tetrad magneto-hydrodynamics via functional formalism. J. Plasma Phys. 81 (6), 495810602.

Materassi, M., Mitchell, C.N., 2007. Wavelet analysis of GPS amplitude scintillation: a case study. Radio Sci. 42 (1), RS1004.

Materassi, M., Alfonsi, L., De Franceschi, G., Mitchell, C.N., Romano, V., Spalla, P., Wernik, A.W., Yordanova, E., 2005. Intermittency and ionospheric scintillations in GPS data. In: Siddiqi, A.H., Alsan, S., Rasulov, M., Oğun, O., Aslan, Z. (Eds.), Proceedings of the International Workshop on Applications of Wavelets to Real World Problems (IWW2005), July 17–18, 2005, Istanbul, Turkey, Istanbul Commerce University Publications.

Materassi, M., Ciraolo, L., Consolini, G., Smith, N., 2011. Predictive space weather: an information theory approach. Adv. Space Res. 47, 877–885. https://doi.org/10.1016/j.asr.2010.10.026.

Materassi, M., Consolini, G., Tassi, E., 2012. Sub-fluid models in dissipative magneto-hydrodynamics. In: Zheng, L. (Ed.), Topics in Magnetohydrodynamics. InTech. https://doi.org/10.5772/36022.

Materassi, M., Consolini, G., Smith, N., De Marco, R., 2014. Information theory analysis of cascading process in a synthetic model of fluid turbulence. Entropy 16, 1272–1286.

Newman Coffey, V., 2009. Oxygen Ion Heating Rate Within Alfvénic Turbulence in the Cusp Near 1 RE Altitude (PhD dissertation). Huntsville, Alabama.

Ohde, T., Bonitz, M., Bornath, T.H., Schlanges, M., 1997. Diffusion and heat transport in a dense partially ionized plasma. Contrib. Plasma Phys. 37 (2–3), 229–238.

Öttinger, H.C., 2005. Beyond Equilibrium Thermodynamics, first ed. Wiley-Interscience. ISBN-10: 0471666580, ISBN-13: 978-0471666585.

Parkinson, M.L., 2008. Complexity in the scaling of velocity fluctuations in the high-latitude F-region ionosphere. Ann. Geophys. 26, 2657–2672.

Phythian, R., 1977. The functional formalism of classical statistical dynamics. J. Phys. A Math. General 10 (5), 777.

Prigogine, I., 1980. From Being to Becoming—Time and Complexity in the Physical Sciences. W.H. Freeman & Co., New York, NY.

Professor Kenneth Davies' Obituary, n.d. Available from: https://www.legacy.com/obituaries/dailycamera/obituary.aspx?n=kenneth-davies&pid=175207849.

Schlegel, K., St.-Maurice, J.P., 1981. Anomalous heating of the polar E region by unstable plasma waves. 1. Observations. J. Geophys. Res. 86, 1447.

Schunk, R., Nagy, A., 2009. Ionospheres. Physics, Plasma Physics, and Chemistry. Cambridge University Press, Cambridge.

Watkins, N.W., Chapman, S.C., Dendy, R.O., Rowlands, G., 1999. Robustness of collective behaviour in strongly driven avalanche models: magnetospheric implications. Geophys. Res. Lett. 26 (16), 2617–2620.

Weinberg, S., 2005. The Quantum Theory of Fields: vol. 1, Foundations. Cambridge University Press, Cambridge.

Wernik, A.W., Alfonsi, L., Materassi, M., 2007. Scintillation modeling using in situ data. Radio Sci 42, 1–21.

Wikipedia, n.d. Definition of "complex system" according to Wikipedia. Available from: https://en.wikipedia.org/wiki/Complex_system.

Zelenyj, L.M., Milovanov, A.V., 2004. Fractal topology and strange kinetics: from percolation theory to problems in cosmic electrodynamics. Phys. Uspekhi 47 (8), 749–788.

Zemansky, M.W., 1957. Heat and Thermodynamics. McGraw-Hill Book Company Inc., New York.

High-resolution approaches to ionospheric exploration

Joshua Semeter

Department of Electrical and Computer Engineering and Center for Space Physics, Boston University,
Boston, MA, United States

1 Introduction

Our view of ionospheric physics has shifted substantially in recent years. Rather than treating the ionosphere as an independent atmospheric layer, the magnetosphere, ionosphere, and thermosphere (M-I-T) are now viewed as a coherently integrated system, exchanging mass, energy, and momentum through processes ranging in scale from thousands of kilometers to centimeters, and from decades to microseconds, with governing physics that can change abruptly across specific spatiotemporal boundaries. This multiscale systems science viewpoint has come to define the modern era of ionospheric investigation. Capturing the interplay among physical processes that span 6+ orders of magnitude is a major observational challenge. Understanding the system-level implications requires sensing strategies that meet the conflicting requirements of broad spatial coverage and high spatial and temporal resolution. Emergent technologies, such as global broadband connectivity, cloud-based storage and computation, inexpensive low-power sensors, and citizen science initiatives are enabling new cost-effective solutions to multiscale ionospheric sensing.

The purpose of this chapter is to offer a mathematical framework for modeling, fusing, and assimilating high-resolution ionospheric data from heterogeneous networked sensors (primarily radio and optical). The sensor landscape continues to evolve rapidly, and so the presentation focuses on the application of universal principles from the fields of discrete inverse theory and data fusion. It is hoped that the use cases covered in this chapter promote creative thinking about new experimental techniques that address a broad range of ionospheric science.

2 Conceptual framework

While all agree that multiscale measurements are important for scientific progress, generally a unified framework is lacking for connecting theoretical predictions to measurements from sensors providing different representations of a physical process. For instance, the transfer of energy from the magnetosphere to the high-latitude ionosphere in the form of Earthward Poynting flux will produce modifications to ionospheric temperatures, resolved by incoherent scatter radar (ISR), transport-induced density enhancements, observable in maps of total electron content (TEC) derived from Global Navigation Satellites System (GNSS) receiver arrays, and a rich spectrum of optical emissions caused by density changes and particle precipitation, quantified through imaging spectroscopy. This leads to the question: how should we combine this complementary information in a way that optimizes physical insight?

Fig. 1 presents one approach to put heterogeneous measurements into a common mathematical framework. This schematic representation shows the forward model connecting theory to measure for four classes of sensors where disruptive advancements in capabilities and coverage are expected to continue: photometric imaging, ISR, digital ionosondes, and GNSS. The forward model is partitioned into four layers: *theory, known physics, projection,* and *instrument*. These layers represent successive abstractions, rather than the actual causal flow from target to instrument. The *theory* layer embodies the hypothesis that we seek to test, which is presumed to be observable in some set of ionospheric state parameters. Here the state parameters are density, temperature, and bulk motion.

The second layer invokes *known physics* (i.e., established physics models) to compute the intrinsic quantities that we seek to estimate with our sensor. For instance, in the left-hand column, the volume emission rate of photons associated with a particular quantum mechanical transition $p_\lambda(\mathbf{r}, t)$ are computed over space and time using a photochemistry model that includes collision cross-sections, reaction rates, and transition probabilities (e.g., Rees, 1989).

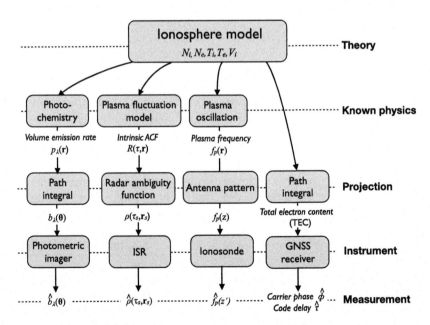

FIG. 1 Forward modeling framework for four classes of ionospheric sensors, organized in four layers representing information flow from theory to measurement.

In the next column, a model of thermal plasma fluctuations is used to compute the intrinsic autocorrelation function (ACF) $R(\tau, \mathbf{r}, t)$ (in the remaining chapter, dependence on t is implied for all parameters). The ACF is estimated using ISR techniques, and is sensitive to density, composition, temperatures, and bulk motions of the plasma (Hysell, 2018). In the third column, the electron plasma frequency is computed, a parameter that may be directly probed with a frequency agile radar called an ionosonde, revealing the bottomside ionospheric density as a function of range (Reinisch et al., 2009).

The third layer is the projection layer. Here the intrinsic quantity is projected according to the sampling modality of the sensor. In a photometric imaging system, the quantity accessible to the sensor is the line-of-site integrated production rate, typically expressed as a brightness in Rayleighs, b_λ (Baker, 1974). For ISR, $R(\tau, \mathbf{r})$ is sensed through a space-time blurring kernel (or ambiguity function) yielding $\rho(\tau, \mathbf{r})$. This blurring kernel depends on the radar pulse pattern, beam width, plasma motion, and integration period (Swoboda et al., 2015). For an ionosonde, the sensed quantity is typically the range-resolved plasma frequency, which is related to the $f_p(\mathbf{r})$ through the antenna pattern. Finally, for GNSS (right-hand column), the projection or sensed parameter is the line-of-site integrated plasma density between satellite and receiver, referred to as the total electron content or TEC.

The final *instrument* layer embodies the mathematical model of the acquisition process. The caret indicates an estimated value, which has a different interpretation for each sensor class. For photometric imagers, the calibrated measurement $\hat{b}_\lambda$ is the expected value of the Gaussian random process describing the photon arrival and the creation of photoelectron events in the detector. The bias and variance depend on source brightness, exposure time, and system parameters such as dark current and electron

read noise (Eq. 5). For ISR, $\hat{\rho}$ is also the expected value of a random variable. This stochasticity arises from the noise-like nature of the scattering process. The variance depends on the number of pulses averaged, the pulse pattern used, and the receiver and sky noise (see Eq. 10). For the ionosonde, the fidelity of the resolved plasma frequency depends on the receiver quality, but assigning range requires correcting for propagation effects through the conducting ionosphere (see Eq. 14). Finally, for GNSS, the measured quantity is the carrier phase and the code delay, which can be resolved rapidly and with high fidelity using modern phase-locked loop technology. These parameters are related to TEC through a model that includes receiver and satellite biases, and uncertainties in establishing group delay (Eq. 11). In the following section, the aforementioned sensor models are explored in more detail.

3 Sensor models

Section 2 described a framework that enables different space-time manifestations of a physical process to be explored simultaneously through different projections and attendant sensors. Such a capability is critical for properly constraining physics models and achieving new insight. For instance, the appearance of a discrete auroral arc in an optical imager represents a new plasma production. The modification of the ionosphere, as detected by ISR, may occur in milliseconds (e.g., temperature changes), seconds (E-region density enhancement), or minutes (F-region enhancements). By formally fusing these diagnostics, we can learn much about how the M-I-T system adjusts to this release of free energy. In the following sections, the model representing each of the four sensor classes in Fig. 1 is expanded upon, in preparation or applications in data fusion.

3.1 Photometric imaging

The M-I-T system reveals much about energy exchange processes through optical emissions. Polar latitudes, for instance, are subject to fluxes of electromagnetic and kinetic energy flux which serve to heat, ionize, and excite the outer atmosphere, producing responses that are readily measured by radar and optical diagnostics (Thayer and Semeter, 2004). The emergence of electron multiplying charge-coupled device (EMCCD) and scientific-grade complementary metal oxide semiconductor (sCMOS) detector technology has enabled high-frame-rate imaging of narrow ephemeral variations in ionospheric airglow and auroras (Semeter et al.,

2008; Dahlgren et al., 2013). Examples of Alfvén wave-induced optical emissions (<1-km in width, 20 ms exposure) are shown in Fig. 2A and B, recorded using an EMCCD-based sensor.

Fig. 2C depicts such a physical system in cross-section. An incident electron flux impinging the atmosphere ϕ_{top} transfers some power into a set of volume production rates $p(\mathbf{r})$ represented by the shaded contours, with z representing field-line distance. The contours could represent the production of excited atomic and molecular states or impact ionization rates. Either may be represented canonically by

$$p_s(\mathbf{r}) = \int n_i(\mathbf{r})\sigma_{ij}(E)\phi(\mathbf{r},E)dE \quad (\text{cm}^{-3}\text{s}^{-1}) \quad (1)$$

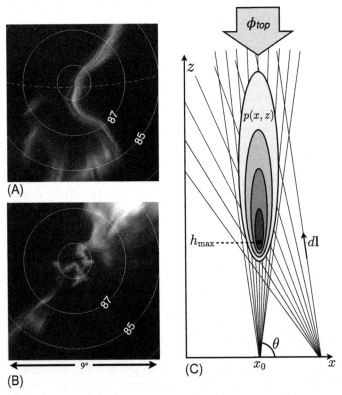

(A)

(B)

(C)

FIG. 2   (A, B) Sample images of narrow (<1 km) ephemeral arcs associated with dispersive Alfvén waves. (C) This physical system depicted in cross-section: A flux of electrons ϕ_{top} impinges the atmosphere which heats, ionizes, and excites the neutral gases. These rates are collectively represented by $p(x, y)$.

where n_i is the density of the ith ground-state neutral species (primarily N_2, O_2, or O), $\phi(\mathbf{r}, E)$ is the net differential energy influx, and $\sigma_{ij}(E)$ is the electron impact cross-section for the jth excitation process acting on the ith species.

One of the triumphs of the field of optical aeronomy has been the development of first-principles models of the optical emission spectrum of the aurora and airglow for a given ionospheric composition and given rate of energy transfer to the gas (e.g., Chamberlain, 1961; Lummerzheim and Lilensten, 1994; Rees, 1989; Strickland et al., 1976). Several models now exist for the reliable computation of Eq. (1) based on established cross-sections. These models form the basis of the *known physics* layer in Fig. 1.

In addition to direct excitation by an external energy source, excited states may also be produced in the release of free energy through chemical reactions. Most notably are the metastable $O(^1D)$ redline emission at 630 nm and $O(^1S)$ greenline emission at 557.7 nm, both produced through dissociative recombination of ionospheric O^+ (Semeter et al., 1996). The multiple excitation mechanisms, long state lifetimes, and dependence on poorly specified quenching rates limits the use of these metastable emissions for high-resolution ionospheric research applications. For prompt emissions, on the other hand, the production rate of photons is proportional to the production rate of excited states, that is,

$$p_\lambda(\mathbf{r}) \propto p_s(\mathbf{r}) \qquad (2)$$

where λ is a wavelength designation for the transition, and with the proportionality constant dependent on known quantum mechanical considerations.

The quantity available for sampling by a ground-based sensor is the path integral of this photon production rate along a line of sight. For optically thin emissions (a good approximation for visible wavelengths), this projection is simply the line-of-sight integral through the emitting region, weighted by extinction factors, determined experimentally.

$$b_\lambda = \int_0^\infty p_\lambda(\mathbf{l})dl \quad (\text{cm}^{-2}\text{s}^{-1}) \qquad (3)$$

where $\mathbf{l}$ is the position vector defining the line of sight. If the emitting region is defined in two-dimensional (2D) Cartesian coordinates (x, z) and the observer's position is described by (x_0, z_0), as in Fig. 2C, this integral may be expressed as an explicit function of elevation angle θ,

$$b_\lambda(\theta;x_0,z_0) = \int_0^\infty \int_0^\infty \delta[z - z_0 \\ -(\tan\theta)(x - x_0)]p_\lambda(x,z)dxdy \qquad (4)$$

where δ is the Dirac delta function which is non-zero only when the argument lies along the line of site. A tomographic inversion problem may be formulated using measurements of b_λ from different ground locations (Semeter et al., 1999). This parameter is captured in Fig. 1 by $b_\lambda(\boldsymbol{\theta})$, where $\boldsymbol{\theta}$ represents a vector representing the direction of integration.

Combining Eqs. (1), (2) suggests a means of estimating the volume energy deposition rate and, hence, particle kinetic energy influx, provided a means of estimating p may be developed. Tomography of the aurora (Frey et al., 1996) and airglow (Semeter et al., 1999) has been explored for many years. Recent advances in detector sensitivity, computer storage, and model-based regularization schemes have enabled the tomographic analysis of features at increasingly smaller scale, using millisecond station synchronization and kilometer-scale baselines (Hirsch et al., 2016). Such a short-baseline common-volume observing scheme is suggested by the two overlapping fan-beam projections depicted schematically in Fig. 2C. More commonly, the integrated intensities are analyzed directly (Rees and Luckey, 1974; Hecht et al., 1989; Strickland et al., 1994; Zettergren et al., 2007).

The *instrument* layer in Fig. 1 represents detection process. For a traditional focal plane array, the image signal-to-noise ratio (SNR) can be expressed

$$SNR = \frac{S_p \cdot QE \cdot t_{int}}{\sqrt{\left(S_p \cdot QE + i_{dark}\right) \cdot t_{int} + \left(\sigma_{read}\right)^2}} \quad (5)$$

Here, S_p is the photon flux incident on a detector pixel (s^{-1}), QE is the quantum efficiency for photon to electron conversion, i_{dark} is the thermal dark current, t_{int} is the integration time, and σ_{read} is the RMS readout noise (Baumgardner et al., 1993). Optical detector technology has undergone disruptive improvements since the year 2000. The most sensitive systems are based on EMCCD technology. In EMCCD, read noise is effectively eliminated through an internal gain which amplifies photoelectron events elevating the electron count above the read noise. Furthermore, the low focal plane temperatures achieved by efficient thermoelectric cooling effectively eliminates dark current. Ignoring the clock-induced charge term (dependent on readout rate), a properly configured EMCCD camera operates effectively at the photon counting limit, such that SNR is determined by Gaussian distributed photon arrival statistics and the detector quantum efficiency,

$$SNR \approx \sqrt{S_p \cdot QE \cdot t_{int}} \quad (6)$$

Under these conditions, the experimenter need not make a priori decisions about exposure time. Memory permitting, images may be recorded at an arbitrarily rapid rate, with optimal image construction left as a postacquisition consideration.

3.2 Incoherent scatter radar

The power spectrum of ion-acoustic fluctuations in the ionosphere is defined as

$$S(\omega, \mathbf{k}_s) = \lim_{V,T \to \infty} \frac{1}{VT} \left\langle \frac{|\Delta N_e(\omega, \mathbf{k}_s)|^2}{n_{e0}} \right\rangle \leftrightarrow \mathcal{F}R(\tau, \mathbf{k}_s) \quad (7)$$

where V is the volume of the plasma, T is the time duration of the observation, the angle brackets represent an average over multiple independent samples of the spectrum, $\Delta N_e(\mathbf{k}_s, \omega)$ is the Fourier transformation in space and time of the small variation of the electron density, n_{e0} is the average electron density within the volume, and $\mathbf{k}_s$ is the scatter wave vector set by the radar system. The theoretical form of S has been derived from different authors and perspectives, all yielding identical results (Dougherty and Farley, 1960; Hagfors, 1961; Rosenbluth and Rostoker, 1962; Evans, 1969; Kudeki and Milla, 2011). This body of work constitutes the "known physics" for the ISR forward model in Fig. 1.

Taking the inverse Fourier transform of S gives the plasma ACF, which we designate R in Eq. (7). Standard ISR processing seeks to estimate spatially resolved ionospheric state parameters by constructing and fitting a discrete representation of R. The individual lags are constructed through appropriate multiplications of gated echoes from the same plasma volume at different times. Many clever techniques have been developed to address the conflicting goals of high range resolution, high time resolution, and high spectral resolution (Gray and Farley, 1973; Lehtinen and Haggstrom, 1987; Lehtinen and Huuskonen, 1996; Hysell et al., 2008).

A practical radar system has a finite beam width and uses waveforms with nonzero time extent such that the measured ACF represents an average over space and time. This blurring effect may be represented by a space-time ambiguity function, L, that operates on the intrinsic ACF,

$$\rho(\tau_s, \mathbf{r}_s, t_s) = \int L(\tau_s, \mathbf{r}_s, t_s, \tau, \mathbf{r}, t) R(\tau, \mathbf{r}, t) \, dV \, dt \, d\tau \quad (8)$$

where the subscript s delineates the variable as the discretized version of its continuous counterpart; that is, the variable τ_s is the sampled version of the lag coordinate τ. This function has the form of a Fredholm integral equation of the first kind, commonly found in blurring, tomography, and other reconstruction problems. In traditional dish-based (mechanically steerable) ISR, L is generally taken as the classic radar ambiguity function, which is a function of range and lag only. Standard ISR analysis does not account for cross-beam plasma drift, as the plasma ACF is assumed stationary during the integration period. This assumption is highly questionable, especially at high latitudes where ionospheric $E \times B$ flows can readily exceed a few kilometers per second.

Electronically steerable arrays such as AMISR and EISCAT-3D can perform pulse-by-pulse repointing, providing a new level of flexibility in ISR analysis. Fig. 3A shows a standard 26-beam sampling pattern used at the PFISR facility. This mode obtains regularly spaced samples of ionospheric parameters over a three-dimensional (3D) volume. Fig. 3B shows an interpolated 3D plasma density constructed from 2-min integration across these samples. If plasma motion is determined by some means, then it becomes possible to track a plasma volume and thus it performs incoherent averaging in the rest frame or of the moving plasma. This reconstruction problem is the 3D analog to 2D blurring encountered in optical imagery of moving objects. Swoboda et al. (2015) showed that the generalized space-time ambiguity function is separable for radar-centered spherical coordinates $\mathbf{r} = (r, \theta, \phi)$,

$$\rho(\tau_s, \mathbf{r}_s, t_s) = \int G(t_s, t) \, F(\theta_s, \phi_s, \theta, \phi)$$
$$W(\tau_s, r_s, \tau, r) \, R(\tau, \mathbf{r}, t) \, dV \, dt \, d\tau \tag{9}$$

where $G(t_s, t)$ is the kernel for the time dimension, $F(\theta_s, \phi_s, \theta, \phi)$ is radar beam shape that acts as a kernel in azimuth and elevation, and $W(\tau_s, r_s, \tau, r)$ is the classic the range-lag ambiguity function. The mathematics of inverting this equation has, as of this writing, only been worked out for simple uniform plasma motion with known velocity (Swoboda et al., 2017).

Having separated out radar ambiguity, what remains in the *instrument* layer in Fig. 1 is a statistical description of the ACF acquisition. The backscattered signal in ISR is "noise-like"— that is, even in the absence of receiver noise and sky background noise, the superposition of signals from the individual electrons creates a signal characterized as a zero-mean Gaussian random variable. The sensing task involves estimating the second-order statistics (i.e., power spectrum) of this random variable. The instrument function in this case represents the acquisition, processing, and incoherent averaging of a set of lag products representing ρ and, hence, R in Eq. (7). The relative error depends on both the density of the scattering medium (affecting the per-pulse signal-to-noise ratio S/N) and the number of pulses averaged K, and can be expressed (Farley, 1969).

$$\frac{\sqrt{\mathrm{Var}\{\hat{\rho}\}}}{\rho} = \frac{1}{\sqrt{K}} \left(1 + \frac{1}{S/N} \right) \tag{10}$$

With electronically steerable ISR, the problem may be expressed in terms of resource allocation. If one defines a time resolution for the desired parameter (e.g., N_e, T_e, T_i), then the total number of pulses is set by the achievable inter-pulse period (IPP) of the radar system. The pulses may be distributed in many angular directions, yielding a large number of resolution elements but high variance in the derived parameters, or in just a few, or even just one, direction, yielding limited spatial information but high confidence. Alternatively, one could begin with a desired spatial sampling pattern, producing a direct tradeoff between confidence and time resolution.

As a numerical example, let us consider the problem of determining plasma density (proportional to backscattered power in the lower ionosphere) for a high S/N case (high plasma density). In this case, the relative error goes as $\sim 1/\sqrt{K}$. Suppose our IPP is 20 ms, such supports 50 pulses in 8 s. This would give a 5% relative error for a single beam, or a 10% relative error for 4 beams, etc. The tradeoff between spatial resolution and error is analogous to a digital camera, in that for dimensions n^2, the relative

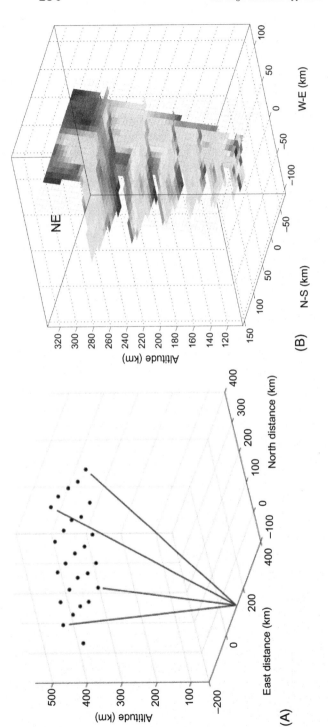

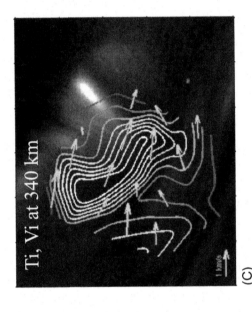

FIG. 3 Examples of PFISR multibeam reconstruction and data fusion. (A) Typical beam pattern (5 × 5 angular grid plus 1 beam in the magnetic zenith). (B) Reconstructed 3D ionospheric density using 2-min integration. (C) Ion temperature (*contour lines*) and $B_\perp$ ion drift (*arrows*) overlaid on auroral image,

error increases by 5% each time we increment n by 1. This results in a linear tradeoff between beam array dimensions and relative error.

3.3 Global navigation satellite systems

ISR and GNSSs provide complementary information about the ionosphere. ISR can be used to construct multidimensional images of the overlying electron density field through analysis of echoes collected in multiple directions (Nicolls and Heinselman, 2007; Semeter et al., 2009), while signals collected by dual frequency GNSS receivers can be used to compute path integrals through this field (Sardon et al., 1994). These two physical parameters are related through the definite integral,

$$\text{TEC} = \int_a^b N(\mathbf{r})ds \qquad (11)$$

where $N(\mathbf{r})$ is the instantaneous electron density field in $\mathbb{R}^3$, and TEC is the "total electron content" for a given transmitter-receiver pair. Position vectors $\mathbf{a}$ and $\mathbf{b}$ could refer to a satellite and ground-receiver, or two satellites (i.e., occultation geometry).

The propagation of the GNSS signal through the conducting ionosphere leads to a group (code) delay and a phase advance that are proportional to TEC,

$$\tau \approx \frac{z}{c} + \frac{40.3}{cf^2}\text{TEC} \quad \phi \approx \frac{2\pi fr}{c} - \frac{80.6\pi}{cf}\text{TEC} \qquad (12)$$

The inverse problem of determining TEC from group delay measured at two frequencies simultaneously yields the TEC and the path correction it produced. Provided the transmitted frequencies are much greater than the plasma frequency, to a good approximation,

$$\text{TEC} = \frac{c(\tau_2 - \tau_1)}{40.3\left(\frac{1}{f_2^2} - \frac{1}{f_1^2}\right)} \quad \frac{d}{dt}\text{TEC} = \frac{cf}{80.6\pi}\left(\frac{2\pi f}{c}\frac{dr}{dt} - \frac{d\phi}{dt}\right)$$

$$(13)$$

3.4 Ionosonde

We included ionosonde in Fig. 1 because it is highly complementary to ISR and GNSS, and because the technology and coverage continue to improve. Modern digital ionosondes (digisondes) can resolve density at ≈ 1 km range resolution as well provide coarse imaging information through angle-of-arrival measurements (Reinisch et al., 2009). For an ionosonde, the *known physics* layer in Fig. 1 refers to the calculation of electron density from measurements of the electron plasma frequency $f_p(\mathbf{r}) = \sqrt{n(\mathbf{r})e^2/4\pi^2\epsilon_0 m_e}$. The estimated quantity (*instrument* layer) is plasma frequency as a function of virtual height, $\hat{f}_p(z')$ (boresight direction of a zenith directed antenna). The intrinsic connection between wave phase velocity and plasma density leads to an Abel equation. The determination of true range h requires numerically solving an integral equation of the form,

$$h(f) = \int_{z_0}^{z_R} n_g dz = \int_{z_0}^{z_R} \frac{dz}{\sqrt{1 - \left(\frac{f_p(z)}{f}\right)^2}} \qquad (14)$$

where z_R is the reflection point and n_g is the group refractive index.

4 Data fusion

In many cases, radio and optical sensors are observing the same physical process through different perspectives. Fig. 3C gives one example. The background auroral image shows the activation of an arc element, producing a region of enhanced electrical conductivity via impact ionization. The contour lines depict the F-region ion temperature and the arrows depict the ionospheric flow field, both constructed using PFISR samples (Fig. 3A, Semeter et al., 2010). The combined data reveal the attenuation of the ionospheric electric field in a region of enhanced conductivity.

Fig. 3 illustrates the scientific insight obtained from a careful space-time coregistration of data from complementary sensors. In other cases, different measurements may have a direct mathematical relationship independent of physical interpretation. For instance, ISR can resolve plasma density as a function of space and time, but the fidelity depends on the integration period (Section 4.1). GNSS-derived TEC provides no range information, but the cross-range sample spacing can be made arbitrarily short. Intuitively, it should be possible to combine these data to obtain a higher resolution representation of N_e. The following sections use the framework developed in Sections 2 and 3 to achieve this.

4.1 Reconstruction filter for ISR

The high-level data product from a multibeam ISR experiment is a set of discrete estimates of ionospheric parameters and their associated uncertainties. A common step in displaying such measurements involves spatial interpolation. Mathematically, interpolation applies a rule to fill in the missing information between samples. In the absence of constraints, this rule must be chosen arbitrarily. Interpolation thus corresponds to an ill-posed inverse problem, since there are an infinite number of solutions consistent with the available data. An example of traditional 3D interpolation was shown in Fig. 3.

In order to incorporate information obtained from other sensors, such as GPS, it is useful to describe the interpolation operation using the formalism of inverse theory (Fathi et al., 2012). Consider the ionospheric plasma density $N(\mathbf{r})$ as an unknown continuous function in $\mathbb{R}^3$, and suppose we have a sensor that obtains discrete samples $N_S(\mathbf{r}_i)$, through an integrating kernel:

$$N_S(\mathbf{r}_i) = \int L_i(\mathbf{r})N(\mathbf{r})dv + \epsilon_i \qquad (15)$$

where ϵ_i is additive noise (which is, in general, dependent on L_i). Note that L was also the parameter, which we used in Eq. (9), and it may also be considered an ambiguity function here. For instance, $L_i(\mathbf{r}) = \delta(\mathbf{r} - \mathbf{r}_i)$ corresponds to point-wise sampling of the continuous field, or the ideal "thumbtack" ambiguity function. For a radar system with finite range and cross-range ambiguity, $L_i(r, \theta, \phi) = F_i(\theta, \phi)W_i(r)$ may be expressed as a separable function embodying the radar beam pattern F and the range ambiguity function W (ignoring for the time being the complexity of the full space-time ambiguity of Eq. 9). However, as we see in the following equation, other choices of L may be more expedient in sensor fusion applications.

The integrating kernel L may be incorporated in the forward model relating the desired reconstruction of the density field to its samples. Let us represent $N(\mathbf{r})$ in a finite basis expansion

$$N(\mathbf{r}) \approx \sum_{j=1}^{J} N_{Rj}\psi_j(\mathbf{r}) \qquad (16)$$

where the discrete coefficients N_{Rj} contain the information needed to reconstruct the density field at some desired fidelity. If ψ represents an orthogonal voxel basis, then the N_{Rj} can represent voxel densities. But other nonstandard bases are also possible. For instance, the $\psi_j(\mathbf{r})$ could represent density profiles corresponding to a particle beam of a particular energy. In that case, the N_R can represent a discrete representation of the differential energy spectrum of the incident particle flux (Semeter and Kamalabadi, 2005).

Assuming N_R is a suitable representation of N, we may combine Eqs. (15), (16) to obtain

$$N_{Si} = \sum_{j=1}^{J} N_{Rj} \underbrace{\int L_i(\mathbf{r})\psi_j(\mathbf{r})d\mathbf{r}}_{\phi_{ij}} \qquad (17)$$

where we have used the shorthand $N_{Si} = N_S(\mathbf{r}_i)$. For now the samples are treated as deterministic ($\epsilon_i = 0$). Uncertainties in ISR sampling arising from the statistical nature of the scattering process (Eq. 10) are treated in the data fusion or reconstruction operation. The term under the bracket in Eq. (17), denoted ϕ_{ji}, represents a mapping between coefficients in our desired representation N_{Rj} and our samples N_{Si}. This function is nonzero only where L_i and ψ_j are both nonzero. Suppose, for instance, the elements W_i define a Voronoi partitioning of the density field (Aurenhammer et al., 2013) and ψ_j are elements of a rectangular orthogonal voxel basis with uniform volumes V. This choice is equivalent to nearest neighbor interpolation. The Voronoi partitioning becomes an orthonormal basis if the elements are weighted by their volume V_i, that is, $L_i(\mathbf{r}) = 1/V_i$ when $\mathbf{r}$ is closer to $\mathbf{r}_i$ than to any other sample. Suppose further that N is represented in a voxel basis with uniform voxel volume v. For this scenario, $\phi_{ji} = v/V_i$ if voxel j lies in cell i, and 0 otherwise. For these choices N_{Si} represents an estimate of the average density in the cell, that is,

$$N_{Si} \approx \frac{1}{V_i} \int_{\mathbf{V}_i} N(\mathbf{r}) dv \qquad (18)$$

Comparing with Eq. (11), we see that Eq. (18) also represents a projection. But rather than an integration along a line of sight, we instead integrate over a spatial region defined by a Voronoi cell.

Expressing Eq. (17) and its inverse in matrix form, we obtain

$$\mathbf{N}_S = \boldsymbol{\phi} \mathbf{N}_R \qquad (19)$$

$$\mathbf{N}_R = \boldsymbol{\phi}^+ \mathbf{N}_S \qquad (20)$$

where $\boldsymbol{\phi}^+$ represents the pseudoinverse of the rank-deficient matrix $\boldsymbol{\phi}$. For the case of the orthogonal voxel basis and Voronoi partitioning described earlier, the Moore-Penrose (MP) pseudo-inverse (the minimum norm solution) yields nearest-neighbor interpolation. This same

solution is also obtained using Tikhonov-type regularization approaches such as total variation (TV) or maximum entropy (ME). This is not too surprising since all of these techniques seek the smoothest possible solution consistent with available data which, in this case, is a constant representing the mean value in the cell.

4.2 Reconstruction filter for TEC

An analogous matrix formulation of the TEC forward model may be obtained by substituting Eq. (16) into Eq. (11) to yield

$$T_{Sk} = \sum_{j=1}^{J} N_{Rj} \underbrace{\int_{S_k} \psi_j(\mathbf{r}) d\mathbf{l}}_{\alpha_{kj}} + b_k + \xi_k \qquad (21)$$

where terms have been included for the bias b_k and uncertainty ξ_k of the measurement. The statistical uncertainty of a TEC measurement is typically small compared to the uncertainty in an ISR density sample. However, TEC bias is a dominant consideration, and its estimation has been treated extensively in the literature (e.g., Sardon et al., 1994; Ciraolo and Spalla, 1998; Coster et al., 2013; Ma et al., 2014). Bias mitigation is facilitated by three key features of the proposed experimental geometry and objectives:

1. The receiver spacing is small compared to expected ionospheric variations.
2. The receivers all view the same satellite, so the contribution from satellite bias is the identical for all measurements.
3. The biases are stable over a period which is long compared to our data fusion sampling window (<1 min). The biases may thus be treated as constant.

Based on these considerations, the differential bias among the receivers may be accurately estimated using techniques described by Otsuka et al. (2002) and Vierinen et al. (2015), which exploit the correlation among the T_{Sk} for long durations. The proximity of the sensors may also make it reasonable to ignore bias variations due to receiver temperature variations (Coster et al., 2013). What remains is a single unknown bias b for the entire array. Expressing Eq. (21) in matrix form with the inclusion of this constant term yields

$$T_S = \alpha N_R + b \qquad (22)$$

where an element α_{jk} represents the length of TEC line-of-sight k in voxel j.

Another challenge with the GNSS-ISR fusion problem is missing density. The limited practical range of PFISR ($\sim$670 km) means that it may miss up to $\sim$25% of the density contributing to the TEC measurements (Makarevich and Nicolls, 2013). This missing density will result in an inconsistency between the ISR and TEC measurements if not accounted for in some way. One approach is to use a basis expansion of N that allows for extrapolation to higher altitudes, such as the use of Chapman functions (Semeter and Mendillo, 1997) or empirical orthogonal functions (Zhao et al., 2005). However, if we make the plausible assumption that this topside contribution does not vary over the limited $\sim$150-km spatial extent of the array, then it will simply appear as an additional contribution to the bias term b. This is equivalent to assuming that all density variations contributing to TEC variations across the receiver array occur within the altitude range observed by the radar (100–600 km). In Section 4.3, we include b as a coefficient for an additional basis function, determined self-consistently with the unknown density field.

4.3 Formulation of the joint reconstruction filter

Through Eqs. (19), (22), we have a discrete forward model connecting our representation of the unknown density field N_R to the combined measurements. Our goal is to combine these measurements to obtain an improved estimate of the $N(\mathbf{r})$ compared with representations

obtained using either measurement technique alone.

Data fusion embodies a broad range of techniques and experimental objectives. Durrant-Whyte (1990) proposed a useful classification based on the relation of the data sources (complementary vs. redundant vs. cooperative). Fig. 4 illustrates schematically the cooperative data fusion problem embodied in Sections 4.1 and 4.2. We seek a discrete representation N_R of the continuous density field N. Both the ISR and TEC observations are treated as projections of N through matrices ϕ and α, respectively. In the ISR case, ϕ describes how elements of N_R are to be combined to form our samples N_S. In the TEC case, α corresponds to numerical integration through N. Our goal is to combine the information to constrain the solution space, thereby obtaining a more reliable representation of N than would be obtained from either modality individually. The products of the fusion step are the estimated representation $\hat{N}_R$ and an associated uncertainty $\hat{\sigma}$.

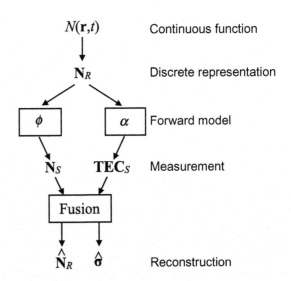

FIG. 4   Flow chart illustrating the framework for data fusion described by Eq. (23).

One approach to data fusion involves concatenation of the observations and projection matrices into a single expression to be inverted. The term b capturing the bias and missing density may be captured in an additional function which affects the TEC observations only. The equation to be solved may be expressed compactly as

$$\begin{bmatrix} \mathbf{N}_S \\ \mathbf{T}_S \end{bmatrix} = \begin{bmatrix} \phi & | & \mathbf{0}_I \\ \alpha & | & \mathbf{1}_K \end{bmatrix} \begin{bmatrix} \mathbf{N}_R \\ b \end{bmatrix} \quad (23)$$

where | denotes concatenation, $\mathbf{0}_K$ is a length-K vector of 0s, and $\mathbf{1}_K$ is a length-K vector of 1s. Defined in this way, b has units of TEC in m^{-2}.

For the voxel basis representation, the solution to Eq. (23) is highly underdetermined, and additional constraints must be incorporated to find a unique solution. One approach that enjoys philosophical and practical justification is the ME method (Mohammad-Djafari, 1993), used in a variety of astronomical (Gull and Daniell, 1978) and geospace (Semeter and Kamalabadi, 2005; Hysell, 2007) applications. Of the infinite solutions for $\mathbf{N}_R$ that are consistent with the available data, the ME approach chooses the solution that maximizes total entropy, $\mathbf{N}_R(\ln \mathbf{N}_R)^T$. ME can be treated as a special case of Tikhonov regularization,

$$\begin{bmatrix} \hat{\mathbf{N}}_R \\ \hat{\mathbf{T}}_R \end{bmatrix} = \underset{\mathbf{N}_R, b}{\arg\min} \left\{ \left\| \begin{bmatrix} \phi & | & \mathbf{0}_I \\ \alpha & | & \mathbf{1}_K \end{bmatrix} \begin{bmatrix} \mathbf{N}_R \\ b \end{bmatrix} - \begin{bmatrix} \mathbf{N}_S \\ \mathbf{T}_S \end{bmatrix} \right\|^2 \right.$$
$$\left. + \lambda^2 \mathbf{N}_R (\ln \mathbf{N}_R)' \right\} \quad (24)$$

where the parameter λ balances the competing interests of data fidelity, expressed as the Euclidean distance between measured and modeled data, and solution entropy. The cost function in Eq. (24) is strictly convex and thus converges for any value of $\lambda > 0$ (Engl and Landl, 1993). A method for determining the optimal value is given by Amato and Hughes (1991) based on the generalized cross-validation (GCV) approach proposed by Wahba (1977). The implementation used in our examples is the parallel log-entropy algorithm described by De Pierro (1991) and used for ionospheric profile inversion by Semeter and Kamalabadi (2005), wherein the Lagrange multiplies are determined iteratively in the solution search.

4.4 Example application: Reconstruction of $B_\perp$ density structure

Let us consider a use case involving reconstruction of a magnetic field-aligned ionospheric feature observed by ISR and a meridional GPS receiver network. For this case, it is insightful to use the actual geometry available by the electronically steerable ISR at Poker Flat, Alaska. Fig. 5A depicts the complete set of standard predefined beam directions used in the design of PFISR experiments, plotted in a radar-centered horizontal coordinate system. The hexagonal boundary of this locus defines the grating lobe-free region, which is defined by the particular antenna array pattern used. The dense set of beams through the center of this locus corresponds to a magnetic meridional elevation scan. The curved lines to the south show a set of GPS satellite tracks projected into this plane. Conveniently, the 50 degrees inclination of the GPS constellation enables them to reach the magnetic zenith of an observer at the radar location.

Consider now a set of radar beams representing a meridional elevation scan and a network of GPS receivers deployed along this meridian. This scenario is depicted schematically in Fig. 5B for a set of radar beams at 5 degrees separation and a colinear set of TEC receivers all observing a single satellite. The spacecraft range is >20,000 km which far exceeds the receiver spacing, and so the receiver-satellite lines can be approximated as parallel.

To prevent artifacts in the density reconstruction process, the region probed by the ISR should sample the entire volume contributing to the TEC density. Stated mathematically, the

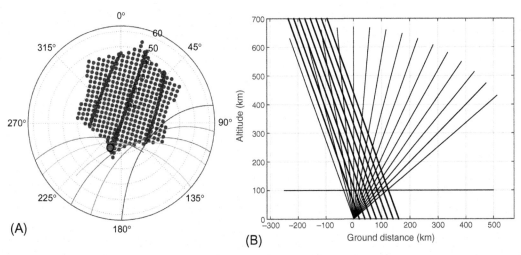

FIG. 5 Illustration of ISR-GNSS observing geometry from Poker Flat, Alaska. (A) Locus of preprogrammed PFISR beam positions (*filled dots*) on a horizontal coordinate system. The *outer circle* depicts the approximate field of view of the colocated all-sky camera. The curved lines show GPS satellite tracks, revealing the regions where co-alignment of radar beams and satellite lines-of-sight is possible. Of particular interest are observations near the magnetic zenith (*largest filled dot*). (B) Fan of radar beams along the magnetic meridian (central cluster of *filled dots* in (A)) and GPS lines of sight corresponding to a coaligned network of TEC receivers deployed to the north of the facility.

ionospheric column contributing to a given TEC measurement should lie entirely within the convex hull defined by the lattice of PFISR samples. At the lower boundary, this criterion sets the maximum northern extent of the receiver network for a given satellite elevation. For instance, suppose we are interested in observing lines of sight along the magnetize zenith (elevation 78 degrees, azimuth 156 degrees W), and suppose we assume the lower ionospheric boundary is 100 km (horizontal line in Fig. 5B). These conditions set the position of the northernmost receiver to ~160 km.

The algorithm used in our simulation cases has the following steps:

1. Define the reconstruction bases ϕ and α based on resolution defined by the GNSS receiver spacing.
2. Compute $\hat{\mathbf{N}}_R$ from $\mathbf{N}_S$ using nearest-neighbor interpolation (equivalent to backprojection, $\hat{\mathbf{N}}_R = \phi^T \mathbf{N}_S$).

3. Solve Eq. (23) for $[\mathbf{N}_R \, b]^T$ using the ME method, using $\hat{\mathbf{N}}_R$ from 2 as the initial guess.

A similar approach has been taken by Chiang et al. (2005) who used ME to refine a direct Tikhonov solution in a study of electron spin resonance.

To assess the efficacy of data fusion to improve representations of dynamic ionospheric structures, we turn to simulation. Fig. 6A shows a phantom ionospheric density function $N(\mathbf{r})$ used for the simulation. The phantom was constructed using two Chapman density profiles, one with peaks density $1 \times 10^{11} \, \mathrm{m}^{-3}$ and the other with peak density of $2 \times 10^{10} \, \mathrm{m}^{-3}$, separated by infinitely steep gradients oriented perpendicular to the ambient magnetic field at Poker Flat (78 degrees). Such structures are commonly observed in the auroral ionosphere, owing to the magnetization of the plasma at F-region altitudes. Fully resolving these $B_\perp$ gradients is an observational challenge.

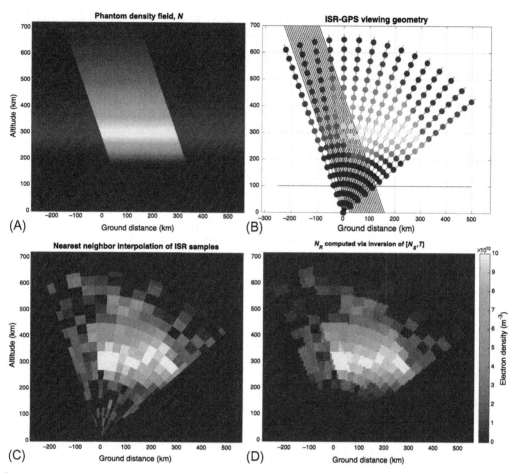

FIG. 6 ISR-GNSS data fusion example. (A) Phantom density structure in the ionosphere, comprised of an enhance density region bounded by steep $B_\perp$ gradients. (B) Simulated PFISR sampling pattern (*filled circles*) and dense GNSS receiver network northward of the facility (*straight lines*). (C) Interpolated density field based on simulated PFISR sampling statistics. (D) Result after fusion with GNSS measurements. Note the improved fidelity of the density gradient.

Panel B illustrates the specific observing geometry through which this phantom is observed. The density field is sampled according to Eq. (15), assuming a 240-µs pulse (36-km range ambiguity) and a 1 degree beam width and 5 degree spacing between beams. The filled circles represent the computed samples $\mathbf{N}_S$. This figure represents the "Projection" layer in Fig. 1, that is, the ambiguity function has been applied but we have not accounted for the statistical uncertainties in the acquisition process.

A network of 20 TEC receivers is placed to the north of PFISR at 7-km separation, following the constraints discussed in the context of Fig. 5. Each receiver observes the same satellite in the magnetic zenith. The data vector $\mathbf{T}_S$ represents a "snapshot" of slant TEC along this meridian. The spacing of TEC receivers corresponds to roughly 20% of the ground-projected distance between ISR samples near the F-region peak for the selected radar mode. The TEC sampling rate can be rapid—1 Hz native sampling rate is typical, although synoptic TEC maps are generally produced at the cadence of several minutes. The TEC measurements thus constitute a high-resolution measurement in the direction perpendicular to the magnetic zenith. The southerly density gradient is fully contained within the common fields of view of PFISR and the receiver array, the northerly gradient is sampled only by PFISR.

For the "Instrument" layer in Fig. 1, we return to Eq. (10). For illustration purposes, let us consider the high-SNR limit, such that the relative error in our ACF samples is $\approx 1/\sqrt{K}$ (i.e., controlled purely by the number of pulses averaged). Let us assume we integrate $K = 100$ pulses per beam direction, corresponding to a 10% relative error in received power. This level of pulse integration represents a minimum reasonable integration for a dynamic high-latitude ionosphere observed using a meridional elevation scan (Semeter et al., 2009). This would correspond to about a 5-s integration period for the 15-beam experiment depicted in Fig. 5.

Panel C depicts the results of applying the ISR reconstruction filter (Eq. 20) to the noise-degraded samples. The filter interpolates the samples according to a Voronoi partitioning, which corresponds identically to nearest-neighbor interpolation. This interpolation scheme has the benefit of clearly conveying the sampling pattern, sample locations, and lack of information between samples. Although this result is generally computed via a pixel assignment algorithm, here we have computed it via inversion of Eq. (19) using the ME regularization approach of Eq. (24). This enables the incorporation of the additional information acquired by the TEC array in a straightforward fashion.

Panel D shows the refined solution obtained from the combined ISR and TEC measurements using Eq. (23). We have assumed that the uncertainties on T_S are much less than the uncertainties on N_S, that is, TEC measurements are essentially assumed to be deterministic to within an additive bias. The sampling period for TEC is also assumed to be comparable to or less than the ISR integration period (indeed, extremely rapid sampling rates are possible; Jiao et al., 2013).

The solution in Panel C was again obtained using the ME approach embodied in Eq. (24), with the nearest-neighbor interpolation (Panel C) as the initial guess. The result represents the maximally uncommitted solution that is consistent with both sets of measurements. In order to avoid large oscillatory behavior in the result, the dimensions of the voxels must be greater than the separation of the TEC receivers. In this case, we have used uniform 10×10-km orthogonal voxels for N_R. Both the location and orientation of the gradient observed by the combined sensors is resolved with greater fidelity in Panel C compared with Panel B. This improvement in fidelity has important physical implications. For instance, the calculation of electromagnetic energy dissipation in the coupled ionosphere-thermosphere system is highly sensitive to errors in the relative position of electric fields and electrical conductance (hence, density) (Cosgrove et al., 2009; Semeter et al., 2016).

5 Summary

The technological landscape is evolving rapidly. ISR's employing electronically steerable arrays and distributed power systems are allowing new lattice sampling modalities and deployment to remote locations. New low-power GNSS receivers are enabling the construction of TEC maps at high sampling density in remote polar locations. New optical detector technology is enabling high-resolution imaging over a wide field with performance approach the photon counting limit. The emerging specter of massive cubesat constellations in low-Earth orbit is not discussed in this chapter, which would work collaboratively with, or provide a complement to, ground-based ionospheric sensor networks.

Current infrastructure trends toward global broadband connectivity, cloud infrastructure, graphics processing unit (GPU)-enhanced computing power, low-power sensors, and artificial intelligence (AI)-based data analytics point toward an ability to "pixelate" the planet with sensors and develop high-resolution representations of ionospheric dynamics with broad spatial coverage. This chapter has sought to develop a framework for the fusing data from different sensors of different types. The approach is based on describing each sensing modality as a "Projection" of the intrinsic physical quantity. Mathematically, this amounts to writing a Fredholm integral equation, with the intrinsic quantity represented in an appropriate basis expansion. The use case presented involved the fusion of TEC and ISR measurements with the goal of increasing the fidelity of dynamic ionospheric gradients.

References

Amato, U., Hughes, W., 1991. Maximum entropy regularization of Fredholm integral equations of the first kind. Inverse Prob. 7 (6), 793.

Aurenhammer, F., Klein, R., Lee, D.-T., 2013. Voronoi Diagrams and Delaunay Triangulations. World Scientific, Singapore.

Baker, D.J., 1974. Rayleigh, the unit for light radiance. Appl. Opt. 13, 2160–2163. https://doi.org/10.1364/AO.13.002160.

Baumgardner, J.L., Flynn, B., Mendillo, M.J., 1993. Monochromatic imaging instrumentation for applications in aeronomy of the earth and planets. Opt. Eng. 32, 3028–3032. https://doi.org/10.1117/12.149194.

Chamberlain, J.W., 1961. Physics of the Aurora and Airglow. Academic Press, New York, NY.

Chiang, Y.-W., Borbat, P.P., Freed, J.H., 2005. Maximum entropy: a complement to Tikhonov regularization for determination of pair distance distributions by pulsed ESR. J. Magn. Reson. 177, 184–196. https://doi.org/10.1016/j.jmr.2005.07.021.

Ciraolo, L., Spalla, P., 1998. Preliminary study of the latitudinal dependence of TEC. Adv. Space Res. 22, 807–810. https://doi.org/10.1016/S0273-1177(98)00102-1.

Cosgrove, R.B., Lu, G., Bahcivan, H., Matsuo, T., Heinselman, C.J., McCready, M.A., 2009. Comparison of AMIE-modeled and Sondrestrom-measured Joule heating: a study in model resolution and electric field-conductivity correlation. J. Geophys. Res. 114, A04316. https://doi.org/10.1029/2008JA013508.

Coster, A., Williams, J., Weatherwax, A., Rideout, W., Herne, D., 2013. Accuracy of GPS total electron content: GPS receiver bias temperature dependence. Radio Sci. 48, 190–196. https://doi.org/10.1002/rds.20011.

Dahlgren, H., Semeter, J.L., Marshall, R.A., Zettergren, M., 2013. The optical manifestation of dispersive field-aligned bursts in auroral breakup arcs. J. Geophys. Res. 118, 4572–4582. https://doi.org/10.1002/jgra.50415.

De Pierro, A., 1991. Multiplicative iterative methods in computed tomography. In: Herman, G.T., Louis, A.K., Natterer, F. (Eds.), Mathematical Methods in Tomography. Springer-Verlag, Berlin, pp. 1441–1450.

Dougherty, J.P., Farley, D.T., 1960. A theory of incoherent scattering of radio waves by a plasma. Proc. R. Soc. Lond. A 259, 79–99. https://doi.org/10.1098/rspa.1960.0212.

Durrant-Whyte, H., 1990. Sensor models and multisensor integration. In: Cox, I.J., Wilfong, G.T. (Eds.), Autonomous Robot Vehicles. Springer, New York, NY, pp. 73–89.

Engl, H.W., Landl, G., 1993. Convergence rates for maximum entropy regularization. SIAM J. Numer. Anal. 30 (5), 1509–1536. https://doi.org/10.1137/0730079.

Evans, J.V., 1969. Theory and practice of ionospheric study by Thomson scatter radar. Proc. IEEE 57, 496–530.

Farley, D.T., 1969. Incoherent scatter correlation function measurements. Radio Sci. 4, 935–953. https://doi.org/10.1029/RS004i010p00935.

Fathi, E., El-Samie, A., Hadhoud, M.M., El-Khamy, S.E., 2012. Image Super-Resolution and Applications. CRC Press, Boca Raton, FL.

Frey, S., Frey, H., Carr, D., Bauer, O., Haerendel, G., 1996. Auroral emission profiles extracted from three-dimensionally reconstructed arcs. J. Geophys. Res. 101 (A10), 21731–21742.

Gray, R.W., Farley, D.T., 1973. Theory of incoherent-scatter measurements using compressed pulses. Radio Sci. 8 (2), 123–131.

Gull, S.F., Daniell, G.J., 1978. Image reconstruction from incomplete and noisy data. Nature 272, 686–690.

Hagfors, T., 1961. Density fluctuations in a plasma in a magnetic field, with applications to the ionosphere. J. Geophys. Res. 66, 1699–1712. https://doi.org/10.1029/JZ066i006p01699.

Hecht, J.H., Christensen, A.B., Strickland, D.J., Meier, R.R., 1989. Deducing composition and incident electron spectra from ground-based auroral optical measurements: variations in oxygen density. J. Geophys. Res. 94, 13553–13563. https://doi.org/10.1029/JA094iA10p13553.

Hirsch, M., Semeter, J., Zettergren, M., Dahlgren, H., Goenka, C., Akbari, H., 2016. Reconstruction of fine-scale auroral dynamics. IEEE Trans. Geosci. Remote Sens. 54, 2780–2791. https://doi.org/10.1109/TGRS.2015.2505686.

Hysell, D.L., 2007. Inverting ionospheric radio occultation measurements using maximum entropy. Radio Sci. 42, RS4022. https://doi.org/10.1029/2007RS003635.

Hysell, D., 2018. Antennas and Radar for Environmental Scientists and Engineers. Cambridge University Press, Cambridge.

Hysell, D.L., Rodrigues, F.S., Chau, J.L., Huba, J.D., 2008. Full profile incoherent scatter analysis at Jicamarca. Ann. Geophys. 26, 59–75. https://doi.org/10.5194/angeo-26-59-2008.

Jiao, Y., Morton, Y.T., Taylor, S., Pelgrum, W., 2013. Characterization of high-latitude ionospheric scintillation of GPS signals. Radio Sci. 48, 698–708. https://doi.org/10.1002/2013RS005259.

Kudeki, E., Milla, M.A., 2011. Incoherent scatter spectral theories—Part I: a general framework and results for small magnetic aspect angles. IEEE Trans. Geosci. Remote Sens. 49 (1), 315–328. https://doi.org/10.1109/TGRS.2010.2057252.

Lehtinen, M.S., Haggstrom, I., 1987. A new modulation principle for incoherent scatter measurements. Radio Sci. 22, 625–634. https://doi.org/10.1029/RS022i004p00625.

Lehtinen, M.S., Huuskonen, A., 1996. General incoherent scatter analysis and GUISDAP. J. Atmos. Terr. Phys. 58, 435–452. https://doi.org/10.1016/0021-9169(95)00047-X.

Lummerzheim, D., Lilensten, J., 1994. Electron transport and energy degradation in the ionosphere: evaluation of the numerical solution, comparison with laboratory experiments and auroral observations. Ann. Geophys. 12, 1039–1051.

Ma, G., Gao, W., Li, J., Chen, Y., Shen, H., 2014. Estimation of GPS instrumental biases from small scale network. Adv. Space Res. 54, 871–882. https://doi.org/10.1016/j.asr.2013.01.008.

Makarevich, R.A., Nicolls, M.J., 2013. Statistical comparison of TEC derived from GPS and ISR observations at high latitudes. Radio Sci. 48, 441–452. https://doi.org/10.1002/rds.20055.

Mohammad-Djafari, A., 1993. Maximum entropy and linear inverse problems. A short review. In: Mohammad-Djafari, A., Demoment, G. (Eds.), Maximum Entropy and Bayesian Methods. In: Fundamental Theories of Physics, vol. 53. Springer, Dordrecht, pp. 253–264.

Nicolls, M., Heinselman, C.J., 2007. Three-dimensional measurements of traveling ionospheric disturbances with the poker flat incoherent scatter radar. Geophys. Res. Lett. 34. https://doi.org/10.1029/2007GL031506.

Otsuka, Y., Ogawa, T., Saito, A., Tsugawa, T., Fukao, S., Miyazaki, S., 2002. A new technique for mapping of total electron content using GPS network in Japan. Earth Planets Space 54, 63–70.

Rees, M.H., 1989. Physics and Chemistry of the Upper Atmosphere. Cambridge University Press, Cambridge.

Rees, M.H., Luckey, D., 1974. Auroral electron energy derived from ratio of spectroscopic emissions: 1. Model computations. J. Geophys. Res. 79, 5181–5186.

Reinisch, B.W., Galkin, I.A., Khmyrov, G.M., Kozlov, A.V., Bibl, K., Lisysyan, I.A., Cheney, G.P., Huang, X., Kitrosser, D.F., Paznukhov, V.V., Luo, Y., Jones, W., Stelmash, S., Hamel, R., Grochmal, J., 2009. New Digisonde for research and monitoring applications. Radio Sci. 44, RS0A24. https://doi.org/10.1029/2008RS004115.

Rosenbluth, M.N., Rostoker, N., 1962. Scattering of electromagnetic waves by a nonequilibrium plasma. Phys. Fluids 5, 776–788. https://doi.org/10.1063/1.1724446.

Sardon, E., Rius, A., Zarraoa, N., 1994. Estimation of the transmitter and receiver differential biases and the ionospheric total electron content from global positioning system observations. Radio Sci. 29, 577–586. https://doi.org/10.1029/94RS00449.

Semeter, J., Kamalabadi, F., 2005. Determination of primary electron spectra from incoherent scatter radar measurements of the auroral E-region. Radio Sci. 40, RS2006. https://doi.org/10.1029/2004RS003042.

Semeter, J., Mendillo, M., 1997. A nonlinear optimization technique for ground-based atmospheric emission tomography. Trans. Geosci. Remote Sens. 35 (5), 1105–1116.

Semeter, J., Mendillo, M., Baumgardner, J., Holt, J., Hunton, D.E., Eccles, V., 1996. A study of oxygen 6300 Å airglow production through chemical modification of the nighttime ionosphere. J. Geophys. Res. 101 (A9), 19683–19699.

Semeter, J., Mendillo, M., Baumgardner, J., 1999. Multispectral tomographic imaging of the midlatitude aurora. J. Geophys. Res. 104 (A11), 24565–24585.

Semeter, J., Zettergren, M., Diaz, M., Mende, S., 2008. Wave dispersion and the discrete aurora: new constraints derived from high-speed imagery. J. Geophys. Res. 113 (A12), 12208. https://doi.org/10.1029/2008JA013122.

Semeter, J., Butler, T., Heinselman, C., Nicolls, M., Kelly, J., Hampton, D., 2009. Volumetric imaging of the auroral ionosphere: initial results from PFISR. J. Atmos. Sol. Terr. Phys. 71, 738–743. https://doi.org/10.1016/j.JASTP.2008.08.014.

Semeter, J., Butler, T.W., Zettergren, M., Nicolls, M., Heinselman, C., 2010. Composite imaging of auroral forms and convective flows during a substorm cycle. J. Geophys. Res. 115 (A14), A08308. https://doi.org/10.1029/2009JA014931.

Semeter, J., Hirsch, M., Lind, F., Coster, A., Erickson, P., Pankratius, V., 2016. GNSS-ISR data fusion: general framework with application to the high-latitude ionosphere. Radio Sci. 51, 118–129. https://doi.org/10.1002/2015RS005794.

Strickland, D.J., Book, D.L., Coffey, T.P., Fedder, J.A., 1976. Transport equation techniques for the deposition of auroral electrons. J. Geophys. Res. 81 (16), 2755–2764.

Strickland, D.J., Hecht, J.H., Christensen, A.B., Kelly, J.D., 1994. Relationship between energy flux Q and mean energy $\langle E \rangle$ of auroral electron spectra based on radar data from the 1987 CEDAR campaign at Sondre Stromfjord, Greenland. J. Geophys. Res. 99 (A10), 19467–19473.

Swoboda, J., Semeter, J., Erickson, P.J., 2015. Space-time ambiguity functions for electronically scanned ISR applications. Radio Sci. 50, 415–430. https://doi.org/10.1002/2014RS005620.

Swoboda, J., Semeter, J., Zettergren, M., Erickson, P.J., 2017. Observability of ionospheric space-time structure with ISR: a simulation study. Radio Sci. 52 (2), 215–234. https://doi.org/10.1002/2016RS006182.

Thayer, J.P., Semeter, J., 2004. The convergence of magnetospheric energy flux in the polar atmosphere. J. Atmos.

Sol. Terr. Phys. 66, 805–817. https://doi.org/10.1016/j.JASTP.2004.01.035.

Vierinen, J., Coster, A.J., Rideout, W.C., Erickson, P.J., Norberg, J., 2015. Statistical framework for estimating GNSS bias. Atmos. Meas. Tech. Discuss. 8, 9373–9398. https://doi.org/10.5194/amtd-8-9373-2015.

Wahba, G., 1977. Practical approximate solutions to linear operator equations when the data are noisy. SIAM J. Numer. Anal. 14 (4), 651–667. https://doi.org/10.1137/0714044.

Zettergren, M., Semeter, J., Blelly, P.-L., Diaz, M., 2007. Optical estimation of auroral ion upflow: theory. J. Geophys. Res. 112 (A11), 12310. https://doi.org/10.1029/2007JA012691.

Zhao, B., Wan, W., Liu, L., Yue, X., Venkatraman, S., 2005. Statistical characteristics of the total ion density in the topside ionosphere during the period 1996–2004 using empirical orthogonal function (EOF) analysis. Ann. Geophys. 23, 3615–3631. https://doi.org/10.5194/angeo-23-3615-2005.

16

Advanced statistical tools in the near-Earth space science

Giuseppe Consolini[a], Massimo Materassi[b]

[a]INAF-Institute for Space Astrophysics and Planetology, Rome, Italy
[b]Institute for Complex Systems of the National Research Council (ISC-CNR), Florence, Italy

1 Introduction

The majority of the real systems are open systems, capable of exchanging energy and matter with the surrounding environment, which evolve outside of equilibrium. This is, for instance, also the case of many space plasma systems, such as solar plasmas, interplanetary media, magnetospheric and ionospheric plasmas, whose dynamics is clearly out-of-equilibrium being characterized by a positive production of entropy, and displaying *dynamical complexity* (Chang et al., 2006; Chang, 2015). Also the Earth's ionosphere has to be considered as an open system capable of exchanging energy, mass, and momentum with the Earth's magnetosphere, on the one side, and the Earth's atmosphere on the other side. As a consequence of such a continuous exchange of energy mass and momentum, and also of the large-scale spatial organization due to the link of the Earth's ionosphere with different regions of the Earth's magnetosphere (due to the geomagnetic field mapping to the different magnetospheric regions), the Earth's ionosphere shows a very

complex dynamics in response to solar wind and interplanetary medium changes. This complex dynamics is particularly marked in the high-latitude polar ionospheric regions, where particle precipitation from remote magnetospheric regions (via a complex system of currents, the *field-aligned currents*, the *Region 0, 1, and 2 current circuits*, etc.; see, for example, Ganushkina et al. (2018) and references therein), as well as directly from the interplanetary medium occurs causing ionospheric plasma energy deposition, heating, and auroral light emission.

Among the various phenomena observed in the high-latitude polar ionosphere, the consequence of its complex dynamics, turbulence is surely one of the most important, being capable of providing an additional plasma heating mechanism and also of causing spatiotemporal sporadic plasma acceleration. Other manifestation of ionospheric spatiotemporal dynamical complexity is related with the scale-invariant features of the magnetospheric energy deposition rate in the auroral regions, as shown by the auroral electrojet (*AE*) indices (Davis and

Sugiura, 1966; Consolini, 1997, 2002; Uritsky and Pudovkin, 1998), which are proxies of Joule heating (Ahn et al., 1983), and auroral ultraviolet imager (UVI) emission blobs (Uritsky et al., 2002), as well as the changes of the ionospheric plasma large-scale convective structure, which is a direct consequence of the dynamical changes occurring in the solar wind and the interplanetary medium. Fig. 1 shows the highly intermittent character of the energy deposition rate in the auroral ionosphere as monitored by the AE index. The behavior of AE index is, furthermore, inherently multiscale, being the fluctuations, associated with this geomagnetic index, not characterized by a characteristic temporal scale, but conversely covering a wide range of timescales from minutes up to hours.

The multiscale nature of fluctuations of ionospheric magnetic and electric fields, and plasma parameters has also been investigated by means of in situ observations via satellites, rockets, etc., showing how such multiscale behavior can be the signature of a turbulent dynamics of ionospheric plasma. For instance, Golovchanskaya and Kozelov (2010) have clearly shown the multiscale nature of the electric field fluctuations in the range from ∼1 km up to ∼250 km in the polar latitude and that these fluctuations

are scale invariant. This evidence of a scale-invariant nature of electric field fluctuations has been shown to be consistent with the occurrence of intermittent turbulence in the polar ionosphere (Golovchanskaya et al., 2006; Kintner, 1976; Tam et al., 2005; Kozelov et al., 2008). Similar studies have been done on the multiscale and turbulent nature of magnetic field fluctuations (Chaston et al., 2008; De Michelis et al., 2015, 2016, 2017, 2019), and on plasma density fluctuations (Spicher et al., 2015) in the high-latitude polar cap.

In this chapter, we will present some statistical and information theory-based approaches and methods to characterize the multiscale and turbulent character of fields and plasma fluctuations in the polar ionosphere, and to infer information on the possible coupling of these fluctuations with some other physical quantities. In particular, in Section 2, we will discuss the meaning of scaling features, and how to investigate and characterize them using methods borrowed from the studies on fluid and plasma turbulence. Section 3 is devoted to an introduction to information theory-based methods to infer information on the coupling and driving of fields and plasma fluctuations. Some applications on simple dynamical models are also shown for clarity.

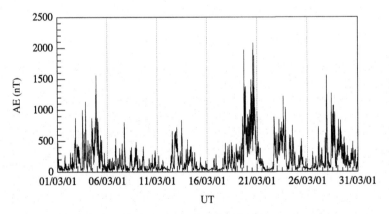

FIG. 1   A sample of the AE index behavior for the period of March 2001.

2 Scale invariance and statistical properties: Analysis methods

Many natural systems and/or signals show a multitude of fluctuations occurring on a wide range of spatial and/or temporal scales; in other words, the dynamics of such systems/signals is inherently multiscale, either in time or in space (or both). This is, for instance, the case of nonequilibrium media/systems whose dynamics is characterized by fully developed (strong) turbulence. In this dynamical regime, a mess of fluctuations (modes) is simultaneously excited so that the dynamics of such systems results in a noisy, chaotic, and poorly predictable evolution (Falkovich, 2008). This mess of fluctuations is the consequence of the fact that in turbulent systems the disturbance/excitation occurs at a macroscopic scale that substantially deviates from the microscopic scales where dissipation takes place.

In spite of the wide range of scales of excited modes/fluctuations, which manifests in very wide *power spectral densities* (PSDs) covering many decades of frequencies/wavenumbers without any peculiar spectral features (such as some characteristic spectral frequency), sometime this mess of fluctuations is characterized by a special kind of symmetry features, namely, *scale invariance*. The scale-invariance concept has been introduced in several different frameworks, from *second-order phase transitions*, to *fractal structures* and *turbulence* (refer to Lesne and Lagües (2012), for an extended discussion) and its direct manifestation is the appearance of spatial and/or temporal structures without a characteristic length or timescale.

The concept of scale invariance can be expressed in mathematical terms by the simple following rationale. Let us consider an observable quantity $f(s)$ depending on the scale s, and let us consider the following *scale transformation*

$$s \rightarrow \lambda s, \tag{1}$$

if the observable quantity $f(s)$ obeys the following equation (i.e., it scales according to the following equation),

$$f(\lambda s) = \lambda^\alpha f(s), \tag{2}$$

then we say that the quantity $f(s)$ is *scale invariant* under the transformation of Eq. (1). In mathematical terms, Eq. (2) is a *first-order homogeneous equation*, whose simple solution is

$$f(s) \propto s^\alpha, \tag{3}$$

that is, a *power-law function*. Thus, the observation of power-law behavior of an observable quantity as a function of the observation scale is an indication of the occurrence of scale invariance.

The existence of scaling relations and scaling laws (such as Eq. 3) is a consequence of the lack of a characteristic *coherence length* and of the validity of central limit theorem and the law of large numbers, and implies the concept of *self-similarity*, that is, the possibility to deduce the behavior of fluctuations at a certain scale, λs, from that at a given one, s.

Another consequence of the existence of scale invariance is that it is possible to construct invariant quantities under scale transformation. For instance, given a quantity $f(s; \ell_0)$ depending on the parameter ℓ_0 (such as the system size, ℓ_0), and scaling according to Eq. (2), then it is possible to construct an invariant quantity $\mathcal{I}(x)$, according to the following transformation:

$$\begin{cases} s & \rightarrow & s/\ell_0 \\ f(s; \ell_0) & \rightarrow & \mathcal{I}(x) = \ell_0^\alpha f(s; \ell_0). \end{cases} \tag{4}$$

In the case of a turbulent signal $f(x)$, such as the observation of velocity fluctuations in a fluid flow at a given spatial point (Eulerian point of view) (Frisch, 1995), a way to investigate the occurrence of scaling features (or in general scale invariance) is to study the scaling properties of the so-called *q*th-*order generalized structure* *functions* $S_q(\delta x)$, which are defined according to the following expression

$$S_q(\delta x) = \langle |f(x+\delta x) - f(x)|^q \rangle, \qquad (5)$$

where δx is a spatial/temporal increment and $\langle \cdot \rangle$ means statistical average, which in the case of stationary time-dependent signals corresponds to a time average under the assumption of ergodicity. Now, for scale-invariant signals the generalized structure functions $S_q(\delta x)$ are expected all to scale according to a power-law over a wide range of δx, that is,

$$S_q(\delta x) \sim \delta x^{\gamma(q)}, \qquad (6)$$

where $\gamma(q)$ is the *scaling exponent* of the moment order q.

According to the definition of Eq. (5), the qth-order generalized structure function, $S_q(\delta x)$, corresponds to the qth-moment of the distribution function, $p(|\delta f|;\delta x)$, of the absolute value of signal increments, δf, at the scale δx,

$$S_q(\delta x) = \int_0^\infty |\delta f(\delta x)|^q p(|\delta f|;\delta x) d|\delta f|. \qquad (7)$$

Thus, the observation of a scaling features of structure functions corresponds to the scaling features of the distribution function $p(\delta f;\delta x)$. Clearly, for symmetric distribution function, $p(-\delta f;\delta x) = p(\delta f;\delta x)$, there is no difference between the scaling features of δf and $|\delta f|$ moments.

The scaling exponents $\gamma(q)$, for simple self-similar distribution functions, are expected to depend linearly on the moment order q, and the corresponding signal is said to be characterized by simple *fractal* features (a *monofractal behavior*). In such a case, the knowledge of a single number $H = \gamma(q)/q$, named *Hurst/Hölder exponent* (see, e.g., Bunde and Havlin, 1991, 1995 for an extensive discussion) is sufficient to characterize the scaling properties of the fluctuations (increments) statistics at different scales. Conversely, if the dependence of the scaling exponents $\gamma(q)$ on the moment order q

is a nonlinear function (generally a convex function), it is not possible to characterize the scaling features of the signal on the basis of a single number H, but we need a hierarchy of numbers (or scaling exponents). According to Mandelbrot (1989), this corresponds to a passage from fractal objects characterized by a single number, to objects whose scaling features are characterized by a function, so that the scaling features are more complex and the signal is said to show *multifractal features*. In the framework of fully developed turbulence such a behavior is related to the occurrence of anomalous scaling features of the structure functions,

$$S_q(\delta x) \neq c(q,p)[S_p(\delta x)]^{\gamma(q)/\gamma(p)}, \qquad (8)$$

where $c(q, p)$ is a constant. This behavior is the signature of the occurrence of *intermittency* in the dissipation field, which is the counterpart of spatially nonhomogeneous scaling features of the dissipation field.

In terms of statistics of fluctuations, the occurrence of anomalous scaling properties, that is, of intermittency, corresponds to the fact that the scaling features of fluctuations acquire a dependence on the fluctuation amplitude. This point can be clearly understood on the basis of Eq. (7), where high (low)-order structure functions correspond to investigate the scaling features of large (small)-amplitude increments due to the exponent factor q which acts as a microscope factor. Using the terminology of multifractal analysis, this corresponds to study those regions where the measure is more concentrated (or rarefied).

Before moving to illustrate some example of signals showing scale invariance, it is worth mentioning that all the previous consideration are strictly valid in the case of *stationary* signals. In the case of real signals, this is clearly an assumption, being not an easy task to verify it. However, if a signal is characterized by a power-law spectral density with a spectral exponent $\beta \in [1, 3]$, it is generally a nonstationary signal with stationary increments so that structure function analysis does not pose any specific critical issue (Davis et al., 1994).

To illustrate an application of the previous methods to the characterization of scaling features of a signal, we will proceed in applying the previous structure function analysis to three different classes of synthetic signals:

(i) A simple monofractal signal, named a *fractional Brownian motion* (FBM) (Mandelbrot and Van Ness, 1968; Bunde and Havlin, 1991, 1995), given by the following modified fractional Weyl integral,

$$x_H(t) = \frac{1}{\Gamma\left(H + \frac{1}{2}\right)} \left[\int_{-\infty}^{t} (t-s)^{H-\frac{1}{2}} dx(s) - \int_{-\infty}^{0} (-s)^{H-\frac{1}{2}} dx(s) \right],$$
(9)

where H is the Hurst exponent ($H \in [0, 1]$).

(ii) A *Lévy flight* motion (Klafter et al., 1987, 1996; Consolini et al., 2005),

$$x_{i+1} = x_i + \eta_i(\alpha),$$
(10)

where η is a δ-correlated noise extracted from a Lévy distribution of characteristic α.

(iii) A *multifractal* motion, generated by means of the p-model (Meneveau and Sreenivasan, 1991),

$$x_{i+1} = x_i + sgn(\eta) * \omega_i(p),$$
(11)

where η is a δ-correlated white noise and ω is an increment generated using the *p-model* with weight p.

Fig. 2 shows three samples of synthetic signals generated using the methods described in Meneveau and Sreenivasan (1991), Benassi et al. (1997), and Consolini et al. (2005, 2013). In particular, the Hurst exponent H for the generation of the FBM is $H = 1/3$, while the Lévy characteristic α is $\alpha = 3/2$ and the p factor relative to the p-model is $p = 0.75$.

Fig. 3 shows the behavior of the generalized structure functions $S_q(\tau) = \langle |\delta x(\tau)|^q \rangle$ as a function of the scale τ (here, $\delta x(\tau) = x(t + \tau) - x(t)$) in the case of the FBM signal. Each structure function

shows a power-law behavior as a function of the scale τ over a wide range of scales (here at least three orders of magnitude) indicating that scaling (self-similarity) is an inherent feature of such a signal. In particular, the scale invariance of the increments with the scale τ as shown by the structure functions is a clear indication of the absence of a characteristic scale in the signal investigated here, suggesting that this signal is statistically self-similar.

The same results are found in the case of the other two signals, which in spite of the different time behavior show scaling features, that is, a power-law dependence of the structure functions on the scale τ for each moment q. However, the nature of the scaling features is different. This can be clearly evidenced by plotting the scaling exponents $\gamma(q)$ of the structure functions as a function of the moment order q.

Fig. 4 shows the scaling exponents $\gamma(q)$ as a function of the moment order q for the three signals. We recover three different behaviors, clearly indicating a different nature of the scaling features of the three signals. The FBM signal shows a linear dependence of the scaling exponent $\gamma(q)$ on q (i.e., $\gamma(q) = Hq$), where $H = 1/3$ is the Hurst exponent. This is a clear signature of the simple fractal character of FBM, whose scaling features do not show any dependence on the amplitude of fluctuations, which are expected to be all characterized by the same statistics. Conversely, the p-model signal shows a clear multifractal character. Indeed, in this case, the scaling exponents $\gamma(q)$ do not linearly depend on the moment order q, but their behavior with the moment order is rather a nonlinear convex function, which implies that one single number is no longer sufficient to characterize the scale invariance of this kind of signal. This means that the scale invariance shown by p-model signal has a more complex nature, being characterized by intermittency, which is the counterpart of the multifractal character of this signal. For such a signal, we should not expect to observe an invariance of the

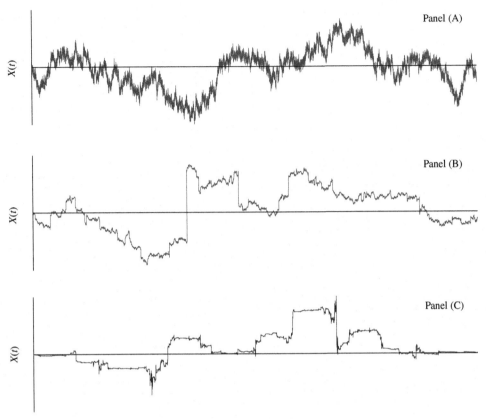

FIG. 2   A sample of the three synthetic signals generated using an FBM model (A), a Lévy motion (B), and a multifractal motion (C).

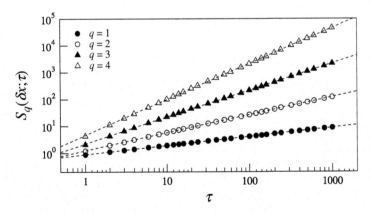

FIG. 3   Generalized structure functions, $S_q(\tau) = \langle |\delta x(\tau)|^q \rangle$, for the FBM signal.

IV. The future era of ionospheric science

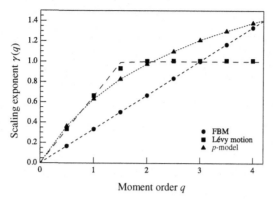

FIG. 4 The scaling exponents $\gamma(q)$ for the three signals are investigated.

distribution function of the increments with the scale τ, being in such a case no longer preserved the stability condition of the corresponding distribution functions at different τ.

A different description is needed for the Lévy flight motion. From the equation defining how this signal is generated, we see that the increments are taken from a Lévy distribution with a characteristic α (here, $\alpha = 3/2$). Lévy distributions are stable distributions (i.e., a linear combination of two independent variables following this distribution still follows the same distribution), whose moments of order higher than α do not exist. The behavior of the scaling exponents $\gamma(q)$ clearly shows a linear trend, that is, a monofractal character for moments orders $q < \alpha$, which is in agreement with the stability condition of the Lévy distribution. Conversely, for moments higher than α the results are not reliable. The flat trend shown in this case is a consequence of the fact that for a limited sample it is not possible to observe infinite increments so that the structure functions at moments higher than α neither diverge nor show a different value of $\gamma(q)$ being not able to sample higher and higher increment values.

To illustrate our previous discussion of different scaling features in terms of distributions of the increments, we show in Fig. 5 the probability density functions (PDFs), $P(\delta x; \tau)$, of signal increments δx at different scales, τ, scaled by the corresponding standard deviation σ_τ of each scale. As expected on the basis of the scaling exponent analysis, in the case of both the FBM and the Lévy flight motion the PDFs collapse onto a single curve (a Gaussian and a Lévy distribution function), while the PDFs of the p-model motion do not show data collapsing as expected, showing significative departures on the distribution tails.

Before moving to a different topic, we would like to discuss another important aspect of the different nature of the observed scale invariance which is related to the concept of *intermittency* and its link to the *multifractality*. Sometimes we read in the literature that because the distribution function departs on a Gaussian shape the signal is expected to be intermittent. However, as already mentioned earlier for a turbulent fluid the intermittency is strictly related to the occurrence of anomalous scaling features of the structure functions, that is, there is no linear relation between the different scaling exponents $\gamma(q)$. In other words, the concept of intermittency is strictly related to that of multifractality. Thus, although the PDFs of the increments of the Lévy flight motion depart on Gaussianity, this is not the signature of an intermittent

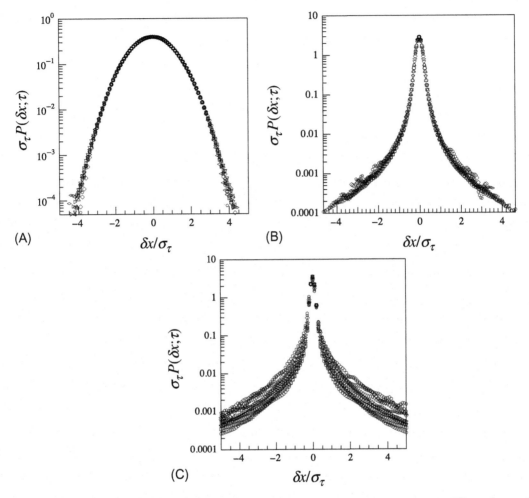

FIG. 5 The PDFs collapsing for the signals: FBM (A), Lévy motion (B), and p-model motion (C). The *different shapes* refer to different scales $\tau = 2^k$ with $k \in [1, 12]$.

character, being the self-similarity essentially associated with a simple monofractal character of this signal.

3 Information theory-based approaches

One of the most striking features of the ionospheric response to interplanetary and magnetospheric dynamics is the nonlinear character of the processes involved in the solar wind-magnetosphere-ionosphere (SMI) coupling. As a consequence of the nonlinear nature of SMI interaction, sometimes it is necessary to make use of approaches different from the traditional linear methods to characterize the coupling between the different dynamical variables considered to be relevant. To this target an alternative approach to estimate the coupling strength and/or driving can be based on information theory quantities, such as the mutual information and the transfer entropy

(TE). In what follows, we will introduce some information theory methods that could be useful to characterize the relevant dynamical variables and their coupling strength with other interplanetary magnetic field and magnetospheric parameters.

The evaluation of uncertainty level in a sequence of characters or in a signal $\{X\}$, which can attain a collection $\{x_i\}$ of N different values/states/characters, can be done by means of the so-called *Shannon entropy* S_X (Shannon, 1948; Jaynes, 2010; Cover and Thomas, 2006). This quantity, also named *information entropy*, is defined via the occupational probabilities p_i associated with each of the N values x_i by the following mathematical expression,

$$S_X = -C \sum_{i=1}^{N} p_i \log_b p_i, \qquad (12)$$

where C is an arbitrary constant which generally is taken to be $C = 1$ if the logarithm base $b = 2$ or e. On the basis of the definition given in Eq. (12) the information entropy, S_X, can attain values in the range $[0, 1]$, where $S_X = 0$ in the case of a null uncertainty ($p = 1$) or of an impossible event ($p = 0$).

As already said, Shannon entropy S_X is a way to estimate the uncertainty level, and thus this quantity is really an *entropic measure*, which gets its maximum value in correspondence with the maximum lack of information. In other words, S_X provides both a measure of our ignorance level and of the amount of knowledge when a measure of the actual value of X is done.

The information entropy is the starting point to define a certain number of quantities capable of estimating the *degree of interaction/coupling* between two interacting stochastic processes. Let us indeed consider two coupled stochastic dynamical processes $\{X\}$ and $\{Y\}$, for which we can write

$$\begin{cases} \dot{X} = f(X, Y; Z, \{\xi\}) \\ \dot{Y} = g(X, Y; Z, \{\xi\}), \end{cases} \qquad (13)$$

where Z is a third external process and $\{\xi\}$ a set of parameters and/or constraints. Let $x(t)$ and $y(t)$ be the corresponding realizations of the two processes $\{X\}$ and $\{Y\}$ in terms of time series.

Moving from the definition of the information entropy, we can introduce a quantity capable to estimate the dependence between the two processes. This quantity is said *mutual information*, $H_{X,Y}$, and is defined by the Shannon entropy as follows:

$$H_{X,Y} = S_X + S_Y - S_{X,Y}, \qquad (14)$$

where $S_{X,Y}$ is the conjugated information entropy, defined as

$$S_{X,Y} = -\sum_i \sum_j p_{i,j}(x_i, y_j) \log_b p_{i,j}(x_i, y_j). \qquad (15)$$

The expression of the mutual information by means of the definition of Shannon entropy and conjugated information entropy can be written in an explicit form as

$$H_{X,Y} = \sum_i \sum_j p_{i,j}(x_i, y_j) \log_b \frac{p_{i,j}(x_i, y_j)}{p_i(x_i) p_j(y_j)}, \qquad (16)$$

where b the logarithm base set the units; *bits* if $b = 2$ or *nats* if $b = e$.

The mutual information provides a measure of the coupling intensity, so that $H_{X,Y} = 0$ means that the two processes $\{X\}$ and $\{Y\}$ are independent, and in the case of processes characterized by bivariate normal distribution we get

$$H_{X,Y} = \frac{1}{2} \log \left(1 - \rho^2(X, Y) \right), \qquad (17)$$

where ρ is the linear correlation coefficient (Palus et al., 2018). This last equation provides the conceptual link between the mutual information and a measure of the degree of correlation between the two processes. Due to this link, the mutual information can be used to get a reliable measure of the global (i.e., both linear, nonlinear and stochastic) correlation between the two processes/signals. Clearly, in order to establish the validity of the results obtained by measuring mutual information, a confidence level should be fixed. A reliable method to do this is to apply some appropriated surrogate data procedure.

To illustrate such an approach, we can consider the following nonlinear map,

$$x_i = 5 \ln (y_i^2 + 1) - \sin (2\pi y_i) + \eta_i, \qquad (18)$$

where η_i is a stochastic δ-correlated Gaussian noise of variance σ_η^2, and y_i is another stochastic process with values in the range $[-1, 1]$ driving the previous one.

Fig. 6 shows a realization of the stochastic process described in Eq. (18) where the standard deviation of the additional noise η has been set $\sigma_\eta = 0.3$. Using Eq. (16), it is possible to evaluate the value of the mutual information $H_{X,Y}$, which is $H_{X,Y} = 1.96$ bits. To assess the significance level, we can evaluate a confidence limit to distinguish between the value we get and the expected value in the case of uncorrelated signals. To this target one can evaluate the corresponding value of the mutual information by shuffling the sequence of $\{x_i\}$ to break the correspondence between the two variables. This process is iterated several times (typically of the order of 10^5) to construct the cumulative statistics of the obtained values of mutual information for the randomized dataset. On the basis of such cumulative function it is possible to establish a critical threshold value H_c to distinguish

between a significant correlation/dependence or less. Generally, one can set this H_c as the value corresponding to a probability $P = .95$ for the cumulative statistics of mutual information for the randomized dataset.

Fig. 7 shows the cumulative distribution relative to the mutual information for the randomized dataset. The corresponding critical threshold H_c checks if a significative correlation is present, which is found to be $H_c = 0.0714$. Since this value is smaller than the obtained $H_{X,Y}$ value, we can conclude that we are in the presence of a very strong correlation between the two signals, especially if we compare this result with the corresponding standard linear correlation analysis which returns a value of the Pearson's coefficient $|\rho| \sim 0.2$, a value indicating a small degree of correlation.

The earlier method can be used in data analysis to unveil possible correlated dynamical quantities that cannot be clearly evidenced using linear approaches, providing also a measure on the possible inference among these dynamical quantities.

The mutual information can be generalized to measure also the correlation delay between the two signals. In this case, we talk of delayed mutual information (DMI), which can be considered as to be equivalent to a cross-correlation measure. The definition of DMI in the case of two signals can be written as

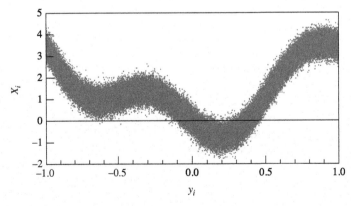

FIG. 6   A realization of the process described in Eq. (18). The standard deviation of the noise η_i is $\sigma_\eta = 0.3$.

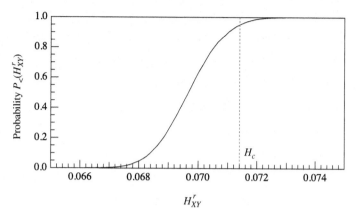

FIG. 7 The cumulative distribution $P_<(H^r_{X,Y})$ relative to the mutual information for the randomized dataset. The *dashed vertical line* indicates the critical threshold value H_c corresponding to $P = .95$.

$$H_{X,Y}(\Delta) = \sum_i \sum_j p'_{i,j}(x_{i+\Delta}, y_j) \log_b \frac{p'_{i,j}(x_{i+\Delta}, y_j)}{p_i(x_i) p_j(y_j)},$$

(19)

where Δ is the delay between the two signals (here considered as sequences), $p'_{i,j}(x_{i+\Delta}, y_j)$ is the probability for shifted variables, and stationarity conditions are assumed, that is, $p(x_{i+\Delta}) = p(x_i)$. The value of the delay Δ where $H_{X,Y}$ has its maximum, defines the delay time, and the corresponding $H_{X,Y}^{max}$ the strength of the correlation (see an example of application in Alberti et al., 2017). Clearly, also in this case a test on the significance of this correlation has to be done.

Although mutual information analysis among a set of variables can help to provide some information on the possible inference between such variables, especially if these are nonlinear coupled, none can be concluded on the occurrence of causal relationships between these variables. Indeed, this quantity cannot provide any indication on the information flow between two variables, that is, there is no way to infer on a possible causality relationship, that is, what is the driver and what is driven.

The identification of causality relations between variables is a crucial point when one would like to attempt a forecast of some quantity. In this context, the information theory is capable of providing some additional element with respect to the more traditional methods based on autoregressive (AR) linear models (Granger, 1969).

With reference to the information theory the functional that better allows the inference of causal relations is the conditional mutual information (CMI), $H(X, Y | Z)$, defined as

$$H(X,Y|Z) = S(X|Z) + S(Y|Z) - S(X,Y|Z) \quad (20)$$

where $S(A|B)$ is the *conditional entropy*

$$S(A|B) = \sum_i \sum_j p_{i,j}(a_i, b_j) \log_b p_{i,j}(a_i|b_j). \quad (21)$$

Moving from the definition of CMI and considering Markovian processes, Schreiber (2000) introduced a functional, named TE, capable of quantifying the information flow from one signal to another, so to provide an information on the occurrence of a causality relation. Given two stochastic processes $\{X\}$ and $\{Y\}$, the information flow from $Y \rightarrow X$ can be estimated by the TE, $T_{Y \rightarrow X}(\tau)$, which is given by the following expression:

$$T_{Y \rightarrow X} = \sum_{x_{i+1}, \mathbf{x}_i^{(k)}, \mathbf{y}_j^{(l)}} p\left(x_{i+1}, \mathbf{x}_i^{(k)}, \mathbf{y}_j^{(l)}\right)$$

$$\log_b \frac{p\left(x_{i+1} | \mathbf{x}_i^{(k)}, \mathbf{y}_j^{(l)}\right)}{p\left(x_{i+1} | \mathbf{x}_i^{(k)}\right)}, \tag{22}$$

where $\mathbf{x}_i^{(k)}$ and $\mathbf{y}_j^{(l)}$ are the state history vectors of the evolution of X and Y (see Kaiser and Schreiber, 2002 for an extensive discussion). This quantity can be generalized to two signals as

$$T_{Y \rightarrow X}(\tau) = \sum_{i,j,k} p(x_i(t+\tau), x_j(t), y_k(t))$$

$$\log_b \frac{p(x_i(t+\tau) | x_j(t), y_k(t))}{p(x_i(t+\tau) | x_j(t))}, \tag{23}$$

where τ is a time lag related to the typical timescale of interaction between $\{X\}$ and $\{Y\}$. In the previous expression, the terms x_i (y_k) refer to the (state) value assumed by the process X_i (Y_k) at a specific time.

As already said the TE $T_{Y \rightarrow X}(\tau)$ provides the information flow from $Y \rightarrow X$, so that if we do not have any funded reason for assuming one specific direction of the causality relation, it could be also useful to compute the reverse quantity, that is, the information flow from $X \rightarrow Y$ via $T_{X \rightarrow Y}(\tau)$. The knowledge of the two directional flow of information allows us to have a direct inference on the driving relation between the two processes. Indeed, if $T_{Y \rightarrow X}(\tau) > T_{X \rightarrow Y}(\tau)$ (or $T_{X \rightarrow Y}(\tau) > T_{Y \rightarrow X}(\tau)$) the process $\{Y\}$ ($\{X\}$) influences (drives) the process $\{X\}$ ($\{Y\}$).

Some notable attempts to apply TE to unveil the occurrence of driving processes in the SMI have been done in the recent past (Materassi et al., 2007, 2011; De Michelis et al., 2011; Wing et al., 2016; Johnson et al., 2018). For instance, De Michelis et al. (2011) have applied the TE analysis to investigate the storm-substorm

causal relationship, unveiling the very complex nature of the storm-substorm coupling. However, the application of TE analysis to real and chaotic signals requires some peculiar attention (Kaiser and Schreiber, 2002; Palus et al., 2018).

4 Conclusions

In this chapter, we have discussed some concepts and methods to investigate the occurrence of multiscale features, complexity and to unveil/infer the underlying causal relations between dynamical quantities and variables. We believe that these approaches can be very useful to get a better description of the complex dynamics of the ionospheric system, which shows nonlinearity, turbulence, and a very complex coupling with the magnetosphere and the neutral atmosphere. We believe that some of the previous described methods can also be useful to understand the ionospheric plasma dynamics and its impact in the framework of Space Weather processes.

Acknowledgments

The authors thank P. De Michelis (INGV, Italy) for the long-standing collaboration. The authors also acknowledge the NSSDC OMNIWeb database (P.I.J.H. King, N. Papatashvilli) for providing some data used in this work.

References

Ahn, B.-H., Akasofu, S.I., Kamide, Y., 1983. The Joule heat-production rate and the particle energy injection rate as a function of the geomagnetic indices AE and AL. J. Geophys. Res. 88, 6275–6287.

Alberti, T., Consolini, G., Lepreti, F., Laurenza, M., Vecchio, A., Carbone, V., 2017. Timescale separation in the solar wind-magnetosphere coupling during St. Patrick's day storms in 2013 and 2015. J. Geophys. Res. 122, A023175. https://doi.org/10.1002/2016JA023175.

Benassi, A., Jaffard, S., Roux, D., 1997. Elliptic Gaussian random processes. Rev. Mat. Iberoamericana 13, 19–90.

Bunde, A., Havlin, S., 1991. Fractals and Disordered Systems. Springer-Verlag, Berlin.

Bunde, A., Havlin, S., 1995. Fractals in Science. Springer-Verlag, Berlin.

Chang, T., 2015. An Introduction to Space Plasma Complexity. Cambridge University Press, Cambridge.

Chang, T., Tam, S.W.Y., Wu, C.-C., 2006. Complexity in space plasmas: a brief review. Space Sci. Rev. 122, 281.

Chaston, C.C., Salem, C., Bonnell, J.W., Carlson, C.W., Ergun, R.E., Strangeway, R.J., McFadden, J.P., 2008. The turbulent Alfvènic aurora. Phys. Rev. Lett. 100, 175003. https://doi.org/10.1103/PhysRevLett.100.175003.

Consolini, G., 1997. Sandpile cellular automata and the magnetospheric dynamics. In: Aiello, S. et al., (Ed.), Cosmic Physics in the Year 2000, Proceedings of VIII GIFCO Conference, SIF, Bologna, p. 123.

Consolini, G., 2002. Self-organized criticality: a new paradigm for the magnetotail dynamics. Fractals 10, 275–283.

Consolini, G., Kretzschmar, M., Lui, A.T.Y., Zimbardo, G., Macek, W.M., 2005. On the magnetic field fluctuations during magnetospheric tail current disruption: a statistical approach. J. Geophys. Res. 110, A07202. https://doi.org/10.1029/2004JA010947.

Consolini, G., De Marco, R., De Michelis, P., 2013. Intermittency and multifractional Brownian character of geomagnetic time series. Nonlinear Process. Geophys. 20, 455–466.

Cover, T.M., Thomas, J.A., 2006. Elements of Information Theory. John Wiley & Sons Inc., Hoboken, NJ

Davis, T., Sugiura, M.J., 1966. Auroral electrojet activity index AE and its universal time variations. J. Geophys. Res. 71, 785–791.

Davis, A., Marshak, A., Wiscombe, W., Cahalan, R., 1994. Multifractal characterizations of non-stationarity and intermittency in geophysical fields, observed, retrieved, or simulated. J. Geophys. Res. 99, 8055–8072.

De Michelis, P., Consolini, G., Materassi, M., Tozzi, R., 2011. An information theory approach to the storm-substorm relationship. J. Geophys. Res. 116, A08225. https://doi.org/10.1029/2011JA016535.

De Michelis, P., Consolini, G., Tozzi, R., 2015. Magnetic field fluctuations at swarm's altitude: a fractal approach. Geophys. Res. Lett. 42. https://doi.org/10.1002/2015GL063603.

De Michelis, P., Consolini, G., Tozzi, R., Marcucci, M.F., 2016. Observations of high-latitude geomagnetic field fluctuations during St. Patrick's day storm: swarm and SuperDARN measurements. Earth Planets Space 68, 105. https://doi.org/10.1186/s40623-016-0476-3.

De Michelis, P., Consolini, G., Tozzi, R., Marcucci, M.F., 2017. Scaling features of high-latitude geomagnetic field fluctuations at Swarm altitude: impact of IMF orientation. J. Geophys. Res. 122, 10548–10562. https://doi.org/10.1002/2017JA024156.

De Michelis, P., Consolini, G., Tozzi, R., Giannattasio, F., Quattrociocchi, V., Coco, I., 2019. Features of magnetic field fluctuations in the ionosphere at Swarm altitude. Ann. Geophys. https://doi.org/10.4401/ag-7789.

Falkovich, G., 2008. Introduction to turbulence theory. In: Cardy, J., Falkovich, G., Gawedzki, K. (Eds.), Non-Equilibrium Statistical Mechanics and Turbulence, In: London Math. Soc. Lect. Note Ser., vol. 355. pp. 1–43.

Frisch, U., 1995. Turbulence: The Legacy of A.N. Kolmogorov. Cambridge University Press, Cambridge.

Ganushkina, N.Y., Liemohn, M.W., Dubyagin, S., 2018. Current systems in the Earth's magnetosphere. Rev. Geophys. 56, 309–332. https://doi.org/10.1002/2017RG000590.

Golovchanskaya, I.V., Kozelov, B.V., 2010. On the origin of electric turbulence in the polar cap ionosphere. J. Geophys. Res. 115, A09321. https://doi.org/10.1029/2009JA014632.

Golovchanskaya, I.V., Ostapenko, A.A., Kozelov, B.V., 2006. Relationship between the high-latitude electric and magnetic turbulence and the Birkeland field-aligned currents. J. Geophys. Res. 111, A12301. https://doi.org/10.1029//2006JA011835.

Granger, C.W.J., 1969. Investigating causal relations by econometric models and cross-spectral methods. Econometrica 47, 424–438.

Jaynes, E.T., 2010. Probability Theory. The Logic of Science. Cambridge University Press, Cambridge.

Johnson, J.R., Wing, S., Camporeale, E., 2018. Transfer entropy and cumulant-based cost as measures of nonlinear causal relationships in space plasmas: applications to DST. Ann. Geophys. 36, 945–952. https://doi.org/10.5194/angeo-36-945-2018.

Kaiser, A., Schreiber, T., 2002. Information transfer in continuous processes. Physica D 166, 43–62.

Kintner, P.M., 1976. Observations of velocity shear driven plasma turbulence. J. Geophys. Res. 81, 5114.

Klafter, J., Blumen, J.A., Schlesinger, M.F., 1987. Stochastic pathway to anomalous diffusion. Phys. Rev. A 35, 3081.

Klafter, J., Shlesinger, M.F., Zumofen, G., 1996. Beyond Brownian motion. Phys. Today 43, 33.

Kozelov, B.V., Golovchanskaya, I.V., Ostapenko, A.A., Federenko, Y.V., 2008. Wavelet analysis of high latitude electric and magnetic fluctuations observed by dynamic explorer 2 satellite. J. Geophys. Res. 113, A03308. https://doi.org/10.1029/2007JA012575.

Lesne, A., Lagües, M., 2012. Scale Invariance. From Phase Transitions to Turbulence. Springer-Verlag, Berlin, Heidelberg.

Mandelbrot, B.B., 1989. Multifractal measures, especially for geophysicist. Pure Appl. Geophys. 131, 5–42. https://doi.org/10.1007/BF00874478.

Mandelbrot, B.B., Van Ness, J.W., 1968. Fractional Brownian motions, fractional noises and applications. Soc. Ind. Appl. Math. Rev. 10, 422–437.

Materassi, M., Wernik, A., Yordanova, E., 2007. Determining the verse of magnetic turbulent cascades in the Earth's magnetospheric cusp via transfer entropy analysis: preliminary results. Nonlinear Process. Geophys. 14, 153.

Materassi, M., Ciraolo, L., Consolini, G., Smith, N., 2011. Predictive space weather: an information theory approach. Adv. Space Res. 47, 877–885. https://doi.org/10.1016/j.asr.2010.10.026.

Meneveau, C.M., Sreenivasan, K.R., 1991. The multifractal nature of turbulent energy dissipation. J. Fluid. Mech. 224, 429.

Palus, M., Krakovska, A., Jakubík, J., Chvostekova, M., 2018. Causality, dynamical systems and the arrow of time. Chaos 28, 075307. https://doi.org/10.1063/1.5019944.

Schreiber, T., 2000. Measuring information transfer. Phys. Rev. Lett. 85, 461–464.

Shannon, C.E., 1948. A mathematical theory of communication. Bell Syst. Tech. J. 27, 379.

Spicher, A., Miloch, W.J., Clausen, L.B.N., Moen, J.I., 2015. Plasma turbulence and coherent structures in the polar cap observed by the ICI-2 sounding rocket. J. Geophys. Res. 120, 10959–10978. https://doi.org/10.1002/2015A021634.

Tam, S.W.Y., Chang, T., Kintner, P.M., Klatt, E., 2005. Intermittency analyses on the SIERRA measurements of the electric field fluctuations in the auroral zone. Geophys. Res. Lett. 32, L05109. https://doi.org/10.1029/2004GL021445.

Uritsky, V.M., Pudovkin, M.I., 1998. Low frequency $1/f$-like fluctuations of the AE-index as a possible manifestation of self-organized criticality in the magnetosphere. Ann. Geophys. 16, 1580.

Uritsky, V.M., et al., 2002. Scale-free statistics of spatiotemporal auroral emissions as depicted by POLAR UVI images: the dynamic magnetosphere is an avalanching system. J. Geophys. Res. 107, 1426.

Wing, S., Johnson, J.R., Camporeale, E., Reeves, G.D., 2016. Information theoretical approach to discovering solar wind drivers of the outer radiation belt. J. Geophys. Res. Space Phys. 121, 9378–9399. https://doi.org/10.1002/2016JA022711.

Ionospheric science in the age of big data

Asti Bhatt[a], Andrew Silberfarb[b]

[a]Center for Geospace Studies, SRI International, Menlo Park, CA, United States
[b]SRI International, Menlo Park, CA, United States

1 Current state of the ionospheric science

The 2012 Solar and Space Physics Decadal Survey has listed determining the dynamics and coupling of Earth's magnetosphere, ionosphere, and atmosphere, and their response to solar and terrestrial inputs a *key science goal* for the current decade.

Unraveling these dynamics requires the system-level view, yet the system itself is incredibly vast containing processes at multiple spatial and temporal scales (Fig. 1). The coupled atmosphere-ionosphere-magnetosphere (AIM) system is inherently complex and nonlinear, with spatial scales ranging from meter-scale to the distances containing multiple earth-radii, and temporal scales from milliseconds to decades.

Our current understanding of the solar—AIM system dynamics comes from physics-based first-principles modeling efforts. These efforts have spanned from whole atmospheric modeling (e.g., Whole Atmosphere Community Climate Modeling—WACCM, Marsh et al., 2013) to general circulation models, including the electrodynamics (e.g., Thermosphere Ionosphere Electrodynamics General Circulation Model—TIEGCM, Qian et al., 2014). In recent years,

owing to the availability of distributed large datasets, assimilative modeling has also allowed us to gain insight (e.g., Assimilative Model of Ionospheric Electrodynamics—AMIE, Cousins et al., 2015). As a result of all these efforts, we have made significant progress in understanding some fundamental properties of the ionospheric plasma, and large-scale processes that drive the plasma circulation across the globe using the available data and modeling efforts. However, similar to the famous adage from Albert Einstein, the more we learn, the more we realize how truly complex the geospace system is. The greater the number of independent variables, the more complex a system. However, not all complex systems are difficult to study. It is the nonlinearity of the system preventing us from breaking it into the sum of its parts that poses a bigger challenge for the complex system. In nature, such systems have complicated behavior in both space and time, and investigating these systems is at the frontiers of scientific inquiry where we often don't even know the questions that we need to ask. Some examples of such systems that have the largest number of variables and highest spatiotemporal complexity are nonlinear waves, plasmas, earthquakes, general relativity, quantum field

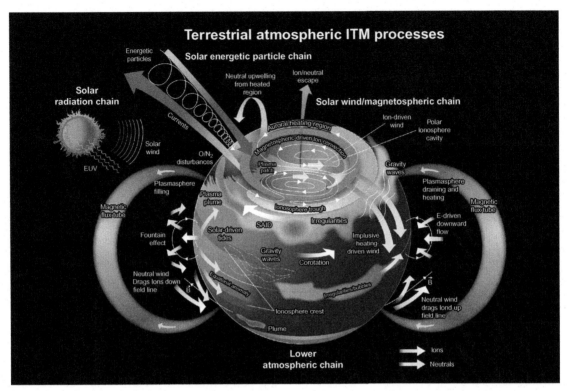

FIG. 1   A figure depicting major geospace processes that occur in the AIM system in response to varied inputs. These processes that involve both charged and neutral gases that tend to evolve in space and time across regions and scale sizes. The complexity and nonlinearity of the system makes it difficult to distinguish, and, in turn, predict sources of variability. *Reproduced from Figure 8.1 of National Research Council, 2012. Solar and Space Physics: A Science for a Technological Society. The National Academies Press, ISBN 978-0-309-38739-2, https://doi.org/10.17226/13060.*

theory, fibrillation, epilepsy, turbulent fluids, and life (Strogatz, 2000). The geospace system is a combination of plasmas, nonlinear waves, and turbulent fluids among other things, and is a perfect example of this, with many known (and several unknown) independent variables at each region from the sun to the lower atmosphere impacting generation and evolution of multiscale processes in the ionosphere.

While the problem at hand is hard and seemingly ill-posed, we have better than ever before tools at our disposal to tackle it. Geospace science is in the midst of the data revolution. Since the advent of radio propagation, several instruments have been developed to study the conducting properties of the upper atmosphere.

The space-based investigations of the solar-magnetospheric-ionospheric connections started with Alouette-1 in 1962 and have only expanded over the years, with over 20 currently active satellite missions across the world (Fig. 2), and the upcoming Geodynamic Constellation mission in the works. The ground-based investigations (Fig. 3) have followed on from using simple chirp sounders worldwide to the modern continuously operating incoherent scatter radars (ISRs) that can make 3D volumetric, high-resolution measurements of the ionospheric electron density, temperatures, and other parameters, with the EISCAT 3D ISR in the works. In addition, several CubeSats, sounding rockets, and campaign-based investigations continue to create high-resolution datasets

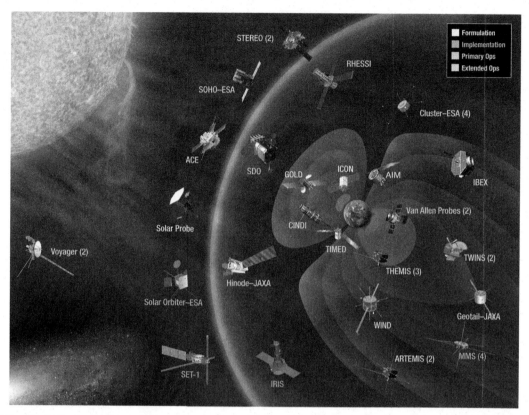

FIG. 2 The Heliophysics System Observatory is a fleet of over 20 active satellite missions continuously measuring key solar-magnetosphere-ionosphere parameters.

spanning short but key time periods and atmospheric-geomagnetic regions. The Global Navigation Satellite System (GNSS) is a great example of ground- and space-based instruments working in conjunction to produce a hugely valuable dataset for characterizing the ionosphere on a large scale.

Combination of all of these has created a data ecosystem for geospace science that is continuously evolving. It is true that while we have sparse measurements of several key quantities such as neutral winds, neutral composition and temperature, and ionospheric conductance, the technological advances are making it easier to measure geophysical parameters across the globe using distributed systems. It is also true that there

have never been more data from geospace system available than ever before measuring other key quantities like the solar wind, geomagnetic currents, ionospheric total electron content, emissions showing aurora and airglow, and others.

Most of these data are continuous, which means that they are measured daily with some periodic consistency. However, arguably we use only a fraction of all the data collected. Most data are used for explosive geomagnetic event-specific analyses or long-term trend studies. Long-term trend studies tend to abstract away the daily variability that can be both an effect and the driver for geospace response to solar and terrestrial inputs. In doing so, we tend to miss important physics.

IV. The future era of ionospheric science

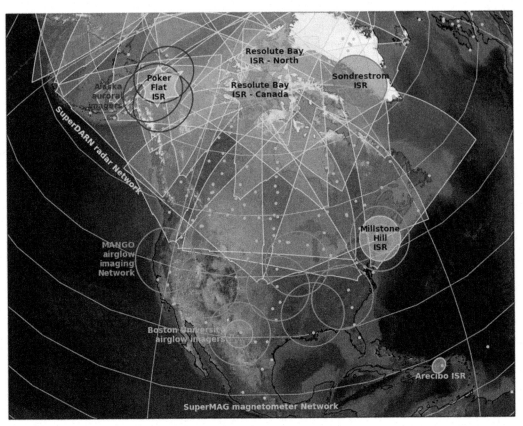

FIG. 3 A map showing fields-of-view of currently operational diverse ground-based geospace instruments in the north America supported by the United States National Science Foundation. These instruments measure ionospheric irregularities, airglow and auroral emissions, and precise ionospheric parameters like electron density and temperature. These instruments operate more-or-less continuously producing more data with each passing day. *Courtesy: Leslie Lamarche, SRI International, USA.*

Characterizing, modeling, and forecasting the complex behavior of the geospace system has also been a longstanding quest from a more operational standpoint. The space environment affects the health of our space assets, communication, navigation, timing accuracy, astronaut health, and the future of space travel. Our ability to predict especially the smaller-scale features is critical to these operational needs as outlined in the National Space Weather Strategy and Action Plan by the United States National Science and Technology Council (2019).

The advent of modern data science approaches has made it possible to use several different approaches in the pursuit of characterizing, modeling, and forecasting the geospace system. In the following section, we will take a look at some of the suitable approaches.

2 Data mining for scientific data landscape

The evolving geospace data landscape has made itself amenable to the approaches related to the so-called big data, where the focus is on the four-V's—*Volume, Velocity, Variety,* and *Veracity.* The ground-based data such as those from the GNSS in its many forms (ground-based receivers, COSMIC, and other satellite missions) are continuously expanding with great *velocity.* With the launch of the Parker Solar Probe, the *volume* of available data is going to increase manifold. To solve any geospace grand challenge, data are needed from a *variety* of disparate sources, and the measured quantities have inherent uncertainties due to the stochastic and continuously evolving nature of many quantities that these instruments measure contributing to the *veracity* of the data.

While the phrases "data science" and "big data" have been made popular in recent times, the statistical methods pertaining to information theory have always been used in geospace science. In the science community, there is a justifiable discomfort in using modern machine-learning methods for finding outcomes. Part of this has to do with the perception that all machine-learning techniques are "black box" that provide results but limited insight into the logic used to arrive at the outcome. This is directly against the purpose of doing science, which is to uncover the rules by which our universe is governed. However, a new paradigm of "Theory-Guided Data Science" aims to leverage the wealth of scientific knowledge for improving the effectiveness of data science models in enabling scientific discovery (Karpatne et al., 2017).

Within the broader space physics community, solar physicists have used machine-learning methods with remarkable success (e.g., Bobra and Couvidat, 2015; Díaz Baso and Asensio Ramos, 2018; Cheung, 2018). A recent book edited by Camporeale et al. (2018) outlines a few examples of applying these methods for specific questions in solar-magnetospheric science. Various agencies like NASA and the European Space Agency (ESA) have taken notice of the potential of these methods and have initiated programs to take advantage of them. For example, the "Frontier Development Lab" (FDL) is an artificial

intelligence (AI) accelerator program supported by both NASA and ESA that brings early-career space and data scientists together to apply machine-learning methods for space science questions in a focused 8-week timeframe. The US FDL program is run by the SETI Institute, and pairs each team of researchers with both domain and machine-learning experts to achieve optimal outcome in a short time period.

In this section, we outline various machine-learning techniques that geospace scientists can use to aid in the scientific discovery process, and describe in detail a couple of examples in geospace science using some of these techniques. This chapter isn't meant to give an overview of *all* data science techniques that geospace scientists can use but rather focus more on how the geospace science can benefit from more modern data science techniques.

Data mining is a broad field to find patterns in datasets using intelligent pattern recognition methods. The field of data mining exists at the intersection of computer science and statistics. The key difference between data *analysis* and data *mining* is that while data analysis is used to test models on datasets, data mining uses machine-learning methods to extract hidden patterns from the available dataset. Most machine-learning methods can be separated into *supervised* or *unsupervised* learning methods. Now prevalent deep-learning techniques have applicability under both of these methods.

2.1 Supervised learning

Supervised learning is a way of inferring a governing function based on prelabeled sets of inputs and outputs. A suite of methods under supervised learning rely on prelabeled *training data* to create a generalized model $Y = m(x)$, use *validation data* to tune model parameters, and then apply the model m to the *test data* and have it perform just as well. A key requirement here is that the data should be distributed identically in each of these three data sets.

Supervised learning is typically separated out in two subfields: Classification and Regression. Classification methods predict the category that the data belong to, while regression methods predict a future numerical value based on prior learning. Modern classification algorithms can be some of the most powerful techniques for scientific data analysis. A popular technique is Support Vector Machines (SVM) (Cortes and Vapnik, 1995), which can be used for both classification and regression, and can nonlinearly map the inputs into high-dimensional feature spaces, making them especially suited for spatiotemporal datasets. SVM is also an *explainable* machine-learning technique, where it is possible to gain insight into the logic of the algorithm.

2.1.1 Application of SVM in ionospheric science

Recently, McGranaghan et al. (2018) used the SVM classification technique to predict the GNSS scintillations projected vertically at Canadian High Arctic Ionospheric Network (CHAIN) receiver stations. The task was to predict whether or not GNSS phase-scintillation occurred (based on a threshold) an hour or more into the future given the information at the current time about solar and geomagnetic drivers and ionospheric state as characterized by the GNSS receivers. The input parameters were selected based on what we understand about drivers of the

high-latitude ionospheric system such as the solar wind and the Interplanetary Magnetic Field (IMF) components obtained from the NASA satellites, geomagnetic indices characterizing active conditions, high-latitude precipitation information using the OVATION Prime model (Newell et al., 2010), and ground-based GNSS parameters like spectral slope and differential Total Electron Content (TEC) to characterize the ionospheric state obtained from the CHAIN stations. The input also contained the current value of GNSS phase scintillation index σ_φ obtained from 17 CHAIN stations, with the objective to create a generalized model for all CHAIN locations.

The SVM technique uses a given data sample, $x_m \in R^N$, where N is the number of input features that defines the problem dimensionality, and corresponding classification label y_m (scintillation or no scintillation in this case), to find a function that separates the data based on their associated label. If the data are linearly separable in the feature space, this function is a hyperplane of the form $f(x) = w^T x + b$, where w is a vector normal to the hyerplane, and b is a constant scalar. x_m belongs to class $y_m = +1$ if $f(x) \geq 0$ and class $y_m = 0$ if $f(x) < 0$ (see Fig. 4). The SVM algorithm then maximizes the distance between the decision hyperplane (also referred to as decision boundary) and the closest data samples, which are known as support vectors. However, the datasets used in this problem of predicting an

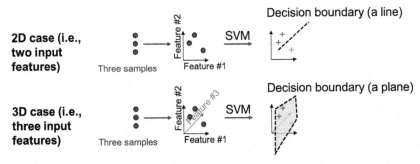

FIG. 4 An illustration of the SVM approach to prediction adopted from McGranaghan et al. (2018) Figure 1 (c). For the 2D case on top, the SVM finds a decision boundary that is a line, while for the 3D case on the bottom, SVM finds a plane that separates the prediction classes.

inherently nonlinear quantity are unlikely to be linearly separable in feature space. McGranaghan et al. (2018) addressed this by mapping the data samples nonlinearly to a higher-dimensional space, and then defining a hyperplane in this space (Cover, 1965). Another challenge in the scintillation dataset is that majority of the data have $\sigma_\varphi < 0.1$, considered nonscintillation, and only a small percentage have $\sigma_\varphi > 0.1$. This is referred to as "class imbalance" in machine learning, and is a major problem causing a strong bias for the majority class. McGranaghan et al. (2018) addressed this challenge by setting up class weight ratios to penalize misclassification. Fig. 5 shows results from a case study for this example for January 20, 2016, with contextual data from B_z, B_y, and geomagnetic indices K_p and AE. The panel (D) showcases the truth, (E) the results from SVM model tested on one of the stations, and (F) shows prediction from the persistence model.

There are a number of ways to evaluate the performance of a prediction model in data science. Many evaluation metrics use True Positives (TP—scintillation predicted and occurred), True Negatives (TN—scintillation not predicted and did not occur), False Positives (FP—scintillation predicted but did not occur), and False Negatives (FN—scintillation not predicted but did occur). Using these four quantities, several metrics can be calculated. One of these is True Skill Score (TSS), which is defined as,

$$\text{TSS} = \frac{TP}{TP + FN} - \frac{FP}{FP + TN}$$

TSS ranges over $[-1, 1]$, where -1 means every event is incorrectly classified, and $+1$ means every event is correctly classified. TSS $= 0$ means that the model predicts consistently with a random chance predictor. TSS is also insensitive to class imbalance. As done in this paper, it is also useful to test against persistence model, which assumes that the predicted value at a future time

is the same as the current value. The TSS for SVM implementation in this example is 0.5 for predicting at 1-min cadence for 1-h prediction, while that of the persistence model is 0.25 for same parameters. However, a better metric calculated to evaluate for GNSS scintillation is TSS over a 15-min mean, which shows high skill with TSS $= 0.51$ for 1-h prediction.

2.2 Other *interpretable* supervised learning techniques

In science, we look for interpretation and understanding of the way an outcome was arrived at. This is hard to do in the case of neural networks, and especially deep learning. However, there are some machine-learning techniques that are classified under "explainable" artificial intelligence (AI), which are transparent in their working. Most classification and regression methods under unsupervised learning fall in this category. In practice, high-level explainable methods are often mixed with lower-level "black box" machine learning. For example, a separate class of algorithms is *Decision Trees* (Rokachand and Maimon, 2007), which can be used to visually gain insight into decision-making processes in data mining. These algorithms also form the basis for ensemble methods of classification, where several supervised learning models are individually trained and the results are merged parallelly using various algorithms to arrive at the final prediction.

Similarly, *Random Forest* decision making, for example, can combine different support vector machine or decision tree models and take the average of all predictions. A random forest is built on several hundred decision trees, each of them performing random extraction of the observations from the dataset and a random extraction of the features, which makes them decorrelated from others. At each decision point, the output is separated into two buckets, each bucket with observations that are more similar within it than with the other bucket. This

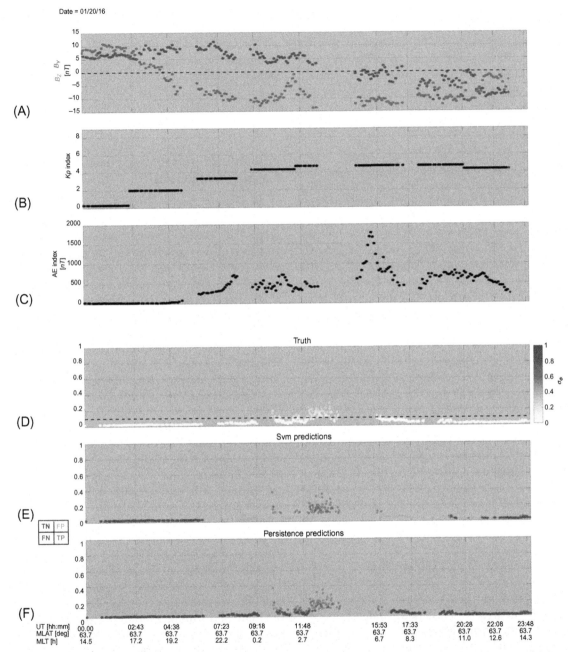

FIG. 5 Results from the SVM model implementation to predict GNSS scintillation, adopted from McGranaghan et al. (2018) figure 9. The top three panels (A–C) show contextual parameters for January 20, 2016, fourth panel (D) shows the truth, fifth (E) the results from SVM implementation against persistence model results in the bottom most panel (F).

property is then used to understand the importance of each input feature for the target outcome.

On the other hand, *Gradient Boosting* algorithm uses a sequential approach to ensemble models, where each model takes output of a previous model as an input. Here, each model tries to correct previous model errors, hence boosting the final prediction.

The foregoing are some of the methods that are *interpretable* in nature, which allow the user to gain an insight into the logic that led to the outcome. This is typically not the case for machine-learning methods that use artificial neural networks and deep learning. However, the neural networks have proven to be extremely useful in understanding patterns in complex data, such as speech or image recognition. In recent years, the advances made in the available software tools, fast-evolving research landscape in the field of deep learning, and broader availability of computational resources has made using these techniques a more feasible endeavor.

2.3 Artificial neural networks and deep learning

Neural networks have been around since at least the 1960s, when the first "perceptron" was developed by Frank Rosenblatt (Rosenblatt, 1958) at Cornell University and used to classify images. Modeled on how biological neurons propagate information in human brains, these methods have been remarkably adept at learning patterns. The usage of neural networks has ebbed and flowed over the years, but in recent times the preponderance of high-end computing and data infrastructure has made it possible to use them to their full potential. Neural network architecture tends to take independent input variables from training data, assign random weights to them, and use an *activation function* to produce an output that is then compared with actual values in the form of a cost function (also known as *"loss function"*), with the goal of minimizing the loss function. Each neural network has one or more "hidden" layers of nodes that use input from the previous layer to learn further. The higher the number of hidden layers, the deeper the network, which is where the name *"deep learning"* is derived from. Many predefined activation and loss functions are available using modern machine-learning libraries, each of which are useful for different applications.

Modern Deep Neural Network (DNN) approaches are particularly suited to dealing with complex systems data as they replace the hand-tuned features of traditional machine-learning approaches with machine-learned replacements. The most straightforward approach to using DNNs is to treat the outputs of complex systems as images and use the large array of DNN image recognition approaches developed over the past decade to identify areas of interest within the images. Auto encoders have been developed using DNNs that can support data compression and feature extraction from large-scale complex systems datasets, for use in transmission over limited channels and data analysis. More ambitious approaches repurpose tools developed for image recognition and language learning to support prediction of the future state of complex systems, either attempting to learn the dynamics directly from historical data, or using convolutional neural networks to approximate known evolution equations in a fraction of the time needed for finite element methods (Fig. 6).

A host of techniques for recognizing interesting phenomena in images is available. Techniques that require specific training based on hand-annotated labels of known object classes include object recognition (Szegedy et al., 2013), activity recognition in video (Vrigkas et al., 2015), predicate recognition in imagery

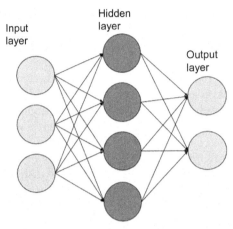

FIG. 6 A simple neural network with three inputs, two outputs, and one hidden layer with four nodes to allow the network to learn connections between various input and output parameters.

(Lu et al., 2016), and automated captioning (You et al., 2016; Vinyals et al., 2015). Other useful techniques require minimal or no annotation, including anomaly detection (Sabokrou et al., 2018) and clustering (Aljalbout et al., 2018). Finally, there are many transfer-learning techniques available that train on one type of data and then can be applied to other data such as vector embedding (Vinyals et al., 2015) and zero-shot learning (Xian et al., 2018). All of these are relevant to finding phenomena in complex systems data.

A recent example of transfer learning is demonstrated by Clausen and Nickisch (2018), who used two linked methods—neural-network-based supervised learning and Ridge classification—to classify auroral images from THEMIS ground-based observatories. They used a pretrained neural network, Inception-v4, provided by Google's TensorFlow open-source library that has been trained on about 1.2 million images of everyday objects like pandas, container ships, and dandelions, as a feature extractor. The extracted features were then paired with labels like *cloudy, moon, clear/ no aurora, arc, diffuse, discrete* and input into a Ridge classifier, which is a linear method for extending and generalizing ordinary linear regression. Using this innovative approach, they

successfully classified aurora by the chosen labels in approximately 82% of cases. When using only a binary classifier of *aurora* and *no aurora*, the classifier was successful in about 96% of cases.

When it comes to complex spatial data representation, a different class of DNNs called *Convolutional Neural Networks* (CNNs) has been quite popular. In a typical neural network implementation for image recognition, the image is flattened into a series of values, which then pass through several hidden layers to derive patterns. However, in this process, the spatial information and correlations between pixels are lost. CNN, on the other hand, uses a spatial filter of the size smaller than the image to scan across the image, the output of which gets convolved with the original pixel values thereby preserving the pattern information in a more robust way. The filter can be customized to detect a specific feature like an edge. CNNs are widely used for image processing and lately where time history of events is involved.

Following the success of the approach by Clausen and Nickisch (2018), a team of researchers at NASA's 2019 FDL program (frontierdevelopmentlab.org) tried to see if it is possible to learn the features of auroral images that classify them into different types using deep learning. This team (Athanasios Vlontzos, Imperial College, London; Kara Lamb, National Oceanic and Atmospheric Administration (NOAA); Edward Wagstaff, Oxford University; Garima Malhotra, University of Michigan, United States) was part of the Living with Our Star challenge at FDL 2019, attempting to improve the results of McGranaghan et al. (2018) to predict high-latitude GNSS phase scintillation by creating a deep learning model that can incorporate information from auroral imagery as observed by the THEMIS all-sky imaging network (Donovan et al., 2006). It has been shown that auroral intensity variations are correlated with GNSS-phase scintillations (e.g., Meeren et al., 2015; Jin et al., 2015); however, the type of irregularities and physical processes producing GNSS-phase scintillations are still being debated. To understand which auroral features related to which type of scintillation events, the team first created a neural-network-based architecture (initially developed by Athanasios

Vlontzos and Benjamin Hu of Imperial College, London, United Kingdom—personal communication), to encode the auroral images (sized 256×256 pixels) to a feature representation of size $32 \times 32 \times 3$. The next step was to see whether it is possible to decode these reduced features to reconstruct original image.

2.3.1 Auto encoders

In computer vision, DNNs are routinely used as auto encoders to support data compression (Theis et al., 2017). Autoencoders learn low-dimension representation from high-dimensional data. This can be used on both single images and image sequences (e.g., video) to compress data in a way that preserves interesting phenomena. In general, one uses a neural network shaped like an hourglass with a small bottleneck layer to learn a compressed representation of the data (Fig. 7), by minimizing the Euclidean distance between the input and the reconstruction. In the case of encoding and decoding auroral images, the network

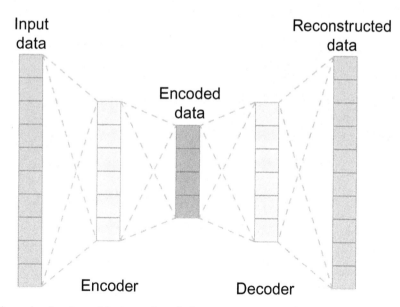

FIG. 7 A typical encoder-decoder architecture, where the latent representation of the complex system being encoded exists in the middle.

architecture developed by Vlontzos and Hu uses a series of convolutional neural networks both for encoding and decoding the complex images.

Using this innovative approach for auroral images, the team was able to reconstruct the auroral images to a high degree of accuracy as shown in Fig. 8A. Fig. 8B shows the result of applying the auto encoder technique to prelabeled auroral data. In this case, since we know the class of the aurora, each dot in the classification diagram corresponds to a specific auroral type. The reconstructed aurora images, while not reproducing every small structure in the image, make it possible to see what class of aurora there may be. The scatter plot in Fig. 8B is from t-distributed stochastic neighbor embedding (t-SNE) analysis (van der Maaten and Hinton, 2008) being applied to the prelabeled images. The scatter plot shows that while it is difficult for the network to learn about the difference between discrete and diffuse, it does remarkably well to separate diffuse aurora and auroral arcs. The dataset used here is from the Fort Simpson THEMIS site for the month of March 2015 between midnight ±3h. Note that the images

with clouds or moon in them have been removed in order to allow the network to learn about smaller-scale structures within the aurora.

The same FDL 2019 team also implemented a DNN-based architecture to improve the prediction for a single CHAIN site, while also reducing the computational load. This was a full data-driven model that used similar inputs as those used in McGranaghan et al. (2018), but dropped the Newell coupling function, TEC, OVATION Prime, and GOES fluxes as inputs to the model. While the details of the model architecture are too extensive to include in this book, it uses several convolutional layers to bring out temporal relationships between different input data (e.g., solar wind, B_z, spectral slope of GNSS signal, etc.), and a custom loss function that accounts for the fact that high scintillation events are rare. The scintillation index is predicted 1h in the future, based on the data from 2h in the past. Using this approach, even with 2months of training data, the GNSS scintillations are predicted with a TSS of 0.64.

DNNs can also be used to directly learn the evolution of complex systems (Kutz, 2017). Here

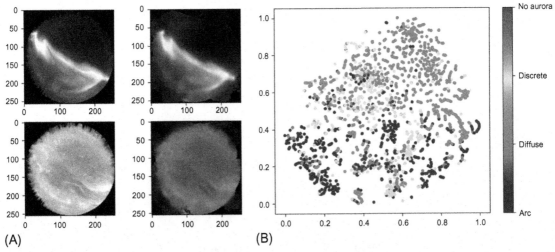

FIG. 8 (A) Two sets of images from an auroral arc *(top)* and discrete aurora *(bottom)*. The left set of two images are original, while the right set of two images are reconstructed 256 × 256 images from the latent feature space of 32 × 32 × 3. (B) A t-SNE diagram consisting of only four classes. The diffuse aurora and arc have the clearer decision boundary than discrete aurora and diffuse, yet, all classes of aurora are seen to occupy a feature space. *Courtesy: Kara Lamb, NOAA.*

the neural network is used to approximate complex system evolution, using its pattern-matching capabilities to skip the intricate details needed to fully model the complex dynamics. Effectively the neural network will recognize patterns in the data (e.g., eddies in a stream) and use past experiences of those same phenomena to predict expected future behavior. The system is trained on pairs of observations, using one observation as input and a later observation as the target output. For a fixed time gap between the two, standard approaches to learning are used. If predictions at multiple time lags are desired, multiple networks can be trained or a single recurrent network can be used to attempt to more closely model the dynamics, though training is more complex due to the need to ensure that the intermediate data are both accurate and carry sufficient information to support future prediction. Recently, libraries that combine neural networks with ordinary differential equation solvers have been developed (Chen et al., 2018). Such techniques are promising for approximating complex dynamics in a way that more accurately reflects the underlying physics and provides a reduction in computational costs over standard neural approximations. Another exciting recent development is context-aware networks (Zhuang et al., 2017) that adjust the neural network weights based on the context of the data. Here, in addition to directly learning weights by minimizing the distance to known outputs, a secondary network learns how to adjust weights based on input context. This technique allows us to learn different dynamic regimes based on classifying the input (e.g., turbulent fluids behave differently than laminar), allowing a more accurate representation of the underlying physics.

2.4 Unsupervised learning

Unsupervised learning methods are a class of algorithms that take input data that we don't already know how to classify, i.e., without pre-existing labels, and find previously unknown patterns in the data. These algorithms are useful in physical sciences where classification of observed phenomena is highly subjective and is often a subject of debate within the scientific community. They use statistical properties of the data to identify clusters of examples and/ or anomalous examples that do not fit patterns.

As an example, Heidrich-Meisner and Wimmer-Schweingruber (2018) used a *k-means* clustering method for solar wind classification. Solar wind classification schemes historically have been heuristic, and relied on expert knowledge primarily. Typically solar wind has been classified as "slow," "fast," and "Coronal Mass Ejection (CME) generated." However, the "decision boundary" that separates these classes is rather fuzzy, and not uniformly accepted. Additionally, within each of these broad labels, subcategories of solar wind have been identified by various researchers. For example, very slow solar wind is often considered a distinct type of solar wind (Sanchez-Diaz et al., 2016), and similarly highly Alfvenic slow solar wind (D'Amicis and Bruno, 2015), and so on for other categories. As a result, all the existing classification schemes don't capture the complexity in the solar wind types. Being able to identify and classify solar wind types correctly is also an essential aspect of understanding the interaction of solar wind with the earth's magnetosphere and space weather impacts. From the data science perspective, different classification labels are available for only a subset of data, which is biased toward data points far away from decision boundaries between classes. Therefore, employing supervised learning for classification would inherit this same bias. In such cases, using unsupervised learning techniques, more specifically "clustering," has merit. The goal is to find a unique cluster for each observation. This depends heavily on the parameters that play a role in identifying decision boundaries and therefore selecting the right parameters (known as "feature selection") is an important process. In this case, the authors created a seven-dimensional feature space consisting of well-known solar wind observables from the Advanced Composition Explorer (ACE) spacecraft, such as the solar wind velocity, density, temperature, magnetic field strength, O charge state ratio, C charge state ratio, and proton-proton collisional age.

As we will detail later in this chapter, most machine-learning methods assume that the training and test datasets are identically distributed, which is usually a challenge in scientific datasets, especially where ground truth is not known. To get around this problem, the authors set aside the equivalent of two Carrington rotations each year from 2001 to 2010 as test data and used the remaining as training data. This arrangement ensures that solar-cycle-dependent changes in the data distributions are captured as well.

Several clustering techniques are prevalent. This example chooses to use a simple clustering method, *k-means*, which assumes that there are k latent clusters and iteratively estimates cluster assignments and centroids using a Euclidean distance metric until a stable solution is obtained. The objective is to assign each observation labeled $n \in \{1...N\}$ to a unique cluster $\{c_1...c_k\}$, where the total number of clusters $K < N$, and the number of data points in the k-th cluster is N_k. The algorithm employed finds a partition $S = \{S_1...S_k\}$ of the training data with N elements and corresponding cluster centers, μ_k, that minimize the inner-cluster distances for all clusters. That is, for $n \in \{1...N\}$, $l = \text{argmin}_{j \in \{1...k\}} ||x_n - c_j||^2$. Then new cluster centers are determined based on current partitions, and the process is repeated until the cluster assignment stabilizes. It is important to note that here k is not determined by the algorithm, rather it has to be predefined to apply the k-means clustering.

Using this approach, the authors were able to cluster the solar wind into seven different types as shown in Fig. 9. The figure shows one of the results from this work displaying the distribution of solar wind velocity and density within each cluster. As can be seen from this figure, the majority of data points are in Cluster 3, which represents slow solar wind, followed by Clusters 2 and 6 that represent interstream solar wind. Clusters 4 and 7 represent coronal hole stream as indicated by high velocities. Analyzing physical properties of all seven clusters of solar wind through domain knowledge, the authors were able to interpret the results of applying this machine-learning algorithm, which is often a difficult task to accomplish.

2.4.1 Other unsupervised learning methods

Anomaly detection

Anomaly detection is a method to identify outliers in the data. This method fits some type of distribution to the data, and then evaluates each data point based on its statistical likelihood. Outliers are selected as possible interesting phenomena to follow up on.

Dimensionality reduction

Many dimensionality reduction algorithms are forms of unsupervised machine learning, relying solely on unannotated data. For example, Principal Component Analysis (PCA) uses the rank analysis of the covariance matrix of the distribution of data points to produce a low-dimensional representation of the data, while more complex methods fit the data to low-dimensional manifolds. Dimensionality reduction is often used as a preprocessing step before supervised methods are applied for classification or regression in order to simplify the problem, and to support more effective generalization.

3 Ionospheric questions that can benefit from machine learning

As noted earlier, several operational needs require accurate forecasting of space weather,

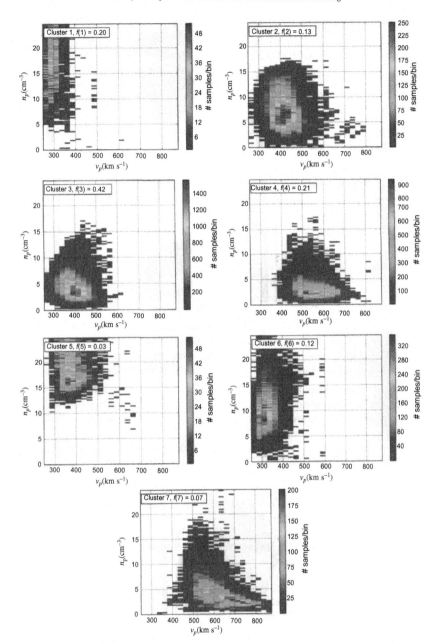

FIG. 9 Clusters of solar wind projected on the solar wind velocity-density plane. Each panel indicates the percentage of data points in that cluster by *f*. *Reproduced from Figure 3 of Heidrich-Meisner, V., Wimmer-Schweingruber, R.F., 2018. Solar wind classification via k-means clustering algorithm. In: Camporeale, E., Johnson, J.R., Wing, S. (Eds.), Machine Learning Techniques for Space Weather. Elsevier (Chapter 16).*

and machine-learning approaches have been tapped to improve forecasting accuracy. However, the scientific community is tasked with a challenge that is much bigger than simply getting the right answer. We need to understand the underlying rules that lead us to the answer. So, we ask ourselves, should we add machine learning to our toolset for scientific inquiry?

As it stands, even after decades of scientific research, we often miss many common phenomena: for example, recent discovery of the subauroral phenomenon, Strong Thermal Emission Velocity Enhancement, or "STEVE" (Gallardo-Lacourt et al., 2018). STEVE appears at subauroral latitudes in the form of an elongated ribbon, sometimes accompanied by auroral beads or "picket fence." Although this phenomenon has always been observed by amateur aurora watchers, it is now a new phenomenon for the upper atmospheric research community. STEVE hid in plain sight for decades, at least since the advent of all-sky auroral imaging, a powerful technique that ionospheric researchers have used for a long time. While there are many reasons for missing such a phenomenon, one reason has to do with is the abundance of data available for human eyes to investigate. The THEMIS ground-based imagers generate one image every 3 s every night. The result is essentially a movie of images for each night, the clarity of which is also dependent on sky conditions—clouds, moon, etc. To analyze each movie, each night, frame by frame, is a difficult task for a human scientist. However, an approach like that of Clausen and Nickisch (2018) can be easily utilized by the ionospheric scientists to classify auroral data free of cloud or moon contamination among various categories already known to science. A further approach would be to use an unsupervised learning algorithm to let the machine identify recurring phenomena that science has yet to identify. As demonstrated earlier, the modern

machine-learning methods are exceptionally good at pattern recognition and classification. By not using these tools at our disposal, we are missing out on essential physics that occurs when we are not looking.

While direct forecasting of space weather can be a complex problem, classification of various structures and phenomena observed by various instruments could be attempted more easily is a low-hanging fruit that modern machine-learning techniques can help us achieve to help further scientific progress. Similarly, as citizen science initiatives such as aurora reporting and ham radio data become more prevalent, machine-learning approaches can be applied to sift through the reports and assign quality flags. Some of the techniques are also suitable for use as intermediate steps in physics-based simulation to arrive at the answer in a computationally efficient manner.

3.1 Notes of caution

It is important however not to lose sight of the scientific process when adding new and seemingly powerful tools to our arsenal for scientific inquiry. When it comes to recognizing patterns for prediction, there are many reasons (or sometimes none) why one independent input variable may affect the output more than others. For example, the geomagnetic drivers may cause ground-based GNSS-phase scintillations and ground-based magnetometers to exhibit variability at the same time. Yet a scientist understands the difference between correlation and causation, and the same rigor must especially be applied to machine-learning problems. It is also important to be clear on the uncertainties inherent within the data and how they propagate through the machine-learning models.

3.1.1 Data distribution

As noted earlier, machine-learning methods assume identical data distributions between *training, validation,* and *test* data sets. If this is not ensured, the models are prone to be severely wrong. Therefore, any machine-learning application calls for a detailed understanding of the data distribution across all the sets. This calls for an especially careful treatment in geospace data that are affected by solar cycle related, climatological, and daily variations. In addition to having a randomly selected test set, it is also recommended to have a time separated test set for data that can have temporal drift. For example, when analyzing data collected from a sensor, constructing a test set from the data collected over the most recent month (or months) can identify if sensor drift and/or secular drift in the observed system can break the machine-learning algorithm. This will provide confidence that algorithms trained on historical data will apply to data collected in the future, and will also help identify how often training needs to be rerun.

3.1.2 Data gaps

Another challenge faced for geospace data is dealing with gaps in continuous datasets. The reasons for these data gaps range from instruments malfunctioning to having snow cover the instrument field-of-view. Any approach that uses machine learning must deal with these data gaps. There are several approaches used by machine-learning practitioners from substituting global mean for missing data to use different interpolation techniques to "fill the data gap"; however, each of these choices must be made consciously.

3.1.3 Hyperparameter tuning

Many of the more powerful machine-learning techniques can be quite sensitive to small changes in model hyperparameters. There are often multiple hyperparameters that need to be tuned, and subtle differences in data can cause catastrophic failure. Best practices for machine learning always include using both

validation and test sets to avoid overfitting, and to measure the generalization ability of the trained systems.

3.1.4 Evaluating the model performance

The data science methods give a new toolkit to science, but in essence, these are a set of mathematical constructs that require the users to understand their inherent limitations and biases. As an example, many standard neural-network-based supervised learning algorithms try to minimize the mean absolute error (MAE) between the input and the output (loss function), and also use it as a metric for how well the model is performing. But as we saw in the case of GNSS scintillation prediction, the scintillation events are comparatively rare; hence, there is a natural "class imbalance." When trying to predict the magnitude of scintillation events using a regression model, simply using MAE as loss or metric doesn't account for the rare, peaky behavior in the data. It is therefore important for a scientific researcher to pay attention to the mathematical constructs behind the loss functions or metrics while using tool bundles like TensorFlow, Keras, or PyTorch that make it easier for anyone to use machine learning.

Finally, it is imperative to not treat the machine-learning methods as standalone answer generators but rather use a combination of techniques and domain science knowledge to fulfill the purpose of scientific inquiry. This is highlighted by the example of the FDL 2019 team mentioned earlier, which was mentored by a domain scientist (the chapter author) and four machine-learning experts from Intel, Oxford University, and ElementAI, in addition to knowledge exchange with various other science and machine-learning experts periodically. Machine learning is a vast field, and requires knowledgeable practitioners to evaluate and implement approaches for effective scientific data analysis.

4 Infrastructure needed to enable big data approaches

Seamlessly applying data science approaches discussed earlier also requires digital data infrastructure to match (McGranaghan et al., 2017). Much of the space weather data are available from disparate sources, which makes it difficult to gather. This is especially important to consider in the age of the American Geophysical Union's FAIR (Findable, Accessible, Interoperable, and Reproducible) data standards (Bhatt et al., 2017). Similar data policies have been adapted by organizations and agencies in various nations dealing with scientific data. Reproducible research is a hallmark of scientific inquiry, and more so with powerful tools that are developed especially to handle big data. Some recent efforts from the geospace community have made headway into making the data accessible using open source approaches and nontraditional partnerships. For example, the Python Satellite Data Analysis Toolkit or pysat allows users to access data from multiple satellite sources on a single platform (Stoneback et al., 2018). Similarly, the Reproducible Software Environment (Resen) toolkit developed under the Integrated Geoscience Observatory project supported by the US National Science Foundation (https://github.com/EarthCubeIn Geo/resen) allows using geospace data and community developed software together to create reproducible results with the use of mainstream technologies like Jupyter and Docker. The Frontier Development Lab program supported by NASA, partners with private industries such as Google, IBM, and Intel to access data and computational infrastructure needed to solve space science challenges. The computational needs for applying data science approaches are significant, and using mainstream approaches is one way to address them.

For developing Deep Neural Networks, it is recommended to have graphical processing units (GPUs) available to substantially reduce training time. While a single GPU is often sufficient and easy to manage, coordinating multiple GPUs may be needed for large datasets. Deployment rarely requires special hardware (e.g., GPUs) as the network only needs to be run once per query, but this also depends on data and data size. The FDL 2019 team for example used 38 NVIDIA GPUs provided by Google Cloud to run over 70 experiments involving 40 datasets with 75,000 points each.

In image processing and speech processing domains, transfer learning has proven particularly useful, where features trained on a large data corpus are later used as the initial layers of a task specific neural network. Training such general purpose features for a given complex systems problem and sharing them among the interested community is a best practice that would dramatically improve accessibility and research speed.

5 Summary and future

Modern data mining approaches provide new tools for scientific inquiry. These are powerful techniques that, if used correctly, can help speed up the process of scientific discovery. As with all tools, machine-learning methods do not replace the scientific insight but rather augment the scientific vision. Geospace science is currently in the epoch of the data revolution. Newer machine-learning techniques are exceptionally suitable for pattern recognition, parameterization, and classification for large datasets representing complex nonlinear phenomena. These methods can help fully exploit the vast continuous data collected from multiple ground- and space-based instruments that are currently underutilized. Significant efforts are being invested into gaining insight into some of the "black box" methods, which are promising in the process of doing science with rigor. A new paradigm of "theory-guided data science" is driving how best to combine physics-driven efforts with machine-learning methods to achieve best results. The scientist of the future may use the continually evolving data landscape and mainstream data science methods to gain insight into the vast, complicated, multiscale and nonlinear nature of the geospace that is slow to reveal its secrets.

References

Aljalbout, E., Golkov, V., Siddiqui, Y., Strobel, M., Cremers, D., 2018. Clustering With Deep Learning: Taxonomy and New Methods. arXiv preprint arXiv:1801.07648.

Bhatt, A., McGranaghan, R., Matsuo, T., Gil, Y., 2017. Essential best practices for the geospace community concerning reproducible research, open science, and digital scholarship. In: White Paper as Part of NSF Earth-Cube Integrative Activity: Integrated Geoscience Observatory. http://sites.nationalacademies.org/cs/groups/ssbsite/documents/webpage/ssb_187012.pdf. Document link:https://goo.gl/wQkb4U. Retrieved 28 February 2019.

Bobra, M.G., Couvidat, S., 2015. Solar flare prediction using SDO/HMI vector magnetic field data with a machine-learning algorithm. Astrophys. J. 798 (2), 11. https://doi.org/10.1088/0004-637X/798/2/135. ID 135.

Camporeale, E., Wing, S., Johnson, J.R. (Eds.), 2018. Machine Learning Techniques for Space Weather. Elsevier. p. iv. ISBN: 9780128117880. https://doi.org/10.1016/B978-0-12-811788-0.09994-7.

Chen, T.Q., Rubanova, Y., Betterncourt, J., Duvenaud, D.K., 2018. Neural ordinary differential equations. In: Advances in Neural Information Processing Systems 31, NIPS Proceedings.

Cheung, M., 2018. Deep learning for heliophysics. In: Graphics Processing Unit (GPU) Technology Conference, 2018, Silicon Valley. ID S8222.

Clausen, L.B.N., Nickisch, H., 2018. Automatic classification of auroral images from the Oslo Auroral THEMIS (OATH) data set using machine learning. J. Geophys. Res. Space Physics 123, 5640–5647.

Cortes, C., Vapnik, V., 1995. Support-vector networks. Mach. Learn. 20, 273. https://doi.org/10.1007/BF00994018.

Cousins, E.D.P., Matsuo, T., Richmond, A.D., 2015. Mapping high-latitude ionospheric electrodynamics with Super-DARN and AMPERE. J. Geophys. Res. Space Physics 120, 5854–5870. https://doi.org/10.1002/2014JA020463.

Cover, T.M., 1965. Geometrical and statistical properties of systems of linear inequalities with applications in pattern recognition. IEEE Trans. Electron. Comput. EC-14 (3), 326–334. https://doi.org/10.1109/PGEC.1965.264137.

D'Amicis, R., Bruno, R., 2015. On the origin of highly alfvenic slow solar wind. Astrophys. J. 805, 84.

Díaz Baso, C.J., Asensio Ramos, A., 2018. Enhancing SDO/HMI images using deep learning. Astron. Astrophys. 614, A5. https://doi.org/10.1051/0004-6361/201731344.

Donovan, E., Mende, S., Jackel, B., Frey, H., Syrjäsuo, M., Voronkov, I., Trondsen, T., Peticolas, L., Angelopoulos, V., Harris, S., Greffen, M., Connors, M., 2006. The THEMIS all-sky imaging array—system design and initial results from the prototype imager. J. Atmos. Sol. Terr. Phys. https://doi.org/10.1016/j.jastp.2005.03.027.

Gallardo-Lacourt, B., Liang, J., Nishimura, Y., Donovan, E., 2018. On the origin of STEVE: particle precipitation or ionospheric skyglow? Geophys. Res. Lett. 45(16). https://doi.org/10.1029/2018GL078509.

Heidrich-Meisner, V., Wimmer-Schweingruber, R.F., 2018. Solar wind classification via k-means clustering algorithm. In: Camporeale, E., Johnson, J.R., Wing, S. (Eds.), Machine Learning Techniques for Space Weather. Elsevier (Chapter 16).

Jin, Y., Moen, J.I., Miloch, W.J., 2015. On the collocation of the cusp aurora and the GPS phase scintillation: a statistical study. J. Geophys. Res. Space Physics 120, 9176–9191. https://doi.org/10.1002/2015JA021449.

Karpatne, A., Atluri, G., Faghmous, J.H., Steinbach, M., Banerjee, A., Ganguly, A., Shekhar, S., Samatova, N., Kumar, V., 2017. Theory-guided data science: a new paradigm for scientific discovery from data. IEEE Trans. Knowl. Data Eng. 29, 10.

Kutz, J.N., 2017. Deep learning in fluid dynamics. J. Fluid Mech. 814, 1–4.

Lu, C., Krishna, R., Bernstein, M., Fei-Fei, L., 2016. Visual relationship detection with language priors. In: European Conference on Computer Vision. Springer, Cham.

Marsh, D.R., Mills, M.J., Kinnison, D.E., Lamarque, J., Calvo, N., Polvani, L.M., 2013. Climate change from 1850 to 2005 simulated in CESM1(WACCM). J. Clim. 26, 7372–7391. https://doi.org/10.1175/JCLI-D-12-00558.1.

McGranaghan, R., Bhatt, A., Matsuo, T., Mannucci, A., Semeter, J., Datta-Barua, S., 2017. Ushering in a new frontier in geospace through data science. J. Geophys. Res. https://doi.org/10.1002/2017JA024835.

McGranaghan, R.M., Mannucci, A.J., Wilson, B., Mattmann, C.A., Chadwick, R., 2018. New capabilities for prediction of high-latitude ionospheric scintillation: a novel approach with machine learning. Space Weather 16 (11), 1817–1846. https://doi.org/10.1029/2018SW002018.

Meeren, C., Oksavik, K., Lorentzen, D.A., Rietveld, M.T., Clausen, L.B.N., 2015. Severe and localized GNSS scintillation at the poleward edge of the nightside auroral oval during intense substorm aurora. J. Geophys. Res. Space Physics 120, 10607–10621. https://doi.org/10.1002/2015JA021819.

National Science and Technology Council (Space Weather Operations, Research, and Mitigation Working Group, Space Weather, Security, and Hazards Subcommittee, and Committee on Homeland and

National Security), 2019. National Space Weather Strategy and Action Plan. Office of Science and Technology Policy, Washington, DC.

Newell, P.T., Sotirelis, T., Wing, S., 2010. Seasonal variations in diffuse, monoenergetic, and broadband aurora. J. Geophys. Res. 115. https://doi.org/10.1029/2009JA014805.

Qian, L., Burns, A.G., Emery, B.A., Foster, B., Lu, G., Maute, A., Richmond, A.D., Roble, R.G., Solomon, S.C., Wang, W., 2014. The NCAR TIE-GCM. In: Huba, J., Schunk, R., Khazanov, G. (Eds.), Modeling the Ionosphere–Thermosphere System (Chapter 7). https://doi.org/10.1002/9781118704417.

Rokachand, L., Maimon, O., 2007. Introduction of decision trees. Data Mining with Decision Trees: Theory and Application. World Scientific Publishing Co. Pte. Ltd. Co., NJ, 5 pp.

Rosenblatt, F., 1958. The perceptron: a probabilistic model for information storage and organization in the brain. Psychol. Rev. 65 (6), 386–408. https://doi.org/10.1037/h0042519.

Sabokrou, M., Fayyaz, M., Fathy, M., Moayed, Z., Klette, R., 2018. Deep-anomaly: fully convolutional neural network for fast anomaly detection in crowded scenes. Comput. Vis. Image Underst. 172, 88–97.

Sanchez-Diaz, E., Rouillard, A.P., Lavraud, B., Segura, K., Tao, C., Pinto, R., Sheeley, N.R., Plotnikov, I., 2016. The very slow solar wind: properties, origin and variability. J. Geophys. Res. Space Physics 121, 2830–2841. https://doi.org/10.1002/2016JA022433.

Stoneback, R.A., Burrell, A.G., Klenzing, J., Depew, M.D., 2018. PYSAT: PYSAT: Python satellite data analysis toolkit. J. Geophys. Res. Space Physics 123, 5271–5283. https://doi.org/10.1029/2018JA025297.

Strogatz, S., 2000. Non-Linear Dynamics and Chaos. CRC Press. ISBN-13: 978-0738204536.

Szegedy, C., Toshev, A., Erhan, D., 2013. Deep neural networks for object detection. In: Advances in Neural Information Processing Systems.

Theis, L., Shi, W., Cunningham, A., Huszár, F., 2017. Lossy Image Compression With Compressive Autoencoders. arXiv preprint arXiv:1703.00395.

van der Maaten, L., Hinton, G., 2008. Visualizing data using t-SNE. J. Mach. Learn. Res. 9, 2579–2605.

Vinyals, O., Toshev, A., Bengio, S., Erhan, D., 2015. Show and tell: a neural image caption generator. In: Proceedings of the IEEE Conference on Computer Vision and Pattern Recognition.

Vrigkas, M., Nikou, C., Kakadiaris, I.A., 2015. A review of human activity recognition methods. Front. Robot. Artif. Intell. 2, 28.

Xian, Y., Lampert, C.H., Schiele, B., Akata, Z., 2018. Zero-shot learning-a comprehensive evaluation of the good, the bad and the ugly. In: IEEE Transactions on Pattern Analysis and Machine Intelligence.

You, Q., Jin, H., Wang, Z., Fang, C., Luo, J., 2016. Image captioning with semantic attention. In: Proceedings of the IEEE Conference on Computer Vision and Pattern Recognition.

Zhuang, B., Liu, L., Shen, C., Reid, I., 2017. Towards context-aware interaction recognition for visual relationship detection. In: Proceedings of the IEEE International Conference on Computer Vision.

Further reading

Covas, Eurico, and Emmanouil Benetos, 2018. Optimal Neural Network Feature Selection for Spatial-Temporal Forecasting. arXiv preprint arXiv:1804.11129.

Greenfeld, I., Yavin, Z., 2016. Method and Apparatus for Aerial Surveillance and Targeting. U.S. Patent Application No. 15/082, 995.

Kulin, M., Kazaz, T., Moerman, I., Poorter, E.D., 2018. End-to-end learning from spectrum data: a deep learning approach for wireless signal identification in spectrum monitoring applications. IEEE Access 6, 18484–18501.

McElwee, S., Heaton, J., Fraley, J., Cannady, J., 2017. Deep learning for prioritizing and responding to intrusion detection alerts. In: MILCOM 2017-2017 IEEE Military Communications Conference (MILCOM). IEEE.

National Research Council, 2012. Solar and Space Physics: A Science for a Technological Society. The National Academies Press, ISBN: 978-0-309-38739-2. https://doi.org/10.17226/13060.

O'Shea, T., 2018. Can a Machine Learn the Fourier Transform. https://cyclostationary.blog/2017/08/03/can-a-machine-learn-the-fourier-transform/.

18

Scintillation modeling

Massimo Materassi[a], Lucilla Alfonsi[b], Luca Spogli[b], Biagio Forte[c]

[a]Institute for Complex Systems of the National Research Council (ISC-CNR), Florence, Italy
[b]Istituto Nazionale di Geofisica e Vulcanologia, Rome, Italy
[c]Department of Electronic and Electrical Engineering, University of Bath, Bath, United Kingdom

1 Introduction

The presence of irregularities in the refraction index of the Earth's ionosphere (EI), due to plasma instabilities (Hysell, 2019) and turbulence (Materassi, 2019), gives rise to random variations of the amplitude and phase of radio waves passing through the medium (Davies, 1990). Such rapid fluctuations are referred to as *ionospheric radio scintillation* (IRS) (Knepp, 2019).

Technological issues related to IRS are remarkable (Wernik et al., 2007): scintillation may affect considerably the performance of the satellite communication and navigation (e.g., reducing the accuracy of the pseudorange and phase measurements in the case of the Global Positioning System (GPS)), and the signal power reduction may be so strong that it may drop below the threshold limit; in this case, the receiver loses lock and GPS positioning becomes impossible. Equally, also strong phase scintillation, resulting in a Doppler shift that is too big, may cause the loss of lock when exceeding the phase lock loop bandwidth.

As IRS is determined by the space and time nature of the ionospheric irregularities, it is not just "an issue" in technological terms, but

also an opportunity in terms of imaging and understanding the irregularities of the EI (Hysell, 2019, Knepp, 2019): on the one hand, there is the necessity of forecasting and mitigating the IRS problems on electromagnetic trans-ionospheric communications; on the other hand, it provides the opportunity to study IRS space-time variability, and hence dynamics, to infer the irregularity dynamics of EI, and learn more about ionospheric turbulence.

For this twofold goal, *IRS modeling is a crucial tool*. Scintillation modeling means putting in relationship the statistics of the fluctuations of the ionospheric medium (IM) with those of fluctuations of the radio signal received past the propagation past the EI: if the irregular electron density is represented as the sum

$$N_e\left(\vec{x},t\right) = N_0\left(\vec{x},t\right) + \Delta N_e\left(\vec{x},t\right) \qquad (1)$$

of a smooth background $N_0\left(\vec{x},t\right)$ and stochastic fluctuations $\Delta N_e\left(\vec{x},t\right)$, where $\vec{x}$ is the position and t is the time, modeling IRS means putting in relationship the ΔN_e statistics (Materassi, 2019), encoded in some probability distribution $P\left[\Delta N_e; \vec{x},t\right]$, with that of the signal

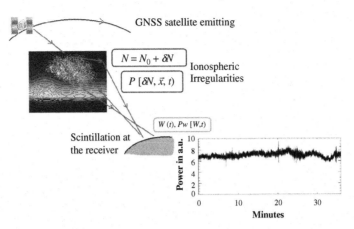

FIG. 1 The origin of ionospheric radio scintillation in a cartoon: the ionospheric medium is the sum of a smooth background $N_0\left(\vec{x},t\right)$ plus stochastic fluctuations $\Delta N_e\left(\vec{x},t\right)$, represented by the statistics $P\left[\Delta N_e; \vec{x},t\right]$. Along its path to the receiver, the electromagnetic wave is randomly scattered, so that its power W is rendered irregular as well, represented by a statistics $P_W(W,t)$ (see text, picture elaborated from Materassi et al., 2008).

received, e.g., its power $W(t)$, encoded in some $P_W(W,t)$ (Materassi et al., 2008; Materassi and Mitchell, 2007) (see Fig. 1).

One can construct a theory in which the statistics of the signal $P_W(W)$ are obtained from that of the medium $P(\Delta N_e)$: *this is properly IRS modeling*, in principle enabling users to assess the likelihood of signal disruption along the radio link at a given time and geometry. Models from scintillation theory focus on *the electromagnetic propagation theory*, and on *simulating the medium statistics*; in these kinds of models, great use is made of wave propagation in random media (Tatarskii, 1971; Yeh and Liu, 1982; Yeh and Wernik, 1993; Bhattacharyya et al., 1992).

On the other hand, it is desirable to infer IM irregularity properties by observing the IRS on Global Navigation Satellite Systems (GNSS) signals collected from ground receivers (e.g., GPS receivers), because nowadays these are tremendously abundant and have a very large coverage: this involves trying to understand something on $P(\Delta N_e)$ from the observation of $P_W(W)$ *and is not quite properly IRS modeling*, but rather is retrieving characteristics of what causes IRS from the observation of it (Wernik

et al., 2008; Grzesiak and Wernik, 2009, 2012). These kinds of studies are based on (possibly large) collections of ground-based scintillation data, sorted according to helio-geophysical conditions (e.g., geomagnetic indices, solar wind speed, interplanetary magnetic field components, local magnetic time, season) (Beniguel et al., 2009; Spogli et al., 2009).

In this chapter, we present some IRS modeling solutions accounting for the complexity and dynamicity of the EI. This chapter *is not intended* as a general review of IRS models available in literature; probably a whole book as long as the present one would not be enough to include them all (see, for instance, Beniguel et al., 2009). Instead, we just focus on a couple of examples, stressing how complexity and dynamicity of the EI must enter the building process of them.

In Section 2 we present the WAM model developed by Wernik, Alfonsi, and Materassi (Wernik et al., 2007; Alfonsi et al., 2017), which is properly a *propagation model*, in which the statistical characteristics of the IM are deduced from real in situ data: as WAM is based on assortments of IM data by geomagnetic activity and season, it is basically a scintillation climatology model, pivoting on a propagation theory, namely Rino's weak scattering theory (Rino, 1979a).

A climatology model of the IRS on received signals is also presented in Section 3, constructed through the use of Ionospheric Scintillation Monitor Receivers, where scintillation and Total Electron Content data from ground-based observations are investigated to describe the recurrent features of the ionospheric irregularities leading to scintillation occurrence at a regional or global scale on a long-term basis.

Critical remarks and scintillation modeling perspectives, according to our vision, are presented in Section 4.

2 Propagation modeling from real in situ measurements: The WAM model

As sketched in the introduction of this chapter, IRS is due to the interplay between the propagating electromagnetic wave (with its own intensity, phase, and frequency spectrum), and the ionospheric irregularities encountered along its path from the emitter to the receiver. Then, IRS is characterized by great spatial and temporal variability. How the irregular IM scatters the electromagnetic wave depends on signal frequency, local time, season, solar activity, and magnetic activity (determining the general aspects of the near-Earth space plasma, as discussed in Lapenta, 2019). IRS also depends on the radio link geometry, i.e., on the satellite zenith angle and on the angle between the raypath and the Earth's magnetic field. At the beginning of the 21st century, all this convinced Wernik that the most faithful way to represent ionospheric fluctuations determining IRS is not to make an abstract theory of them, which must necessarily be approximate, but rather deducing their characteristics from *real local measurements*, collected by satellites traveling across the EI.

The Wernik-Alfonsi-Materassi (WAM) model, presented in Wernik et al. (2007) and further extended in Alfonsi et al. (2017), is a *wave propagation model*, i.e., a tool to predict the radio scintillation on an assigned link, from the knowledge of random medium crossed by the wave: the link assignment includes a specific time, so that season information on the IM contained in the WAM model may be used.

2.1 Theoretical framework

As stated in Yeh and Liu (1982), and anticipated here in Section 1, in scintillation theories, both the hypotheses on the IM irregularities and their consequences on the IRS along the link are intended in terms of statistics of fluctuations, of the IM, and of the received wave proxies, respectively.

The WAM model is based on the following points:

- The necessary IM fluctuation statistical quantities are deduced from *real in situ measurements*.
- The wave propagation is mimicked by a *single phase screen* technique (Yeh and Liu, 1982; Rino, 1979a).
- The IM is represented as a smooth ionosphere, given by the empirical model IRI95 (Bilitza, 1997), with electron density irregularities all concentrated in a slab.

Those are the crucial points of WAM "philosophy," making reference to the propagation geometry sketched in Fig. 2.

The electromagnetic wave is emitted by a GNSS satellite, supposed to fly along a trajectory $\vec{x}_{GNSS}(t)$, and is received at the position $\vec{x}_R$ of the receiver. The radio link is represented by the segment from $\vec{x}_{GNSS}(t)$ to $\vec{x}_R$; the signal is a plane wave of wave vector $\vec{k}$. The characteristics of the ionospheric irregularities ΔN_e are obtained from real in situ measurements collected by an LEO satellite flying along a trajectory $\vec{x}_S(t)$: these data are collected during a time interval embracing a period long enough to collect a good diversity of helio-geophysical conditions. The time at

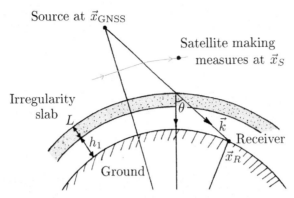

FIG. 2   The general geometry of the ionospheric propagation adopted in WAM (see text. Picture freely elaborated from Singh et al., 1997).

which scintillation is simulated along the radio link between $\vec{x}_{GNSS}(t)$ and $\vec{x}_R$ is generally different from the historical period in which real IM measurements were collected. For example, the version of the WAM model in Wernik et al. (2007) was based on the local measurements of the retarding potential analyzer of the Dynamic Explorer 2 (Hanson et al., 1981), collected from 1981 to 1983: this determines the climatology of ΔN_e on which the model is based, but in principle the IRS may be simulated along links of any epoch.

Other details reported in Fig. 2 are explained in what follows.

As reported, in WAM, one models the electromagnetic wave propagation through the irregular IM: in order to construct such a model, it is necessary to choose both the *propagation equation* for the electromagnetic signal and the *type of medium*.

"Choosing the equation," or "choosing the type of medium," may sound a little funny, as an electromagnetic wave just travels according to Maxwell Equations, while the medium we are discussing is "simply" the EI with irregularities: still, *approximations* may be chosen, according to the degree of accuracy desired, with respect to what situation one deals with. The *logical path* of the chain of approximations leading to WAM is reported in Fig. 3 (from Materassi

et al., 2008): the steps to follow to construct the model are indicated by Roman numbers in the insets, from I to VI. Here they are revised, and their physical meaning is highlighted and commented.

One starts by considering the irregular EI as a random medium, i.e., a medium with a refraction index treated as a stochastic variable due to the assumption (Eq. 1). The statistics of ΔN_e will enter WAM equations: as anticipated, the point of view of WAM is to obtain this statistics *from real in situ measurements* of the EI, inset I in Fig. 3. Attempts to use in situ data in IRS modeling before WAM formulation included those in Basu et al. (1976, 1981): the important novelty in the WAM model with respect to those two examples was the calculation of *all* the irregularity parameters out of real in situ measurements, which included the *spectral index p* and the *turbulence strength C_{sr}*.

As far as wave propagation is concerned, one is interested in IRS on a GNSS radio link, so it is safe to assume that only the phase of the wave is distorted as the wave propagates through the irregular ionosphere; the *phase screen approximation* is chosen (inset II), in which backscattering of the wave is neglected, as well as any loss, while the random part of the medium is concentrated in just a thin *slab of irregularities* (SoI); see also Fig. 4.

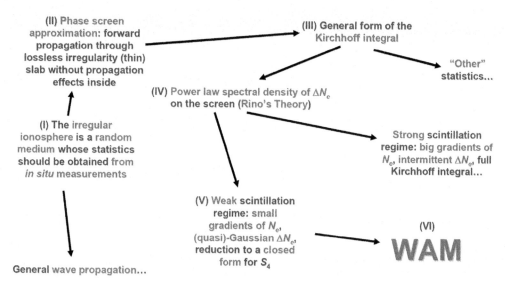

FIG. 3 The theoretical "choices" selecting the WAM model among possible IRS models based on wave propagation through a random medium (see text. Picture from Materassi et al., 2008).

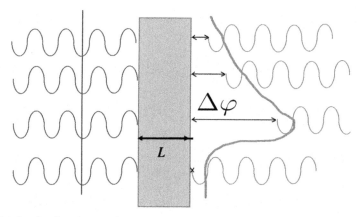

FIG. 4 A cartoon explaining the electromagnetic wave propagation through an irregular phase screen (the path of the wave goes from left to right): when crossing the (thin) screen, the plane wave (with bluish plane wavefront) is phase-altered in a local, "empirically" casual way. The wavefront is deformed (reddish deformed shape right to the screen), and a casual interference pattern forms downstream "according" to the irregularity in the phase screen. At the receiver, radio scintillation appears.

The SoI is depicted in Fig. 2 and its construction is "explained" in Fig. 5: the background ionosphere N_0 is supposed to be proportional to the International Reference Ionosphere IRI95 (Bilitza, 1997):

$$N_0\left(\vec{x},t\right) = \alpha N_{\text{IRI95}}\left(\vec{x},t\right), \qquad (2)$$

where α is fixed so that $\alpha N_{\text{IRI95}}\left(\vec{x}_S(t),t\right)$ coincides with the background $N_0\left(\vec{x}_S(t),t\right)$

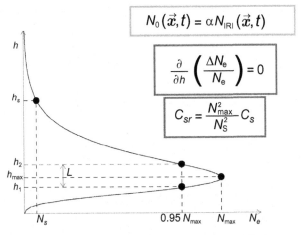

FIG. 5 The ionospheric profile adopted in WAM, with the important heights and the important ionospheric density values highlighted. In the insets, the choice of the background and the mathematical assumption leading to the relationship between C_{rs} and C_s (see text below, picture freely elaborated from Wernik et al., 2007 and Materassi et al., 2008).

obtained via detrending from the time series $N_e\left(\vec{x}_S(t), t\right)$ of in situ data. The SoI is centered at the height $h_{\max}$ of the ionospheric density peak according to N_{IRI95}, while its thickness L is the difference between the two quotes h_1 and h_2, so that $h_2 > h_{\max} > h_1$, at which the ionospheric density is 95% of the peak value:

$$N_{\mathrm{IRI95}}(h_1) = N_{\mathrm{IRI95}}(h_2) = \frac{95}{100} N_{\mathrm{IRI95}}(h_{\max}); \quad (3)$$

all in all, $L = h_2 - h_1$ (see Fig. 5. Of course, Eq. (3) holds both for N_{IRI95} and for the adapted background $N_0 = \alpha N_{\mathrm{IRI95}}$). In the case of the IRI95 model background, L ranges between 60 and 80 km: considering that the whole distance from $\vec{x}_{\mathrm{GNSS}}(t)$ to $\vec{x}_R$ is of the order of 2×10^4 km, so that $\frac{|\vec{x}_{\mathrm{GNSS}}(t) - \vec{x}_R|}{L} \gg 1$, the SoI may really be considered a thin layer, and approximated as a screen altering just the phase of the wave (the absence of losses in the medium is guaranteed by the high frequency of the GNSS signal under investigation).

Once the SoI is located as described, the values of fluctuations ΔN_e and their statistics in it are calculated from those measured along

the $\vec{x}_S(t)$ trajectory, at height h_S, by assuming the condition that the ratio $\frac{\Delta N_e}{N_e}$ is height independent (Wernik et al., 2007):

$$\frac{\partial}{\partial h}\left(\frac{\Delta N_e}{N_e}\right) = 0 \Rightarrow \Delta N_e(h_{\max})$$
$$= \frac{N_{\mathrm{IRI95}}(h_{\max})}{N_e(h_S)} \Delta N_e(h_S) \quad (4)$$

(In Eq. (4) the quantities $N_e(h_S)$ and $\Delta N_e(h_S)$ are real data, while $N_{\mathrm{IRI95}}(h_{\max})$ is calculated from the IRI95 model.) A consequence of Eq. (4) is the rescaling condition (8), see later; moreover, the SoI location determines the value of the zenith angle θ as in Fig. 2: it is the angle the radio link forms with the vertical direction at the pierce point with the SoI.

Going back to the "path-of-approximations going to WAM," the phase screen approximation choice leads to obtaining the electromagnetic wave propagated from the emitting satellite to the ground receiver as a phasor u in the general form of Kirchhoff integral (inset III in Fig. 3, see Yeh and Liu, 1982). The wave u depends on the scattering process through the

Fourier spectrum of the electron density fluctuations ΔN_e at the phase screen: as this should mimic ionospheric *turbulence*, a power law spectrum is assumed for ΔN_e, according to Rino (1979a, b); in Fig. 2 then one moves down to the inset (IV), in which Rino's theory from the two papers just mentioned is chosen.

The Kirchhoff integral giving the statistics of the power $W = |u|^2$ at the receiver reduces to a rather simple form, as one assumes the gradients of irregularities fluctuations ΔN_e to be small enough, so that ΔN_e is a quasi-Gaussian stochastic variable: this allows for the calculation of the scintillation index S_4 in a closed form as a function of the variance of ΔN_e and other few quantities characterizing the irregularity slab and the radio link geometry (see Eq. 5 later): this is going to the inset V of Fig. 3, i.e., working in the *weak scattering regime* described in Rino (1979a).

A few more choices, in particular that of the ionospheric background already made in Eq. (2), and on the irregularity forms (determining the factor F in Eq. 5, see later), lead to the WAM model (the final inset, VI).

The statistics of the signal power at the receiver is described by the power scintillation index $S_4 \overset{\text{def}}{=} \sqrt{\frac{\langle (W - \langle W \rangle)^2 \rangle}{\langle W \rangle^2}}$: according to the theory described thoroughly in Rino (1979a), and applied in Wernik et al. (2007) and Alfonsi et al. (2017), an expression of S_4 can be given in terms of quantities describing the statistics of the medium. In particular, the relationship that WAM is based on reads as follows:

$$S_4 = r_e^2 \lambda^2 \sec\theta \, L Z^{\frac{p}{2}} C_{sr} \frac{\Gamma\left(1 - \frac{p}{4}\right)}{\sqrt{\pi} p \Gamma\left(\frac{p}{4} + \frac{1}{2}\right)} F: \quad (5)$$

This is the power scintillation index attributed to a monochromatic plane wave crossing the ionosphere. In Eq. (5) the properties of the received signal S_4 emerge *as a match between those of the*

electromagnetic wave and those of the irregularities: it contains properties of the wave, properties of the wave-medium relative geometry, and properties of the medium.

Besides r_e, which is the electron classical radius, the only wave property is the wavelength λ.

The quantities describing the relative geometry of the wave propagation with respect to the medium are the propagation zenith angle θ and the Fresnel zone parameter $Z = \frac{\lambda \sec\theta z_S z}{4\pi(z_S + z)}$, being z_s the satellite-receiver distance, and z the distance between the receiver and the phase screen.

The quantities describing the medium are: the thickness of the irregularity layer L, the 1D spectral index p of the ionospheric fluctuations, the turbulence strength parameter at the height of the phase screen C_{sr}, and the irregularity form factor F. These parameters are all constructed both from real in situ measurements and from suitable theoretical assumptions.

As anticipated, one starts from time series $\Delta N_e(t) \overset{\text{def}}{=} \Delta N_e\left(\vec{x}_S(t), t\right)$ of electron density fluctuations collected along the LEO satellite DE2 trajectory $\vec{x}_S(t)$; these broadband time series are elaborated, and their power spectrum $P_{\Delta N_e}(f)$ is calculated every 8 s long data segment: the log-log slope of this power spectrum is the p invoked in Eq. (5), obtained by the least squares fit over 1–20 Hz (Wernik et al., 2007). From $P_{\Delta N_e}(f)$, one may arrive at the parameter C_{sr} in Eq. (5) from the local turbulence strength C_s as follows: first of all, the effective satellite-medium velocity V_{eff} is needed, a parameter depending on the real velocity of the satellite $\vec{v}_S$ and on the geometry of the ionospheric irregularities; see again Wernik et al. (2007) and the references therein. Then, one assumes a certain relationship between the "outer scale" ℓ_0 of the ionospheric turbulence (Rino, 1979a; Yeh and Liu, 1982), this effective velocity V_{eff}, and the frequency f at which $P_{\Delta N_e}(f) \propto f^{-p}$; all in all, one may state:

$$\left(\frac{\ell_0 f}{V_{\text{eff}}}\right)^2 \gg 1 \Rightarrow P_{\Delta N_e}(f) = \frac{\Gamma\left(\frac{p}{2}\right) C_s V_{\text{eff}}^{p-1}}{(2\pi)^{p+2}\Gamma\left(\frac{p}{2}+1\right)} f^{-p}. \tag{6}$$

This equation states that, provided we consider a frequency so that $\ell_0^2 f^2 \gg V_{\text{eff}}^2$ (i.e., for high enough frequencies f, a large enough outer scale of the turbulence ℓ_0, or small enough effective velocities V_{eff}), the power spectrum of $\Delta N_e(t)$ follows *precisely* a power law, and for each frequency a local turbulence strength can be calculated as follows:

$$C_s(f) = \frac{(2\pi)^{p+2}\Gamma\left(\frac{p}{2}+1\right)}{\Gamma\left(\frac{p}{2}\right) V_{\text{eff}}^{p-1}} P_{\Delta N_e}(f) f^p. \tag{7}$$

In the WAM model, an f-independent C_s is obtained as the average of the values predicted by Eq. (7) at the three different frequencies $f = 1\,\text{Hz}$, $f = 6\,\text{Hz}$, and $f = 20\,\text{Hz}$:

$$\overline{C}_s = \frac{C_s(1\text{Hz}) + C_s(6\text{Hz}) + C_s(20\text{Hz})}{3}.$$

Then, this $\overline{C}_s$ still gives a measure of how strong the ionospheric turbulence is along the satellite trajectory. However, we need this information in the irregularity slab, where one locates the phase screen: this is the renormalized turbulence strength C_{sr} appearing in Eq. (5), easily obtained from Eq. (7) as follows:

$$C_{sr} = \frac{N_{\text{max}}^2}{N_S^2} \overline{C}_s. \tag{8}$$

The last element in equation (5) needing some further explanation is the *form factor F*, taking into account the shape of the ionospheric irregularities, and of how they are oriented with respect to the geomagnetic field. This is given in the formulae in Rino (1979a), considering the shape of the irregularities determined by the following assortment of the parameters a, b, and c defined in the same paper:

$$\begin{cases} \Lambda \leq 65° \Rightarrow a:b:c = 10:10:1, \text{Sheet}-\text{like}, \\ 65° < \Lambda \leq 70° \Rightarrow a:b:c = 10:10:1, \text{Rod}-\text{like,long}, \\ \Lambda > 70° \Rightarrow a:b:c = 3:1:1, \text{Rod}-\text{like,short} \end{cases} \tag{9}$$

In Eq. (9) the Λ is magnetic latitude of the link-SoI pierce point. A cartoon of the shapes suggested in Eq. (9) quantitatively is reported in Fig. 6.

One can then calculate F depending on the type of irregularity that populates the ionospheric medium at the link-SoI pierce point, reading $F(a,b,c)$ to be placed in Eq. (5) from Rino (1979a) according to the classification (Eq. 9).

2.2 Predictions of the WAM model

The capability of WAM in modeling the climatological features of the ionospheric scintillation has been validated before at high latitude (Wernik et al., 2007); the model has subsequently been extended to the low latitudes and validated in the Brazilian region (Alfonsi et al., 2017).

We note here that the model is based on the in situ plasma density data taken by the DE2

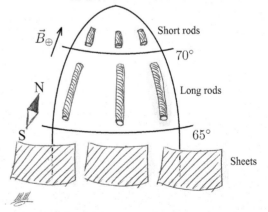

FIG. 6 The irregularity shapes adopted in WAM, according to Eq. (9) and determining the form factor F in Eq. (5) (see text. Picture from Materassi et al., 2008).

mission between August 1981 and February 1983, under solar maximum conditions.

The main outcome at the WAM modeling at high latitude is the maps of the so-called overhead scintillation index S_4, provided in terms of magnetic local time (MLT) and invariant geomagnetic latitude (ILAT). The overhead S_4 is the value the scintillation index would have if the ray-path of the trans-ionospheric were exactly at the zero zenith angle as seen from the receiver. Thus, a map of overhead S_4

provides the modeled scintillation independently on the geometry of the transmitter(s)-receiver(s) system. Fig. 7 reports the aforementioned maps, sorted in quiet (panels A and C) and disturbed geomagnetic (panels B and D) activity conditions according to the K_p index ($K_p \leq 3$ and $K_p > 3$) and for winter (panels A and B) and summer (panels C and D) seasons. Maps are for a signal wavelength of 1.2 GHz, simulating the behavior of L-band signals like those emitted by GNSS.

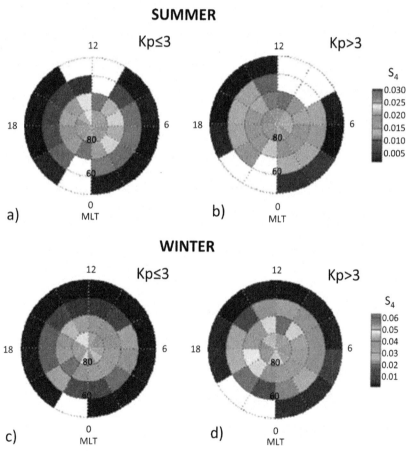

FIG. 7 Maps of the mean amplitude scintillation index S_4 at 1.2 GHz under overhead propagation geometry for the irregularity anisotropy model described in the text. The map coordinate systems are invariant latitude and magnetic local time. Maps are sorted according to different geomagnetic disturbance levels and seasons. The four subplots indicated with labels (A), (B), (C), and (D) refer to the different range of K_p considered: (A) low K_p in summer months, (B) high K_p in summer months; (C) low K_p in winter months, (D) high K_p in winter months.

In general, scintillation is found to be weaker in summer than in winter, and it intensifies according to the geomagnetic disturbance level. In correspondence with the increased geomagnetic activity, the found scintillation tends to expand equatorward, following the displacement of the auroral oval under disturbed conditions of the geospace. This is also in agreement with what was found climatologically by Spogli et al. (2009) in the European high-latitude sector in the period October to December 2003, covering extreme conditions of the geomagnetic activity. During summer and quiet magnetic times, the scintillation is larger around the magnetic noon and midnight, and in correspondence with the expected position of the cusp and of the boundaries of the auroral oval. During winter and at low geomagnetic disturbance level, the bulk of scintillation is found within the polar cap. This

models well the effect of the polar cap patches in hosting scintillation-effective irregularities (see, e.g., Spogli et al., 2010; Alfonsi et al., 2011) and references therein). In winter, the expansion of the scintillation zone is less defined in the midnight sector, and the polar cap scintillation intensity weakens.

To focus better on the dependence of the amplitude scintillation level on the geomagnetic disturbance level, Fig. 8 reports the dependence of the average modeled S_4 as a function of the K_p index. The plots are sorted according to the season (winter/summer) and to the sector of invariant magnetic latitude (ILAT lower/>70 degree). As already highlighted for Fig. 7, the larger values of S_4 are found in winter and for ILAT>70 degree, although they tend to decrease as the K_p becomes larger. This decrease is found also in the same ILAT sector for

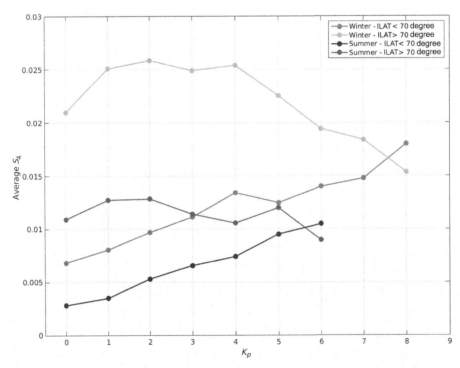

FIG. 8 Average S_4 as a function of K_p in summer and winter and by considering two sectors of invariant magnetic latitude (ILAT): <70 and >70 degree.

summer time. This is likely due to the displacement toward the equator of the boundaries of the auroral oval during disturbed times. The latter leads to an enhanced probability of finding scintillation-effective ionospheric irregularities at lower magnetic latitudes. The values of S_4 tend to increase as a function of K_p for ILAT < 70 degree for both summer and winter times.

The WAM model has been also compared with scintillation found on UHF data by Basu et al. (1985), in which scintillation at 250 MHz observed at Thule, Greenland (76.5°N, 68.7°W, 86° ILAT), is studied in terms of occurrence of fading levels. The result of such a comparison is provided in Fig. 9. The left-hand panel of this figure reports the modeled occurrence of S_4 above 0.12 (blue), 0.22 (red), 0.32 (black), and 0.40 (green) as a function of the month from August 1981 to February 1983, by considering 400 MHz signals. Such thresholds for the S_4 occurrence calculation at 400 MHz correspond to the fading excess threshold of 5, 10, 15, and 20 dB reported in Fig. 7 of Basu et al. (1985) at

250 MHz. The right-hand panel of Fig. 9 is inspired by Fig. 7 of Basu et al. (1985): occurrence of signal fading exceeding 5 dB (blue), 10 dB (red), 15 dB (black), and 20 dB (green) is shown as a function of the month from August 1981 to December 1982.

The capability of the model to reproduce the seasonal variation found with real data, especially when considering the intensity fading in excess of 10 dB, is shown in Fig. 9. In the specific, both panels of this figure show maxima in fall 1981 and 1982 and spring 1982, and a deep minimum in summer 1982. This minimum is likely due to the presence of a highly conducting ionospheric E layer that results in a reduced probability of formation of small-scale ionospheric irregularities.

The WAM model for low latitude has been tuned to work in the region for which geographic latitude is in the range [−40°N; 0°N] and longitude in [−70°E; −30°E], to focus on the southern crest of the Equatorial Ionospheric Anomaly (EIA) in Latin America. One of the main differences with respect to the high-latitude

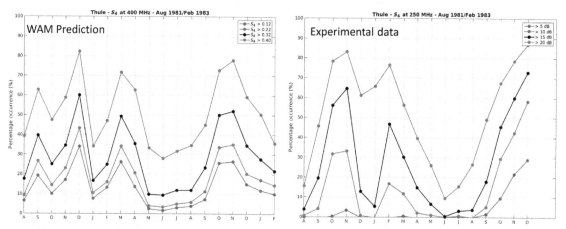

FIG. 9 (*Left*) Occurrence of S_4 above 0.12 (*blue*), 0.22 (*red*), 0.32 (*black*), and 0.40 (*green*) as a function of the month from August 1981 to February 1983 modeled over Thule, Greenland (76.5°N, 68.7°W, 86° ILAT), by considering 400 MHz signals (for *black and white figure* in the print versions: *blue, red, black*, and *green* plots are found from bottom to top). (*Right*) Occurrence of signal fading exceeding 5 dB (*blue*), 10 dB (*red*), 15 dB (*black*), and 20 dB (*green*) as a function of the month from August 1981 to December 1982, as reported by Basu et al. (1985) (for *black and white figure* in the print versions: *blue, red, black*, and *green* plots are found from bottom to top). The levels of fading in the right plot correspond to the threshold for the occurrence in the left plot.

formulation of the model is the use of the NeQuick (Coïsson et al., 2008) model instead of IRI95 model to rescale the C_s parameter at the phase screen altitude. In addition, the low-latitude scintillation is characterized by reaching often the strong regime ($S_4 > 0.7$). This poses a serious limitation in the use of Rino's formulas for the weak scattering approximation. To overcome this limitation, the S_4 in the strong scattering regime, S_{4s}, is calculated by correcting the corresponding weak regime index S_{4w} according to the equation provided by Secan et al. (1995):

$$S_{4s}^2 \approx 1 - e^{-S_{4w}^2}.$$

Some of the model prediction capabilities are presented in Fig. 10, in which the S_4 occurrence at 1.2 GHz above the 0.1 threshold is reported as a function of the local time and day of the year in the geographic sector reported earlier. The 0.1 threshold identifies the weak to strong scintillation regimes. In the figure, the left plot is for quiet ($K_p \leq 3$) geomagnetic conditions and the right plot is for disturbed ($K_p > 3$) geomagnetic conditions. The black line in the two plots

represents the solar terminator. The WAM model is able to draw the scintillation patterns due to small-scale irregularities embedded in the plasma bubbles, whose probability of formation maximizes immediately after the sunset hours (Balan et al., 2018). In addition, the seasonality of the scintillation occurrence is well drawn, having a peak in the equinoctial months and a minimum in the solstice months. Moreover, the right panel shows how the scintillation occurrence is severely modified when disturbed geomagnetic conditions exist. The prompt penetration electric fields (PPEF, Wei et al., 2015) and the disturbance dynamo electric fields (DDEF, Scherliess and Fejer, 1997) processes likely following a geomagnetic storms change the electrodynamics of the low-latitude ionosphere, altering the regular formation of the small-scale irregularities embedded in the post-sunset equatorial plasma bubbles. This results in suppression or intensification and in retard or anticipation of the scintillation occurrence (see, e.g., Spogli et al., 2016), well drawn by the WAM model.

The low-latitude formulation of WAM has been also tested against GNSS data collected by

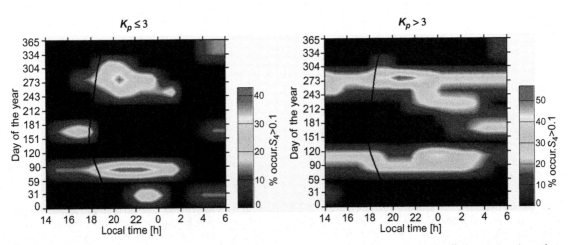

FIG. 10 Occurrence of amplitude scintillation index S_4 at 1.2 GHz above 0.1 (weak to strong scintillation regimes) as a function of the local time and day of the year over Brazil. The left plot is for quiet geomagnetic conditions ($K_p \leq 3$) and the right plot is for disturbed ($K_p > 3$) geomagnetic conditions. The black line indicates the solar terminator.

a SCINTMON (SCINTillation MONitor, http://gps.ece.cornell.edu/realtime.php) receiver in Presidente Prudente in Brazil (22.12°S, 51.41°W; geomagnetic coordinates: 12.62°S, 19.73°E) and managed by Instituto Nacional de Pesquisas Espaciais (INPE). Data refer to the period from January to October 2009 and to night-time hours (from 21:00 UT to 09:00 UT, corresponding to 18:00 LT to 06:00 LT). In addition, all-day-long data from a PolaRxS receiver (Bougard et al., 2011) in Presidente Prudente have been considered for the same months.

The result of the comparison is reported in Fig. 11, which shows the maps in latitude versus universal time (LT=UT-3) of mean S_4 during the equinoxes predicted by WAM (top) and obtained by the S_4 measurements on GPS L1 frequency at Presidente Prudente by PolaRxS (middle) and by SCINTMON (bottom) receivers.

The comparison between modeled and measured S_4 presents a qualitatively good agreement, as in all three panels the expected post-sunset increase of S_4 (LT=UT-3) appears, clearly located in the southern crest of the EIA. A lack of measurements from the SCINTMON receiver (bottom panel) is due to the SCINTMON settings, as already mentioned, and limits the comparison between 10:00 and 20:00 UT.

Found discrepancies are likely to be due to many factors, which include the following:

- Different solar phase conditions of the DE2 data are used for WAM formulation (1981–1983, solar maximum) and PolaRxS/SCITMON data (2009, solar minimum).
- The quantity of DE2 data in the Brazilian region is not sufficient to get a WAM modeling able to reproduce the complexity of the low-latitude electrodynamics triggering L-band scintillation.
- The assumption of isotropic irregularities may not be suitable for all cases.
- Rino's theory for weak scattering regime has significant limitations to reproduce low-latitude ionospheric irregularities.

3 From semi-empirical to empirical climatology

As scintillation are caused by random variations in the refractive index of the upper atmosphere, their observation provides valuable information about the structuring of the electron density and about the plasma dynamics. The sizes of the irregular electron density structures, termed as "ionospheric irregularities" or simply "irregularities," can be inferred from the amplitude scintillations. This is because the amplitude response of an electromagnetic wave to ionospheric structuring is determined by the Fresnel filtering mechanism, which strongly suppresses the contribution due to irregularities larger than about the radius of the first Fresnel zone. This quantity, d_F, is given by:

$$d_F = \sqrt{\lambda \left(z - \frac{L}{2} \right)},$$

where λ is the frequency of the electromagnetic wave, and z and L are the position and the thickness of the irregularity slab, respectively (Yeh and Liu, 1982). Therefore, by knowing λ and by assuming the position z and the thickness L, it is possible to derive d_F and, consequently, to learn the upper limit of the size of the irregularities producing amplitude scintillations. Assuming, realistically, a slab thickness negligible with respect to its position and posing the irregularities z at 350 km (the average position of the F-layer), on the VHF/UHF frequency range d_F is of the order of kilometers, and on the L-band d_F is of the order of 100 m (Aarons et al., 1995 and references therein).

Until the 1990s, scintillation climatology was derived from sparse ground-based VHF-UHF campaigns or from semi-empirical modeling based on electron density in situ data (Aarons et al., 1995; Mendillo et al., 1992; Aarons, 1993; Basu et al., 1994; Kersley et al., 1995; Pryse et al., 1996). GPS opened a

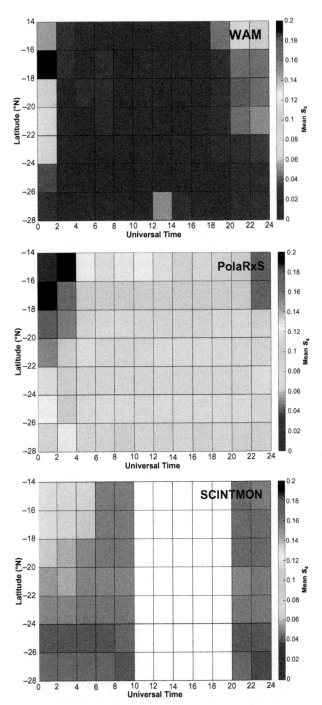

FIG. 11 Maps of mean S_4 during the equinoxes, based on predictions by WAM (*top*) and on measurements on GPS L1 frequency at Presidente Prudente by PolaRxS (*middle*) and by SCINTMON (*bottom*) receivers.

new era in the assessment of scintillations on a long-term basis (Aarons et al., 1995; Pi et al., 1997). The GPS constellation, visible globally and always available, was suddenly exploited for ionosphere monitoring and study (Mendillo, 2006). The GPS receivers, in fact, can provide indirect information on the ionospheric changes, as identified by TEC (Total Electron Content), from the delay induced by the ionosphere on the signals received on the ground. A further elaboration of TEC allows derivation of the TEC spatial gradients, thus identifying the presence of plasma irregularities that can produce scintillations. An example of TEC gradients climatology achieved over 1 year of GPS and GLONASS data is given by Cesaroni et al. (2015). This analysis made use of the GNSS (Global Navigation Satellite Systems) networks, nowadays widely used globally, to retrieve information usable to recognize irregularities triggering scintillations. Nevertheless, as the Fresnel filtering the size of TEC gradients and of irregularities producing scintillations at L-band does not match. In the scintillation production mechanisms, the amplitude scintillation is biased by irregularities' probing size: amplitude scintillations at L-band are sensitive to irregularities of 100-m scale sizes. Alternatively, TEC gradients can identify irregularities of scales that vary by an order of magnitude (a few kilometers), due to the pierce point velocity variation with the zenith angle and its dependence on the spectral index of phase. For instance, as reported by Alfonsi et al. (2011), if the TEC gradients are inferred from the Rate of TEC changes (ROT) computed over 1 min, calculating the difference between the relative (slant) TEC values provided by the receiver, at high latitudes the irregularities scale length span from a few to tens of kilometers (Basu et al., 1999). In addition, the ROTI (Rate of TEC Index) based on the standard deviation of ROT, generally calculated over 5 min, is considered to identify smaller-scale irregularities

(Pi et al., 1997). Nevertheless, the use of TEC-derived parameters can support the detection of the irregularities producing scintillation, but it cannot provide the full scenario. In this framework, a real advancement in the monitoring and, then, in the study of the scintillation was provided by the analysis of the data acquired by new kinds of receivers able to track the GNSS constellation by sampling the received signals at a high frequency rate (typically from 50 to 100 Hz). The pioneers were the GSV 4004 (Van Dierendonck et al., 1993, 1996) and the SCINTMON (http://www.gps.ece.cornell.edu/realtime.php#scintmon) receivers. Thanks to internal oscillators that prevent the pollution of the received signals with electronic noise, these receivers provide intensity, I, and phase high-rate samplings, generally indicated as raw data. Firmware embedded in the receiver computes the scintillation indices S_4 and σ_Φ, the TEC and the ROT. Such devices are termed TEC and scintillation monitors.

As introduced earlier in the chapter, S_4 is the amplitude scintillation index defined as the normalized variance of the signal intensity I, typically computed over 60 s, and σ_Φ is the phase scintillation index defined as the standard deviation of the detrended carrier phase, typically computed over 60 s on 50 Hz measurements at L1 frequency. The detrending is usually made adopting a cut-off frequency of 0.1 Hz (Van Dierendonck et al., 1993). Nevertheless, some recent papers support the definition of scintillations as being due to the sole diffractive (stochastic) effects, claiming that the refractive (deterministic) effects can be mitigated through an adaptive choice of the cut-off frequency. Such a choice would lead to a reduction of the "phase-without-amplitude scintillation occurrence" (Mushini et al., 2012). Some promising attempts to find the optimal choice have been recently made by Wang et al. (2018). The challenge now is to catch the "best" cut-off frequency to filter long-term (years) data series: the authors of

De Franceschi et al. (2019) have demonstrated that this is not an easy task.

The availability of the scintillation indices on a long-term basis, usually years, ideally over one or more solar cycle(s), enables the description of the scintillation climatology. The coverage and the duration of the climatology depend on the networks' lifetime and location. Several authors exploited this possibility; among the pioneers, we cite (Spogli et al., 2009; Prikryl et al., 2011).

3.1 The ground-based scintillation climatology

The Ground-Based Scintillation Climatology (GBSC) is a method proposed by Spogli et al. (2009) to describe the ionospheric scintillations and TEC on a regional or global scale on a long-term basis. The GBSC is input by TEC and scintillation indices provided by networks of Ionospheric Scintillation and TEC Monitor Receivers, and releases maps of occurrence of the scintillation indices over a given threshold and maps of mean and rms values of scintillation indices, TEC, and TEC-derived parameters. The representation of TEC and scintillation climatology can help to infer when and where the ionosphere becomes irregular, producing scintillations. By-products of the GBSC are maps of mean and rms values of the signal to noise ratio, C/N_0, on the different carrier frequencies (L1 and L2 for GPS, for instance) and Code Carrier standard deviation. These quantities may help to characterize the site with respect to the multipath sources supporting the assessment of the quality of the data provided by each station. Fig. 12 reports a sketch of the GBSC method.

The GBSC can release information on single sites or can describe the ionospheric behavior on selected regions and time intervals, merging the data coming from several stations. The threshold chosen to distinguish between the weak to moderate and the moderate to strong scintillation

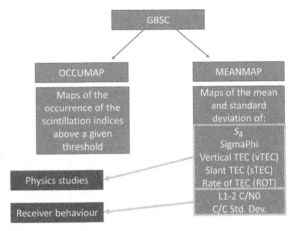

FIG. 12   A sketch of the GBSC method.

regimes are 0.1 and 0.25 for S_4 and 0.1 (rad) and 0.25 (rad) for σ_Φ. The mapping is achieved by computing the percentage of occurrence for each bin identified by the chosen coordinates (geomagnetic/geographic; UT/LT/MLT).

The threshold chosen to distinguish between the weak to moderate and the moderate to strong scintillation regimes are 0.1 and 0.25 for S_4 and 0.1 and 0.25 rad for σ_Φ. The adoption of a given threshold depends on the need of a good compromise between the large statistics required in each bin and a meaningful fragmentation of the map. Data acquired at elevation angles lower than a certain value (typically 30 degree) are filtered out to reduce the impact of the multipath on the assessment of the level of scintillation. To minimize the impact of the observation geometry on the evaluation of the scintillation indices, the indices considered in the GBSC are projected to the vertical, as follows:

$$\sigma_\varphi^{\mathrm{vert}} = \sigma_\varphi^{\mathrm{slant}}\left(\frac{1}{F(\alpha_{\mathrm{elev}})}\right)^a,$$

$$S_4^{\mathrm{vert}} = S_4^{\mathrm{slant}}\left(\frac{1}{F(\alpha_{\mathrm{elev}})}\right)^b,$$

$$F(\alpha_{\text{elev}}) = \frac{1}{\sqrt{1 - \left(\dfrac{R_e \cos \alpha_{\text{elev}}}{R_e + H_{IPP}}\right)^2}},$$

where α_{elev} is the elevation angle S_4^{vert} and σ_Φ^{vert}, R_e is the Earth radius, and H_{IPP} is the height of the ionospheric pierce point, generally fixed at 350 km, the average position of the electron density maximum (F layer). The exponents a and b are equal to:

$$a = 0.5,$$

$$b = \frac{p+1}{4}$$

in which p is the spectral slope of detrended phase, which is usually estimated by the firmware of last generation receivers. If p is not measured, some average values can be assumed. According to Rino (1979a) and as described in Spogli et al. (2009), the exponent p can be reasonably chosen to be $p = 2.6$ at high latitude, from which $(p+1)/4 = 0.9$.

The use of the vertical index brings then some assumption, as the angular dependence of the scintillation indices in the earlier formulas is strictly valid only under the conditions of weak scattering and when single phase screen approximation is suitable. However, the nature of the GBSC approach is statistical and its aim is to identify climatological patterns of scintillation in terms of occurrence above selected thresholds. The risk of applying the weak formula to the moderate/strong regime is reduced in GBSC to a slight underestimation of the occurrence. The strength of the use of vertical indices is a scintillation picture that is almost independent of the locations of the receivers from which it is drawn.

The GBSC outputs can be sorted according to geomagnetic, Interplanetary Magnetic Field (IMF) conditions and any other parameters characterizing the geospace environment. A schematic summary of the GBSC is given in Table 1.

3.2 Examples of GBSC outputs

The GBSC outputs can be a powerful tool to interpret the physical conditions that make the ionosphere a plasma characterized by a highly uneven electron density distribution.

The maps in Fig. 13 describe the climatology of the occurrence of the phase and amplitude scintillation indices above 0.25 (rad), being the threshold to identify moderate to strong scintillation

TABLE 1   Summary of the GBSC main assumptions.

Quantity	Description	Typical assumption
Elevation angle	Reduce the impact of large values of the scintillation indices not related to ionospheric scintillation (e.g., multipath)	30
Vertical/slant	Scintillation indices can be projected to the vertical to minimize the impact of the geometry	Vertical
System of coordinates	These can be geographical coordinates, geomagnetic coordinates, universal and local time, magnetic local time, azimuth, and elevation	N/A
Geomagnetic conditions	A select of the geomagnetic behavior of each day based on K_p/Dst/AU/AL indices	Quiet/disturbed/all
IMF conditions	Selection of the IMF components and total field	B_x, B_y, $B_z < 0$ or > 0 Ranges of $|B_{\text{tot}}|$

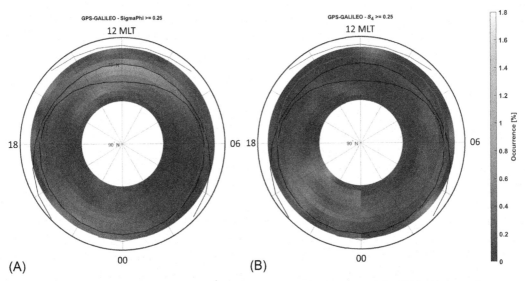

FIG. 13 Scintillation climatology maps over Ny Ålesund (Svalbard Islands, Norway) for 2016: (A) maps of σ_Φ occurrence 0.25 rad; (B) maps of occurrence above 0.25. Both thresholds identify moderate to strong scintillation regimes.

regimes. Phase scintillations are derived by detrending the signal with a cut-off frequency at 0.1 Hz. The climatology is based on the GPS e GALILEO L1 data acquired in 2016 at Ny Ålesund (Svalbard Islands, Norway). The field of view is peculiar because it allows one to observe the auroral, cusp, and polar cap regions of the ionosphere, depending on the MLT. The outputs are presented in the form of polar maps in MLT and Altitude Adjusted Corrected Geomagnetic Latitude (AACGMLat) (Baker and Wing, 1989), superimposing also the quiet and disturbed (IQ=0 and IQ=6, respectively) auroral oval boundaries position provided by the Feldstein, Holzworth, and Meng model (Feldstein, 1963; Holzworth and Meng, 1975).

The most severe phase scintillations are mainly confined in the auroral oval around noon, with an asymmetry favoring the morning sector, and in the pre-midnight polar cap. The most intense amplitude scintillations occur during the evening hours within the polar cap.

The maps in Fig. 14 describe the climatology of the occurrence of the amplitude (and phase) scintillation indices above 0.1 (rad) according

to the auroral indices AU and AL. Also in this case, the cut-off frequency to detrend the signal is 0.1 Hz. The climatology is based on the GPS L1 data acquired from September 2003 to December 2016 at Ny Ålesund (Svalbard Islands, Norway).

To highlight the effects of the currents systems flowing in the lower ionosphere over our scintillation climatology, we use the AU and AL indices. These have been employed extensively and widely since their introduction in space weather (Davis and Sugiura, 1966), being able to characterize the intensity of the eastward and westward auroral electrojects, respectively. Data are then sorted according to low and high levels of the AU and AL indices; the low level identifies when the absolute value of the indices is below 400 nT (Troshichev and Janzhura, 2009).

At first glance, the maps of amplitude and phase scintillation occurrence result very different. Fig. 14 shows the results obtained for the percentage occurrences of S_4 (>0.1) and σ_Φ (>0.1 rad), and for the mean ROT and its standard deviation according to the four possible permutations of the indices level (low AU/low AL, low AU/high AL, high AU/low AL, high AU/high AL).

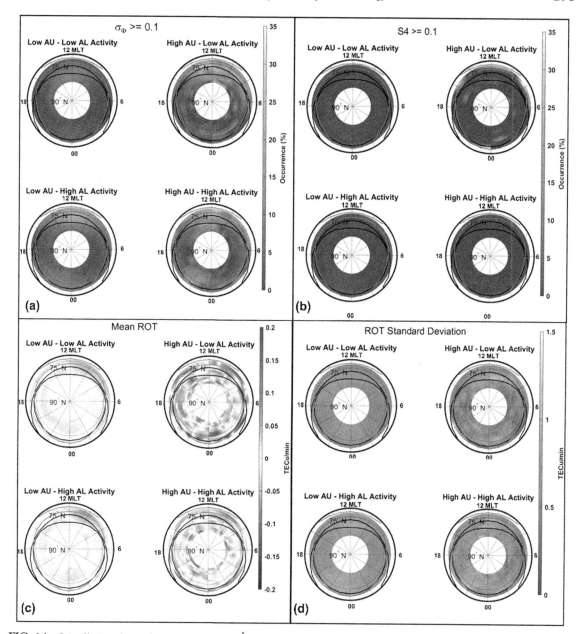

FIG. 14 Scintillation climatology maps over Ny Ålesund (Svalbard Islands, Norway) for 2016: (A) maps of σ_Φ scintillation and ROT climatology over Ny Ålesund (Svalbard Islands, Norway) for 2003–2016 sorted according to AU/AL levels: maps of occurrence σ_Φ (panel A) and (panel B) above the weak scintillation threshold; mean (panel C) and standard deviation (panel D) values of ROT.

Fig. 14A indicates that for low auroral activity (low AU/low AL), the only peak (even if weak) in the σ_Φ occurrence is recorded around noon at auroral and cap latitudes. This slight enhancement is not visible in the amplitude scintillation occurrence (Fig. 14B), possibly meaning that the irregularities present in noon are producing only weak refractive effects identifying changes in the ionospheric plasma velocity and/or the presence of irregularities with scale sizes greater than the first Fresnel radius (at L-band hundreds of meters). When AU is high and AL is low (prevalence of the eastward electroject), the bulk of scintillations is around noon, with an asymmetry in the morning, mainly concentrated at auroral latitudes (Fig. 14A and B). The GBSC reveals also a weaker increase of scintillations during the night in the polar cap and at the poleward edge of the auroral oval. The opposite condition (low AU/high AL), which is indicative of a prevalence of the westward electroject, shows an increase of refractive effects around noon and in the pre-midnight sector within the auroral oval. When AU and AL are both high, the areas mainly affected by a higher occurrence of refractive effects are: (i) auroral and cap regions around noon and (ii) the auroral sector at pre- and post-midnight.

Fig. 14C shows that positive ROT values are confined in the auroral and subauroral regions for low AU (regardless AL) and maximize in the pre-noon hours. Under such conditions, weak negative TEC gradients are located within the cap. When AU is high, ROT shows quite a patchy distribution, especially within the cap: coupled with low/high AL, positive gradients appear at very high latitudes during afternoon and evening time, at almost all the time-sectors.

The ROT standard deviation shows quite a homogeneous distribution under low auroral activity (low AU, low AL) (Fig. 14D). When AL increases, ROT standard deviation peaks around noon in the quiet auroral oval and in the evening/pre-midnight hours in the polar cap. The patchy ROT behavior found for the combination high AU/low AL reflects on its standard deviation that is uneven in the pre-noon sector at auroral latitudes, and from dusk to dawn at auroral latitudes and at the polar cap. When both AU and AL are high, the peaks appear around noon at all the latitudes and in the evening/pre-midnights hours within the polar cap.

4 Critical review and perspectives

The modeling of radio wave scintillation is complicated by the interplay of two main aspects (Priyadarshi, 2015). First, the knowledge of the medium properties and particularly the distribution of small-scale irregularities is rather limited in both space and time; this aspect leads to assumptions around the specification of the medium in propagation models. Second, while low-to-moderate scintillation is well modeled on the assumption of a phase-changing screen along the propagation path, high scintillation is likely to be the result of scattering from several phase screens. The knowledge of the precise distribution of phase screens occurring along scintillating propagation paths is limited by the experimental resolution in both space and time.

The spatial distribution of ionospheric irregularities together with their temporal evolution determines the source term in the partial differential equation describing the propagation problem.

The problem with current experimental measurements of scintillation is that they offer an indication of the scale of irregularities on the assumption of a single phase screen occurring along the propagation path.

It is necessary to sample their distribution (in space and time) in order to refine the accuracy of any scintillation model. For example, it is reasonable to expect that the study of ionospheric irregularities and their effect on radio wave propagation will improve in future with the increase in spatial and temporal resolution of experimental techniques (for example, likely to be offered by instruments such as EISCAT-3D).

The importance of the refinement of the accuracy of scintillation models lies in the increasing need of scintillation forecast from end users: for example, those concerned with real-time safety-critical applications.

In order to meet this demand, future models will have to migrate from a climatologic approach to a weather approach. Hence, the challenge ahead is that of refining the resolution in the knowledge of the spatial distribution of ionospheric irregularities and their temporal evolution. This represents a critical aspect in any weather-like predictive model as the spatial spectrum of the irregularities (i.e., from large to small scales) together with its temporal evolution would need to be determined. The challenge in the experimental determination of the irregularities spectrum currently lies in too-sparse measurements being available. In this regard, ionospheric prediction models appear still very far from weather prediction models.

References

Aarons, J., 1993. The longitudinal morphology of equatorial F-layer irregularities relevant to their occurrence. Radio Sci. 63, 209.

Aarons, J., Kersley, L., Rodger, A.S., 1995. The sunspot cycle and auroral F-layer irregularities. Radio Sci. 30, 631.

Alfonsi, L., Spogli, L., De Franceschi, G., Romano, V., Aquino, M., Dodson, A., Mitchell, C.N., 2011. Bipolar climatology of GPS ionospheric scintillation at solar minimum. Radio Sci. 46(3).

Alfonsi, L., Wernik, A.W., Materassi, M., Spogli, L., 2017. Modelling ionospheric scintillation under the crest of the equatorial anomaly. Adv. Space Res. 60 (8), 1698–1707.

Baker, K.B., Wing, S., 1989. A new magnetic coordinate system for conjugate studies at high latitudes. J. Geophys. Res. Space Physics 94 (A7), 9139–9143.

Balan, N., Liu, L., Le, H., 2018. A brief review of equatorial ionization anomaly and ionospheric irregularities. Earth Planet. Phys. 2 (4), 257–275.

Basu, S., Basu, S., Khan, B.K., 1976. Model of equatorial scintillations from in situ measurements. Radio Sci. 11, 821–832.

Basu, S., Basu, S., Hanson, W.B., 1981. The Role of In Situ Measurements in Scintillation Modelling, Pap. 4A-8. Naval Research Laboratory, Washington, DC.

Basu, S., Basu, S., MacKenzie, E., Whitney, H.E., 1985. Morphology of phase and intensity scintillations in the auroral oval and polar cap. Radio Sci. 20 (3), 347–356.

Basu, S., Basu, S., Chaturvedi, P.K., Bryant, C.M., 1994. Irregularity structures in the cusp cleft and polar-cap regions. Radio Sci. 29, 195.

Basu, S., Groves, K.M., Quinn, J.M., Doherty, P., 1999. A comparison of TEC fluctuations and scintillations at Ascension Island. J. Atmos. Sol.Terr. Phys. 61 (16), 1219–1226. https://doi.org/10.1016/S1364-6826(99)00052-8.

Beniguel, Y., Romano, V., Alfonsi, L., Aquino, M., Bourdillon, A., Cannon, P., De Franceschi, G., Dubey, S., Forte, B., Gherm, V., Jakowski, N., Materassi, M., Noack, T., Pożoga, M., Rogers, N., Spalla, P., Strangeways, H.J., Warrington, E.M., Wernik, A., Wilken, V., Zernov, N., 2009. Ionospheric scintillation monitoring and modelling. Ann. Geophys. 52, 391.

Bhattacharyya, A., Yeh, K.C., Franke, S.J., 1992. Deducting turbulence parameters from transionospheric scintillation measurements. Space Sci. Rev. 61, 335–386.

Bilitza, D., 1997. International reference ionosphere—status 1995/96. Adv. Space Res. 20 (9), 1751–1754.

Bougard, B., Sleewaegen, J.-M., Spogli, L., Veettil Vadakke, S., Galera Monico, J.F., 2011. CIGALA: challenging the solar maximum in Brazil with PolaRxS. In: Proceedings of the 24th International Technical Meeting of the Satellite Division of the Institute of Navigation (ION GNSS 2011), Portland, OR, pp. 2572–2579.

Cesaroni, C., Spogli, L., Alfonsi, L., De Franceschi, G., Ciraolo, L., Monico, J.F.G., … Bougard, B., 2015. L-band scintillations and calibrated total electron content gradients over Brazil during the last solar maximum. J. Space Weather Space Clim. 5, A36.

Coïsson, P., Nava, B., Radicella, S.M., Oladipo, O.A., Adeniyi, J.O., Gopi Krishna, S., Rama Rao, P.V.S., Ravindran, S., 2008. NeQuick bottomside analysis at low latitudes. J. Atmos. Sol. Terr. Phys. 70, 1911–1918.

Davies, K., 1990. Ionospheric Radio. Peter Peregrinus Ltd.

Davis, T.N., Sugiura, M., 1966. Auroral electrojet activity index AE and its universal time variations. J. Geophys. Res. 71 (3), 785–801. https://doi.org/10.1029/JZ071i003p00785.

De Franceschi, G., Spogli, L., Alfonsi, L., Romano, V., Cesaroni, C., Hunstad, I., 2019. The ionospheric irregularities climatology over Svalbard from solar cycle 23. Sci. Rep. https://doi.org/10.1038/s41598-019-44829-5.

Feldstein, Y.I., 1963. On morphology and auroral and magnetic disturbances at high latitudes. Geomagn. Aeron. 3, 138.

Grzesiak, M., Wernik, A.W., 2009. Dispersion analysis of spaced antenna scintillation measurement. Ann. Geophys. 27, 2843–2849. www.ann-geophys.net/27/2843/2009/.

Grzesiak, M., Wernik, A.W., 2012. Ionospheric drifts estimated using GPS scintillation data during magnetic storm on 5–6th of April 2010. In: Contribution Presented at IGS Workshop 2012, Olsztyn (Poland). http://www.igs.org/assets/pdf/Poland%202012%20-%20P03%20Grzesiak%20PO74.pdf.

Hanson, W.B., Heelis, R.A., Power, R.A., Lippincott, C.R., Zuccaro, D.R., Holt, B.J., Harmon, L.H., Sanatani, S., 1981. The retarding potential analyzer for dynamics explorer-B. Space Sci. Instrum. 5, 503–510.

Holzworth, R.H., Meng, C.-I., 1975. Mathematical representation of the auroral oval. Geophys. Res. Lett. 2, 377–380. https://doi.org/10.1029/GL002i009p00377.

Hysell, D.L., 2019. From Instabilities to Irregularities. Chapter [Hysell] of this book.

Kersley, L., Russel, C.D., Rice, D.L., 1995. Phase scintillation and irregularities in the northern polar ionosphere. Radio Sci. 30, 619.

Knepp, D.L., 2019. Scintillation Theory. Chapter [Knepp] of this book.

Lapenta, G., 2019. Complex Dynamics of the Sun-Earth Interaction. Chapter [Lapenta] of this book.

Materassi, M., 2019. The Complex Ionosphere. Chapter [CompIono] of this book.

Materassi, M., Mitchell, C.N., 2007. Wavelet analysis of GPS amplitude scintillation: a case study. Radio Sci. 42(1).

Materassi, M., Wernik, A.W., Alfonsi, L., 2008. The WAM scintillation model: overview and application to ionospheric studies. In: Talk given at the COST Workshop "Ionospheric Scintillation: Scientific Aspects, Space Weather Application and Services", Nottingham, UK, 20–22 February.

Mendillo, M., 2006. Storms in the ionosphere: patterns and processes for total electron content. Rev. Geophys. 44. https://doi.org/10.1029/2005RG000193.

Mendillo, M., Baumgardner, J., Pi, X., Sultan, P., 1992. Onset conditions for equatorial spread F. J. Geophys. Res. 97.

Mushini, S.C., Jayachandran, P.T., Langley, R.B., MacDougall, J.W., Pokhotelov, D., 2012. Improved amplitude-and phase-scintillation indices derived from wavelet detrended high-latitude GPS data. GPS Solutions 16 (3), 363–373.

Pi, X., Mannucci, A.J., Lindqwister, U.J., Ho, C.M., 1997. Monitoring of global ionospheric irregularities using the worldwide GPS network. Geophys. Res. Lett. 24 (18), 2283–2286.

Prikryl, P., Jayachandran, P.T., Mushini, S.C., Chadwick, R., 2011. Climatology of GPS phase scintillation and HF radar backscatter for the high-latitude ionosphere under solar minimum conditions. Ann. Geophys. 29(2).

Priyadarshi, S., 2015. A review of ionospheric scintillation models. Surv. Geophys. 36, 295. https://doi.org/10.1007/s10712-015-9319-1.

Pryse, S.E., Kersley, L., Walker, I.K., 1996. Blobs and irregularities in the auroral ionosphere. J. Atmos. Terr. Phys. 58, 205.

Rino, C.L., 1979a. A power law phase screen model for ionospheric scintillation, 1. Weak scattering. Radio Sci. 14, 1135–1145.

Rino, C.L., 1979b. A power law phase screen model for ionospheric scintillation: 2. Strong scatter. Radio Sci. 14, 1147–1155.

Scherliess, L., Fejer, B.G., 1997. Storm time dependence of equatorial disturbance dynamo zonal electric fields. J. Geophys. Res. Space Physics 102 (A11), 24037–24046.

Secan, J.A., Bussey, R.M., Fremouw, E.J., Basu, S., 1995. An improved model of equatorial scintillation. Radio Sci. 30, 607–617.

Singh, A.K., Narayan, D., Singh, R.P., 1997. Weak scattering theory for ionospheric scintillation. Il Nuovo Cimento. 20(4).

Spogli, L., Alfonsi, L., De Franceschi, G., Romano, V., Aquino, M.H.O., Dodson, A., 2009. Climatology of GPS ionospheric scintillations over high and mid-latitude European regions. Ann. Geophys. 27, 3429–3437. https://doi.org/10.5194/angeo-27-3429-2009.

Spogli, L., Alfonsi, L., De Franceschi, G., Romano, V., Aquino, M.H.O., Dodson, A., 2010. Climatology of GNSS ionospheric scintillations at high and mid latitudes under different solar activity conditions. Il Nuovo Cimento B. https://doi.org/10.1393/ncb/i2010-10857-7.

Spogli, L., Cesaroni, C., Di Mauro, D., Pezzopane, M., Alfonsi, L., Musicò, E., …Linty, N., 2016. Formation of ionospheric irregularities over Southeast Asia during the 2015 St. Patrick's day storm. J. Geophys. Res. Space Physics 121(12).

Tatarskii, V.I., 1971. The Effects of the Turbulent Atmosphere on Wave Propagation. NOAA Rep. TT-68-50464, U.S. Department of Commerce, Springfield, VA.

Troshichev, O., Janzhura, A., 2009. Relationship between the PC and AL indices during repetitive bay-like magnetic disturbances in the auroral zone. J. Atmos. Sol. Terr. Phys. 71 (12), 1340–1352.

Van Dierendonck, A.J., Klobuchar, J.A., Hua, Q., 1993. Ionospheric scintillation monitoring using commercial single frequency C/A code receivers. In: Proc. ION ITM, Salt Lake City, UT.

Van Dierendonck, A.J., Hua, Q., Fenton, P., Klobuchar, J.A., 1996. Commercial ionospheric scintillation monitoring receiver development and test results. In: Proc. ION Annual Meeting, pp. 573–582.

Wang, Y., Zhang, Q.H., Jayachandran, P.T., Moen, J., Xing, Z.Y., Chadwick, R., ... Lester, M., 2018. Experimental evidence on the dependence of the standard GPS phase scintillation index on the ionospheric plasma drift around noon sector of the polar ionosphere. J. Geophys. Res. Space Physics 123 (3), 2370–2378.

Wei, Y., Zhao, B., Li, G., Wan, W., 2015. Electric field penetration into Earth's ionosphere: a brief review for 2000–2013. Sci. Bull. 60 (8), 748–761.

Wernik, A.W., Alfonsi, L., Materassi, M., 2007. Scintillation modelling using in situ data. Radio Sci. 42. https://doi.org/10.1029/2006RS003512.

Wernik, A.W., Pożoga, M., Grzesiak, M., Rokicki, A., Morawski, M., 2008. Monitoring ionospheric scintillations and TEC at the polish polar station on spitsbergen: instrumentation and preliminary results. Acta Geophys. 56 (4), 1129–1146. https://doi.org/10.2478/s11600-008-0060-8.

Yeh, K.C., Liu, C.H., 1982. Radio wave scintillations in the ionosphere. Proc. IEEE 70, 324–360.

Yeh, K.C., Wernik, A.W., 1993. On ionospheric scintillation. In: Tatarski, V.I. et al., (Ed.), Wave Propagation in Random Media (Scintillation). Int. Soc. for Opt. Eng., Bellingham, pp. 34–49.

19

Multiscale analysis of the turbulent ionospheric medium

Paola De Michelis, Roberta Tozzi

Istituto Nazionale di Geofisica e Vulcanologia, Roma, Italy

1 Introduction

The ionospheric environment is a complex system where dynamic phenomena, such as plasma instabilities and turbulence, generally occur as a consequence of the coupling processes between solar wind, the magnetosphere, and the ionosphere. The analysis of fluctuations of plasma density, electrostatic potential, and magnetic and electric fields have suggested the existence of a turbulent state in the ionospheric environment and shown that it is at the base of some crucial dynamical processes and is capable of influencing the cross-scale coupling in an essential way. Therefore, turbulence is a fundamental element in the ionospheric research, as also evidenced by the growing number of scientific papers that recognize the importance of turbulence phenomena in the ionospheric environment (Dyrud et al., 2008; Pécseli, 2016; Grach et al., 2016; De Michelis et al., 2017). For instance, in the last few years it has been suggested that the turbulent character of the ionospheric plasma density also enters into the formation and dynamics of ionospheric inhomogeneities and irregularities (see, e.g., Basu et al., 1984, 1988; Earle et al., 1989; Giannattasio et al., 2018), which essentially characterize the active equatorial and polar regions. These plasma density structures can be responsible for the delay, distortion, or total loss of electromagnetic signals while passing through the ionosphere. This means that turbulence is also able to compromise seriously the performance of the global navigation satellite system (GNSS), e.g., GPS and Galileo. These technologies rely on the propagation of signals through the ionosphere and are vital for precision positioning, navigation, and timing applications; these, in turn, are nowadays extensively used for commercial as well as for defense purposes. Thus, ionospheric turbulence indirectly plays a key role also in the framework of space weather. Indeed, the arrival of solar perturbations can deeply modify the plasma, the energetic particle distributions, and the electric and magnetic fields within the magnetosphere and ionosphere, thus paving the way for an increase of ionospheric turbulence and, as a consequence, for the formation of inhomogeneities and irregularities. Furthermore, progress in the knowledge of the origin of turbulence, as well as of its spatial

The Dynamical Ionosphere
https://doi.org/10.1016/B978-0-12-814782-5.00019-4

and temporal evolution, is certainly beneficial for more reliable and accurate mathematical modeling of the spatial and temporal characteristics of the ionosphere on both a global and a local scale.

The paper by Kintner and Seyler (1985), published in the mid-1980s, is one of the first review papers giving high-latitude ionospheric plasma turbulence considerable attention. The authors not only present their observations and theories on plasma turbulence in the high-latitude ionosphere but also propose possible connections between the different theories and models with the aim of describing the ionospheric turbulent dynamics, thus establishing the experimental evidence. In this paper, the authors analyze large collections of data relative to the electron density fluctuations recorded by rockets and satellites at different altitudes (from 400 to 8000 km) with the purpose of characterizing the turbulent state of the ionospheric environment on a wide range of spatial scales: from meters to a few thousand kilometers. In the mid-1980s, however, several other studies pointed out the capability of the ionospheric turbulence to generate coherent structures and plasma inhomogeneities (see, e.g., Basu et al., 1984, 1988; Earle et al., 1989), highlighting the important role played by turbulence in driving plasma irregularities. Indeed, it was at the end of 1980s that, analyzing in situ plasma density measurements below 1000 km of altitude, it was found that one-dimensional spectra of the density irregularities obey power-law behavior, suggesting the existence of turbulence phenomena. It was also immediately clear that the origin of the irregularities is actually a complex process, which can only partly be associated with turbulent processes that are not easy to discern. It is important to mention that in the time since the work of Kintner and Seyler (1985), great attention has been paid also to the analysis of turbulent properties of electric and magnetic fluctuations observed by ground stations,

rockets, and satellites at different altitudes and latitudes, in order to characterize the ionospheric turbulence as a whole and to try to understand its driving mechanisms. In this context, many interesting features of electric field fluctuations and their possible origin have been discovered. For example, using data from the low-altitude Dynamics Explorer 2 (DE 2) spacecraft during its several crossings of the auroral zone and polar cap under different interplanetary magnetic field orientations, it has been found (Golovchanskaya et al., 2006; Golovchanskaya and Kozelov, 2010) that the scaling features of electric and magnetic field fluctuations on small-scales (0.5–256 km) are basically the same in the different regions of the high-latitude ionosphere and are due to the development of intermittent turbulent processes (Basu et al., 1988).

Further analyses have permitted relation of the ionospheric turbulence directly to solar wind turbulence (e.g., Parkinson, 2006; Abel et al., 2009) in the polar cap and to the occurrence of shear flow instabilities developing in the regions of the large-scale field-aligned currents in the auroral zone (see, e.g., Tam et al., 2005; Kozelov et al., 2008). Thus, different theories and models have been developed where the observed turbulent behavior of the electric field fluctuations, or equivalently of the plasma velocity fluctuations, has been related to solar wind turbulence and/or to structures of magnetospheric origin (Golovchanskaya et al., 2006). Some of these results have been successively confirmed by Cousins and Shepherd (2012), who analyzed the electric field fluctuations on small spatial (45–450 km) and temporal scales (2–20 min) measured by the high-frequency (HF) radars of the Super Dual Auroral Radar Network (SuperDARN), in both hemispheres. The authors found that the shapes of distribution functions of the electric field fluctuations and the dependence of the power of these fluctuations on the scale size are both in agreement with what was expected for a turbulent flow and previously observed by other authors (e.g., Heppner et al., 1993; Parkinson, 2006). Moreover, Cousins and Shepherd (2012) suggested that both the plasma instabilities and the gradients in the ionospheric electrical conductance can be important sources of the small-scale variability of the electric field, while other parameters, related to the interplanetary magnetic field orientations and geomagnetic activity, seem to influence it only little.

Recently, the statistical turbulent properties of the magnetic field fluctuations on the mesoscale (8–300 km) have been studied also through measurements of the vector magnetometers on board the low-altitude polar-orbiting Swarm constellation. De Michelis et al. (2015, 2016, 2017, 2018) found that different turbulence regimes exist in the regions crossed by the satellites and that the scaling features of magnetic field fluctuations seem to be influenced by the interplanetary magnetic field and its orientation. These findings have suggested that geomagnetic activity can be a significant driver for the different turbulence regimes, which have been revealed in the F-region of the ionosphere (De Michelis et al., 2017). The characterization of the properties of the fluctuations of the magnetic field and, in particular, of their scaling features is an important element in the overall picture of the turbulent nature of the high-latitude ionosphere. Indeed, the results acquired over the years analyzing the fluctuations of the electric field cannot be necessarily considered valid also for the fluctuations of the magnetic field. The correlation between electric and magnetic field is not trivial; a simple linear relationship seems to exist only concerning the auroral oval when the East-West component of the magnetic field (B_y) and the North-South component of the electric field (E_x) are considered, as reported by Sugiura et al. (1982) and confirmed by Weimer et al. (1985) using DE 2 observations in the auroral oval.

2 Multiscale nature of the ionospheric medium

The occurrence of turbulence in the ionosphere is one of the signatures of the complex nature of the ionosphere. The processes occur on a large range of temporal and spatial scales and they interact with each other, at least in the so-called inertial range, i.e., at least in the range of length scales over which the transfer of energy occurs with a dissipation due to molecular viscosity that is negligible. Usually, it is difficult to relate the properties of a system observed at a particular scale to those at larger or smaller scales. This, however, is necessary when investigating the mechanisms of coupling between different scales and, more generally, when trying to understand complex physical systems, whose description is based on the key concepts of scale invariance and universality.

Scale invariance occurs when a certain property of the physical quantity under investigation does not change when a scale transformation is performed, i.e., when the values of the physical quantity are multiplied by a constant. In the case of a fractal, for example, by choosing a small part of it and scaling it up to the original fractal dimensions, we obtain a perfect copy of the latter: its geometric properties are therefore scale invariant. Among the most significant examples of systems displaying scale invariance are the fractal aggregations, the large-scale structure of the universe, turbulence in fluids, the spatio-temporal distributions of earthquakes, and the fluctuations in the field of economics and finance (Mandelbrot, 1997). In a fractal, volume and length are connected by a nontrivial relationship characterized by a single exponent named the "fractal dimension," this relationship corresponds to a scaling law with respect to the metric properties of the structure, from which the concept of fractal dimension derives. However, it is possible to identify scale invariance also in relation to properties that are different from metric properties. This property has important conceptual and practical consequences on the nature of the fluctuations that are intrinsic to each scale. An important consequence is the absence of a characteristic scale, apart from the minimum and maximum ones between which self-similarity develops. Scale invariance involves characteristics of intrinsic irregularities that cannot be described by traditional mathematical methods. It is possible to demonstrate that the power laws constitute the natural mathematical structure corresponding to the property of scale invariance. Indeed, from a mathematical point of view the invariance to scale transformations implies that, by changing the scale of the variable from x to $x' = \lambda x$, the function $f(x')$ is identical to $f(x)$ up to a constant factor, which does not depend on the variable x. Thus, in the case of scale invariance, the following holds:

$$\left\{ \begin{array}{l} x \to \lambda x \\ f(x) \to \lambda f(x) \end{array} \right\} \qquad (1)$$

Consequently, if for example $f(x) = x^n$, then it will result $f(\lambda x) = (\lambda x)^n = \lambda^n f(x)$. In the systems characterized by scale invariance, the scaling laws acquire particular importance. They characterize the change of the system under the effect of a transformation of the length scale, and represent an essential element for understanding the complexity of the system. The scaling laws are indeed a fundamental property of those systems where the knowledge of the individual elements is not sufficient to characterize the structure as a whole, and where individual elements interacting with each other in a nonlinear way give rise to complex structures, whose properties cannot be traced back to those of the individual constituents. Consequently, it is easy to realize that there are well-defined laws of scale not only in many fields of physics but also in disciplines such as economics, biology, mathematics, and social sciences.

Finally, an important feature of a system characterized by scale invariance properties is its universality. Systems that are apparently different at a certain scale can converge to the same statistical properties. In this case, such systems belong to the same class of universality and will have precisely the same critical exponents, which means that the corresponding scaling functions will be identical. The concept of universality is also very important from an experimental point of view. Indeed, if a real system and its simplified theoretical model belong to the same class of universality, then the model can be considered a faithful representation of the real system. To recognize and investigate the multiscale nature of a physical system, different methods can be used, for example: Fourier analysis, methods based on time-frequency analysis partly originated from Fourier analysis, wavelet transform, structure function analysis, and detrended fluctuation analysis. There are also new methods including, for example, the empirical mode decomposition and the associated Hilbert spectral analysis (Huang and Wu, 2008). Certainly, each method has its own advantages and limitations so that for processes with different dynamics, different methodologies might have different performances. Among the methodologies introduced in recent years to characterize the multiscale statistics, we focus our attention on structure function analysis. A possible way to investigate the scaling features of a time series (B_H) is in fact based on the estimation of the qth-order structure function $S_q(\tau)$ for different scales τ. This method, proposed by Kolmogorov in 1941 in his famous paper relative to the characterization of turbulence phenomena (Kolmogorov, 1941), consists in the evaluation of the following quantities:

$$S_q(\tau) = \langle |B_H(t + \tau) - B_H(t)|^q \rangle_T \qquad (2)$$

where τ is a time delay belonging to the inertial range and which quantifies the scale of interest, $\langle \rangle$ denotes the statistical average of the increments of the variable (B_H) taken over all pairs of points separated by a time delay τ, and T is the time interval over which the average is calculated. $S_q(\tau)$, as defined in Eq. (2), is known as qth-order generalized structure function. In the approach used in fluid turbulence, the physical quantity usually used to describe turbulence is the fluid velocity; the structure functions of the velocity field are computed in respect to spatial differences along the flow direction and do not contain the absolute value (Frisch, 1995). In our case, the term generalized refers to the fact that the structure function is evaluated using the absolute value of the increments given by $[B_H(t + \tau) - B_H(t)]$ and the computation is done in the temporal domain. If the analyzed time series describes a system characterized by scale invariance properties, the qth-order structure function $S_q(\tau)$ will be expected to scale as follows:

$$S_q(\tau) \sim \tau^{\xi(q)} \qquad (3)$$

where $\xi(q)$ is the scaling exponent of the structure function $S_q(\tau)$. This expression permits the definition of a hierarchy of scaling exponents $\xi(q)$, which can depend on q either linearly or nonlinearly. When the relationship between $\xi(q)$ and q is linear, i.e., $\xi(q) = cq$, and c is a constant, the scale invariance of the analyzed time series can be described by only one scaling exponent, i.e., the fluctuations of the analyzed signal are statistically self-similar with a single scaling exponent. Conversely, when the signal is more complex, the relationship between $\xi(q)$ and q can be nonlinear. The departure from a linear trend is a signature of multifractality. When this property is observed in turbulent media, in quantities as velocity and/or magnetic field, it is named intermittency and reflects the nonhomogeneous distribution of energy at the smaller scales (Frisch, 1995).

Interesting information on the analyzed signal, and consequently on the dynamical processes which generate the signal, can be obtained considering the structure functions of the first ($S_1(\tau)$) and second order ($S_2(\tau)$). By estimating the scaling exponent of the first-order structure function $S_1(\tau)$, it is possible to characterize the temporal (or spatial) structure of the analyzed signal. Indeed, this scaling exponent, which is considered equivalent to the Hurst exponent (H) (Hurst, 1956), is a measure of the long-term memory of time series and quantifies the relative tendency of a time series either to regress strongly to the mean or to cluster in a direction. A value of H in the range $0.5 < H < 1$ indicates a persistent behavior of the time series, meaning that a high value in the time series will probably be followed by another high value; similarly a low value in the time series will probably be followed by another low value. In this case the time series will tend to cluster in a direction and H will provide information on the memory of the analyzed time series and, consequently, on the dynamics of the system it describes. A value of the Hurst exponent in the range $0 < H < 0.5$ indicates an antipersistent behavior of the time series, meaning the tendency of the time series to switch between high and low values. Lastly, $H = 0.5$ is expected for a completely uncorrelated time series as for example in the case of a Brownian random motion; $H = 0$ is expected for a white noise and $H = 1$ for a linear trend. Some interesting information on the features of the magnetic fluctuations can also be obtained by analyzing the scaling exponent of the second-order structure function $S_2(\tau)$. In this case it is possible to have information on the spectral properties of the analyzed signal. Indeed, the second-order structure function scaling exponent $\xi(2)$ can be associated with the Fourier power spectrum exponent (β) via the Wiener-Khinchin relationship, according to which

$$\xi(2) + 1 = \beta. \qquad (4)$$

Lastly, some insights on anomalous scaling features and the occurrence of intermittency can be obtained from the first- and second-order structure function scaling exponents, $\xi(1)$ and $\xi(2)$. Indeed, $\xi(1)$ and $\xi(2)$ can allow defining the parameter I, named "degree of intermittency," defined as $I = \xi(2)/\xi(1)$. The degree of intermittency describes the existence of fluctuations in the analyzed signal that occur sporadically, in bursts, and a nonhomogeneous distribution of energy from the larger to the smaller scales.

3 Application: Case study in the ionosphere

In this section, an application of the methods based on the concepts explained earlier is applied to a real dataset of magnetic measurements recorded in the ionosphere by the Swarm constellation. Swarm is an European Space Agency (ESA) mission, which consists of three identical satellites that measure the Earth's magnetic field, electron density, and temperature with high precision and high resolution (Friis-Christensen et al., 2006). All three satellites fly on circular and almost polar orbits at two different altitudes. Two of them (Swarm A and Swarm C) fly at an altitude that, in mid-2016, was close to 460 km while Swarm B flies at a mean altitude of 510 km. Thus, using data recorded by these satellites, it is possible to explore the statistical scaling features of the fluctuations of the magnetic field and of some plasma parameters in the F-region of the ionosphere. A possible way will be illustrated to describe the multiscale nature of the ionospheric medium by analyzing the features of the low-resolution (1 Hz) magnetic field fluctuations in the F-region of the ionosphere using measurements recorded by the Swarm A satellite during a period of 2 years (April 2014–March 2016). The CHAOS-6 geomagnetic field model (Finlay et al., 2016) is used to remove, from the observed magnetic field, the core and crustal fields to obtain the external magnetic field of magnetospheric and ionospheric origin. Successively, the intensity of the horizontal magnetic field component (B_H) is evaluated, focusing on the Northern high-latitude regions during geomagnetically quiet periods. These periods are selected resorting to the use of geomagnetic activity indices, in detail AE (Davis and Sugiura, 2012) and Sym-H indices (Wanliss and Showalter, 2006).

Indeed, the AE index monitors the level of geomagnetic disturbance resulting from the auroral electrojets, being a proxy of the electric currents in the auroral zone, while the Sym-H index gives information about the strength of the ring current around the Earth, which increases during disturbed periods. Only data satisfying simultaneously the following conditions have been used for the analysis: $AE < 60$ nT and -5 nT$ < Sym - H < 5$ nT. The values selected for these two geomagnetic indices permit minimization of the presence in the data of magnetic perturbations introduced by the occurrence of storm and substorm events, and consequently to select geomagnetic quiet periods.

Fig. 1 displays an example of the values of the horizontal intensity of the geomagnetic field due to sources external to the Earth during one of the Swarm A high-latitude crossings of the Northern Hemisphere. B_H large fluctuations of Fig. 1 indicate nonstationarity in the data but also that they could be produced by a physical nonlinear mechanism. This means that the governing equation for this magnetic field component must be nonlinear. The nonstationarity and nonlinearity are typical features of data obtained from the real world, which display the tendency to have multiscale properties with fluctuations over a large range of scales. The detrended structure function analysis is used to estimate the first-order scaling exponent; this technique is commonly used for multiscale analysis and has the purpose of minimizing the effect of nonstationarity by detrending the time series. According to this method, before calculating the increments, it is necessary to remove from the original data their local trends, which are evaluated within selected windows. In detail, a time window of $T=400$ s is considered and in each time window the time series is detrended. This detrended time series is then used to calculate the increments needed for the estimation of the standard structure function defined in Eq. (2). The next step is to set the range of the scales to investigate; a general rule is to choose the

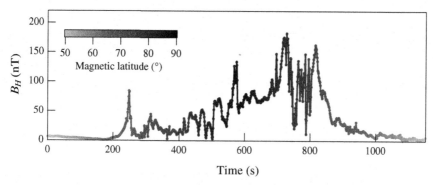

FIG. 1 An example of the used dataset. Horizontal intensity of the geomagnetic field obtained after the removal of the magnetic field's internal contribution by CHAOS-6; the plot shows data covering a time interval of about 20 min. Data have been recorded by Swarm A satellite during one of its high-latitude crossings of the Northern Hemisphere (magnetic latitude >50°N). Data refers to April 6, 2014 from 13:38:40 to 13:57:55.

largest scale around 10 times smaller than the size of the moving window; this guarantees that also the largest scale can rely on good statistics and hence provide a reliable estimation of the structure function for that scale. In the case presented here, scales τ between 1 and 40 s will be investigated.

Fig. 2 reports an example of the first-order structure function $S_1(\tau)$ evaluated using the dataset obtained as explained earlier.

A power-law behavior is evident within the selected range of time scales, and the corresponding first-order scaling exponent defined in Eq. (3) can be estimated. This behavior assesses the scale invariance of the analyzed time series in the selected range of time scales. We note that the data analysis is performed in the temporal domain but, under the assumption of Taylor's hypothesis (Taylor, 1938), the temporal scales can be related to the spatial scales.

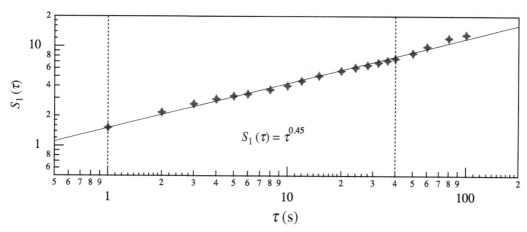

$$S_1(\tau) = \tau^{0.45}$$

FIG. 2 Values of the first-order structure function $S_1(\tau)$ *(blue diamonds)* *(dark gray* in print version) during one of the Swarm A satellite high-latitude crossings of the Northern Hemisphere (April 6, 2014 from 13:38:40 to 13:57:55). A power-law behavior *(blue solid line)* *(light gray* in print version) can be observed in the selected range of time scales, i.e., between 1 and 40 s *(black vertical dashed lines)*.

Thus, taking into account the orbital velocity of the Swarm satellite (~7.6 km/s), the range of time scales τ corresponds to the range of spatial fluctuations between ~8 and ~300 km. Thus, investigating time scales between 1 and 40 s means analyzing the features of the magnetic fluctuations at mesoscales.

Fig. 3 reports the values obtained for the first-order scaling exponent $\xi(1)$ in the Northern high latitudes (50°N–90°N) obtained by analyzing the fluctuations of the horizontal intensity of the Earth's magnetic field. The obtained values exhibit a strong asymmetry in the magnetic local time (MLT) character of the fluctuations, which present an antipersistent ($H < 0.5$) character in the sunlit hemisphere (05:00 < MLT < 18:00) and a persistent ($H > 0.5$) one in the dark sector. We recall that the Hurst exponent is equivalent to the first-order scaling exponent.

Fig. 4 displays the values of the second-order scaling exponent $\xi(2)$ obtained evaluating the second-order structure function for the analyzed dataset. Using the Wiener-Khinchin theorem (see Eq. 4), $\xi(2)$ can provide information on the values of the Fourier power spectral density exponent β. It is worth underlining that the regions characterized by high values of the H exponent, are the same where the second-order structure function assumes higher values. That happens in the part of the hemisphere that is not sunlit (from MLT = 18:00 to MTL = 06:00). The different values assumed by this scaling exponent correspond to different values of the Fourier power spectral exponent β, which according to Eq. (4) result in being often less than 2, being $\xi(2)$ less than 1 as shown by Fig. 4. A plasma shear flow turbulence characterized by an inverse energy cascade or a turbulence due to current convective or strong

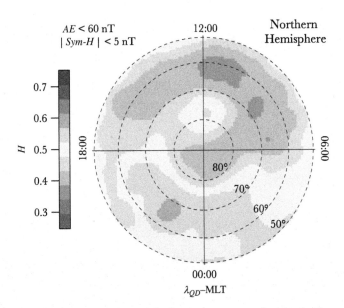

FIG. 3 The Hurst exponent values obtained evaluating the first-order structure function for the considered dataset (i.e., the horizontal intensity of the magnetic field due to external sources recorded by Swarm A during a period of 2 years and quiet geomagnetic conditions). The Hurst exponent values are relative to the Northern high latitudes and are reported in magnetic local time (MLT) and quasidipole magnetic latitude (λ_{QD}) in a polar representation. *Dashed circles* are drawn at magnetic latitudes of 50°, 60°, 70°, and 80°.

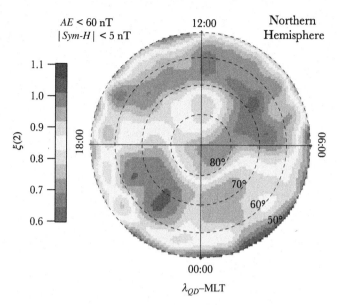

FIG. 4 Second-order structure function scaling exponents evaluated for the selected dataset. The values are relative to the Northern high latitudes and are reported in magnetic local time (MLT) and quasidipole magnetic latitude in a polar representation. *Dashed circles* are drawn at magnetic latitudes of 50°, 60°, 70°, and 80°.

gradient drifts may explain the Fourier power spectral exponent being less than 2 (Kintner, 1976). Alternatively, spectral slopes near −5/3 could also be the result of strong turbulence of shear Alfven waves under the hypothesis of critical balance (Sridhar and Goldreich, 1994). Conversely, values of the Fourier power spectral exponent β greater than 2 may be, for example, a consequence of both a strong plasma shear flow turbulence and a weak turbulence regime where the energy cascade occurs on a plane due to the fluctuations, which develop essentially in a perpendicular direction to the strong mean magnetic field (Galtier et al., 2000, 2005).

Lastly, Fig. 5 shows some information regarding the anomalous scaling features and the intermittency that characterize the analyzed data. In the case of a turbulent regime where a homogeneous transfer of energy between scales happens and the relationship between $\xi(q)$ and q is linear, the scaling exponent of the second-order structure function is related to the scaling exponent of the first-order structure function by the relation $\xi(2) = 2\xi(1)$. As already mentioned, when that does not occur, the turbulence is termed intermittent. In this case, the fluctuations of the analyzed signal occur sporadically in bursts and there is a nonhomogeneous distribution of energy from the larger to the smaller scales. The spatial distribution of intermittency is reported in Fig. 5. The values where intermittency is high, i.e., $\xi(2) = 2\xi(1)$ far from 2, seem to characterize the areas immediately outside the auroral oval, while inside the auroral oval, the level of intermittency is low. In general, evidence for a low level of intermittency can be found in correspondence with those regions where the power spectrum of the geomagnetic field fluctuations is around 2 (that means $\xi(2)$ near 1), while evidence for a high level of intermittency can be found in correspondence with those regions where the power spectrum of the geomagnetic field fluctuations is around

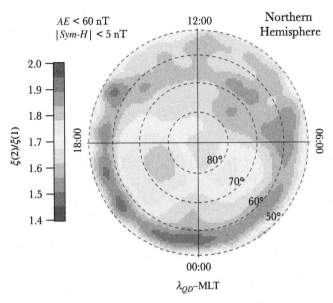

FIG. 5 Degree of intermittency of the considered dataset (i.e., the horizontal intensity of the magnetic field due to external sources recorded by Swarm A during a period of 2 years in a quiet geomagnetic condition). The values are relative to the Northern high latitudes and are reported in magnetic local time (MLT) and quasidipole magnetic latitude in a polar representation. *Dashed circles* are drawn at magnetic latitudes of 50°, 60°, 70°, and 80°.

1.6–1.7 ($\xi(2)$ between 0.6 and 0.7) which means it is around 5/3. This supports the hypothesis that the origin of intermittency may be researched in the occurrence of secondary instabilities requiring the presence of local large-scale gradients (Hallatschek and Diamond, 2003; Sudan, 1983) associated with a long wavelength component.

This case study shows how it is possible to describe the multiscale nature of the ionospheric medium using data recorded by the Swarm constellation. The evaluation of the local scaling indices of the first- and second-order structure functions of the magnetic field fluctuations permits information to be provide on the ionospheric turbulence and new insights to be given about the processes due to the ionosphere-magnetosphere coupling.

Acknowledgments

The results presented rely on data collected by one of the three satellites of the Swarm constellation. We thank the European Space Agency (ESA), which supports the Swarm mission. Swarm data can be accessed online at http://earth.esa.int/swarm. The authors acknowledge use of NASA/GSFC's Space Physics Data Facility's OMNIWeb service, and OMNI data. The authors acknowledge financial support from European Space Agency (ESA contract No. 4000125663/18/I-NB "EO Science for Society Permanently Open Call for Proposals EOEP-5 BLOCK4" (INTENS)).

References

Abel, G.A., Freeman, M.P., Chisham, G., 2009. IMF clock angle control of multifractality in ionospheric velocity fluctuations. Geophys. Res. Lett. 36, L19102. https://doi.org/10.1029/2009GL040336.

Basu, S., Basu, S., MacKenzie, E., Coley, W.R., Hanson, W.B., Lin, C.S., 1984. F-region electron density irregularity spectra near auroral acceleration and shear regions. J. Geophys. Res. 89, 5554.

Basu, S., Basu, S., MacKenzie, E., Fougere, P.F., Coley, W.R., Maynard, N.C., Winningham, J.D., Sugiura, M., Hanson, W.B., Hoegy, W.R., 1988. Simultaneous density and electric field fluctuation spectra associated with velocity shears in the auroral oval. J. Geophys. Res. 93, 115–136.

Cousins, E.D.P., Shepherd, S.G., 2012. Statistical maps of small-scale electric field variability in the high-latitude ionosphere. J. Geophys. Res. 117, 12304. https://doi.org/10.1029/2012JA017929.

Davis, T.N., Sugiura, M., 2012. Auroral electrojet activity index AE and its universal time variations. J. Geophys. Res. 71, 785–801. https://doi.org/10.1029/JZ071i003p00785.

De Michelis, P., Consolini, G., Tozzi, R., 2015. Magnetic field fluctuation features at Swarm's altitude: a fractal approach. Geophys. Res. Lett. 42, 3100–3105. https://doi.org/10.1002/2015GL063603.

De Michelis, P., Consolini, G., Tozzi, R., Marcucci, M.F., 2016. Observations of high-latitude geomagnetic field fluctuations during St. Patrick storm: Swarm and SuperDARN measurements. Earth Planets Space 68, 105. https://doi.org/10.1186/s40623-016-0476-3s.

De Michelis, P., Consolini, G., Tozzi, R., Marcucci, M.F., 2017. Scaling features of high latitude geomagnetic field fluctuations at Swarm altitude. J. Geophys. Res. 122, 10548–10562. https://doi.org/10.1002/2017JA024156.

De Michelis, P., Consolini, G., Tozzi, R., Giannatasio, F., Quattrociocchi, V., Coco, I., 2018. Features of magnetic field fluctuations in the ionosphere at Swarm altitude. Ann. Geophys. 61. https://doi.org/10.4401/ag-7789.

Dyrud, L., Krane, B., Oppenheim, M., Pécseli, H.L., Trulsen, J., Wernik, A.W., 2008. Structure functions and intermittency in the ionosphere plasma. Nonlinear Processes Geophys. 15, 847–862.

Earle, G.D., Kelley, M.C., Ganguli, G., 1989. Large velocity shears and associated electrostatic waves and turbulence in the auroral F region. J. Geophys. Res. 94, 15321–15333. https://doi.org/10.1029/JA094iA11p15321.

Finlay, C.C., Olsen, N., Kotsiaros, S., Gillet, N., Tffner-Clausen, L., 2016. Recent geomagnetic secular variation from Swarm and ground observatories as estimated in the CHAOS-6 geomagnetic field model. Earth Planets Space 68, 112. https://doi.org/10.1186/s40623-016-0486-1.

Friis-Christensen, E., Lühr, H., Hulot, G., 2006. Swarm: a constellation to study the Earth's magnetic field. Earth Planets Space 58, 351.

Frisch, U., 1995. Turbulence: The Legacy of A.N. Kolmogorov. Cambridge University Press.

Galtier, S., Nazarenko, S.V., Newell, A.C., Pouquet, A., 2000. Weak turbulence theory for incompressible magnetohydrodynamics. J. Plasma Phys. 42, 447–488.

Galtier, S., Pouquet, A., Mangeney, A., 2005. On spectral scaling laws for incompressible anisotropic magnetohydrodynamic turbulence. Phys. Plasmas 12, 092310.

Giannattasio, F., De Michelis, P., Consolini, G., Quattrociocchi, V., Coco, I., Tozzi, R., 2018. Characterising the electron density fluctuations in the high-latitude ionosphere at swarm altitude in response to the geomagnetic activity. Ann. Geophys. 61. https://doi.org/10.4401/ag-7716.

Golovchanskaya, I.V., Kozelov, B.V., 2010. On the origin of electric turbulence in the polar cap ionosphere.

J. Geophys. Res. 115, A09321. https://doi.org/10.1029/2009JA014632.

Golovchanskaya, I.V., Ostapenko, A.A., Kozelov, B.V., 2006. Relationship between the high-latitude electric and magnetic turbulence and the Birkeland field-aligned currents. J. Geophys. Res. 111, A12301. https://doi.org/10.1029/2006JA011835.

Grach, S.M., Sergeev, E.N., Mishin, E.V., Shindin, A.V., 2016. Dynamic properties of ionospheric plasma turbulence driven by high-power high-frequency radiowaves. Phys. Usp. 59, 1091–1128.

Hallatschek, K., Diamond, P.H., 2003. Modulation instability of drift waves. New J. Phys. 5, 29.1–29.9.

Heppner, J.P., Liebrecht, M.C., Maynard, N.C., Pfaff, R.F., 1993. High-latitude distributions of plasma waves and spatial irregularities from DE2 alternating current electric field observations. J. Geophys. Res. 98, 1629.

Huang, N.E., Wu, Z., 2008. A review on Hilbert-Huang transform: methods and its applications to geophysical studies. Rev. Geophys. 46, RG2006. https://doi.org/10.1029/2007RG000228.

Hurst, H.E., 1956. Methods of using long-term storage in reservoirs. ICE Proc. 5 (704), 519.

Kintner Jr, P.M., 1976. Observations of velocity shear driven plasma turbulence. J. Geophys. Res. 81, 5114.

Kintner Jr, P.M., Seyler, C.E., 1985. The status of observations and theory of high latitude ionosphere and magnetospheric plasma turbulence. Space Sci. Rev. 41, 91–129.

Kolmogorov, A., 1941. The local structure of turbulence in incompressible viscous fluid for very large Reynolds' numbers. Dokl. Akad. Nauk SSSR 30, 301.

Kozelov, B.V., Golovchanskaya, I.V., Ostapenko, A.A., Fedorenko, Y.V., 2008. Wavelet analysis of high-latitude electric and magnetic fluctuations observed by the dynamic explorer 2 satellite. J. Geophys. Res. 113, A03308. https://doi.org/10.1029/2007JA012575.

Mandelbrot, B.B., 1997. Fractals and Scaling in Finance: Discontinuity, Concentration, Risk. Springer. https://doi.org/10.1007/978-1-4757-2763-0.

Parkinson, M.L., 2006. Dynamical critical scaling of electric field fluctuations in the greater cusp and magnetotail implied by HF radar observations of F-region Doppler velocity. Ann. Geophys. 24, 689–705.

Pécseli, H., 2016. Turbulence in the ionosphere. In: Pécseli, H. (Ed.), Low Frequency Waves and Turbulence in Magnetized Laboratory Plasmas and in the Ionosphere. IOP Science (Chapter 24).

Sridhar, S., Goldreich, P., 1994. Toward a theory of interstellar turbulence: I. Weak Alfvénic turbulence. Astrophys. J. 432, 612–621.

Sudan, R.N., 1983. Unified theory of type-I and type-II irregularities in the equatorial electrojet. J. Geophys. Res. 88, 4853–4860.

Sugiura, M., Maynard, N.C., Farthing, W.H., Heppner, J.P., Ledley, B.G., Cahill Jr, L.J., 1982. Initial results on the correlation between the magnetic and electric fields observed from DE 2 satellite. Geophys. Res. Lett. 9, 985–988.

Tam, S.W.Y., Chang, T., Kintner, P.M., Klatt, E., 2005. Intermittency analyses on the SIERRA measurements of the electric field fluctuations in the auroral zone. Geophys. Res. Lett. 32, L05109.

Taylor, G.I., 1938. The spectrum of turbulence. Proc. R. Soc. London 164, 476.

Wanliss, J.A., Showalter, K.M., 2006. High-resolution global storm index: Dst versus SYM-H. J. Geophys. Res. 111, A02202. https://doi.org/10.1029/2005JA011034.

Weimer, D.R., Goertz, C.K., Gurnett, D.A., 1985. Auroral zone electric fields from DE1 and 2 at magnetic conjunctions. J. Geophys. Res. 90A, 7479.

The future of the ionosphere (according to us...)

Massimo Materassi[a], Anthea J. Coster[b], Biagio Forte[c], Susan Skone[d], Michael Mendillo[e]

[a]Institute for Complex Systems of the National Research Council (ISC-CNR), Florence, Italy
[b]MIT Haystack Observatory, Westford, MA, United States
[c]Department of Electronic and Electrical Engineering, University of Bath, Bath, United Kingdom
[d]The University of Calgary|HBI Department of Geomatics Engineering, Calgary, AB, Canada
[e]Department of Astronomy, Boston University, Boston, MA, United States

After going through the chapters of this book, the reader has come to the point at which we would like to give a sketch of what we expect, and wish for, the future development of ionospheric science to be.

The main take-home message of this book, which also inspired the title, is that *the Earth's ionosphere (EI) is a highly dynamical system.*

This is not "just an expression," emphasizing the great ionospheric variability: a significant improvement is expected in understanding ionospheric physics, as the most recent tools, developed in dynamical system theory and science of complexity, are employed to define new analysis techniques for ionospheric data, and to construct new physical models of the ionospheric processes.

In particular, two aspects of complex dynamical systems are of relevance for the study of the EI: the need for *huge databases* with many time resolutions; and the need for *highly advanced mathematical tools* to make models of such systems.

These two needs represent a great opportunity for huge innovation in ionospheric science.

Nowadays, huge ionospheric databases are available, providing unprecedented opportunities: reference must be made to GNSS transmitter and receiver networks, arrays of magnetometers, radar and ionosonde infrastructure, and new fleets of satellites performing in situ measurements.

The study of such a wide collection of information about a unique dynamical system is timely, because new tools that did not previously exist, such as machine learning and big data science, can now be applied.

This all offers the opportunity of investigating aspects of ionosphere's nature never investigated before: the choice of these questions should be considered very carefully. There are phenomena, traditionally referred to as "worst cases," in which ionospheric processes appear that cannot be predicted by the present

The Dynamical Ionosphere
https://doi.org/10.1016/B978-0-12-814782-5.00020-0

models, based on classical fluid dynamics and electromagnetism.

As far as new mathematical tools to model the EI "worst cases" are concerned, there is a need, and opportunity, for borrowing such tools from different theoretical sectors and bridging the ionospheric physics with other branches of science and technology: new alliances will bring unexpected cross-fertilizations. WE mention here just a few of those new theoretical approaches, and mathematical tools: valuing the role of information in EI dynamics; using the theory of phase transitions; employing the original fractional and fractal calculus; and importing the techniques of stochastic dynamics to take into account "sub-grid" elements of dynamics.

We argue that the data analysis tools and physical modeling techniques explored in the book, together with many others, may be important examples of future ionospheric science aspects, as previously suggested.

Before taking our readers' leave, we would like to focus briefly on three significant future challenges of ionospheric science, which could profit from what has been reported in this volume.

The first challenge is that of space weather for human infrastructure protection.

Focusing the value of space weather awareness, science, and technology to modern society (electricity-dependence, transport, communication in a global world) is definitely a must for the stability and prosperity of our global society. Even those issues brought by our global lifestyle as challenging problems plaguing our planet (e.g., conflicts and environmental issues) can be dealt with only in a stable and prosperous society with adequate technological infrastructure (Buzulukova, 2017). Technology will not be rendered more human and human-centered if it is severely harmed by space weather "big events": technological disruption due to natural causes can only lead to a regression in welfare, human rights, and the "history clock."

The second important challenge that EI science of the future must face is the study of space weather for space exploration, both manned and unmanned.

The space environment is extremely dangerous both for Earth's living creatures and for human technology, due to the high level of ionizing radiations, which are carcinogenic for living beings and harmful for electric and electronic devices (Schrijver et al., 2015). The abundance of ionizing radiation, in terms of both background values and peak occurrence in "extreme" events (e.g., particle ray bursts), is definitely part of the effects of the helio-geospace dynamics, given by fields, plasma motion, and electric currents. Constructing effective models of the different regions of the geospace will enable us to travel through them safely, both for unmanned missions and, more importantly, for manned flights.

Last but not least, the third challenge we expect ionospheric science will have to face in future is a deeper understanding of the role of an ionosphere, like the EI, in the development of life on a planet. This branch of research will be important both in terms of trying to establish permanent human colonies on other celestial bodies of our solar system, and also for diagnosing the possibility of a biosphere on exoplanets.

The habitable zone of a star is defined by the range of orbits around a star within which a planetary surface can support liquid water given a sufficient atmospheric pressure. It does not guarantee that the planet is habitable. Understanding the impact of stellar activity and the space environment around a planet and its moons is critical to determining the key factors that can control, through a complicated coupled chain of physical processes, planetary habitability. Despite the wealth of measurements from Earth, Mars, and Venus, we still do not understand the critical factors that determine the ultimate loss of an atmosphere to space, yet sustaining an atmosphere over long periods of time is one of the key factors in planetary habitability. One important question here is whether a global planetary magnetic field is necessary to sustain an atmosphere.

About this, it is worth mentioning that the complexity of our near-Earth plasma environment is unique in the solar system. This arises from the fact that the maximum plasma density (N_{max}) of every other planet with an atmosphere has the electron density matched by molecular ions. This occurs on Venus, Mars, and all four of the gas giant planets. Earth has atomic oxygen ions (O^+) for its N_{max} and h_{max}, which is due to the fact that ongoing oxygenic photosynthesis produces so much O_2 that it leads to O atoms as the dominant gas in the upper atmosphere. Atomic O exists on other planets—caused by photo-dissociation of CO_2 or H_2O—but it is never the dominant gas ionized by solar EUV at h_{max}. That criterion thus defines a biomarker for our solar system and beyond. If an exoplanet is discovered to have O^+ ions at its ionospheric peak, then that planet has thriving global life (Mendillo et al., 2018; Mendillo, 2019). Ionospheres and life are thus linked complex systems.

In conclusion, we would like to leave the reader with the expectation, and the wish, that future ionospheric science will be that of a complex dynamical system, studied through the big database that the community has collected and is collecting, and modeled with the most advanced mathematical tools of theoretical physics. Moreover, ionospheric science renewed and regenerated in this way will protect our infrastructures and safety from space weather hazards, and help us to explore, and possibly colonize, space.

References

Buzulukova, N. (Ed.), 2017. ISBN: 9780128127001. Extreme Events in Geospace, first ed. Elsevier.

Mendillo, M., 2019. The ionospheres of planets and exoplanets. Astron. Geophys. 60 (1), 25–30.

Mendillo, M., Withers, P., Dalba, P.A., 2018. Atomic oxygen ions as an ionospheric biomarker on exoplanets. Nat. Astron. https://doi.org/10.1038/s41550-017-0375-y.

Schrijver, C.J., et al., 2015. Understanding space weather to shield society: a global road map for 2015–2025 commissioned by COSPAR and ILWS. Adv. Space Res. 55 (12), 2745–2807.

Index

Printed in the United States
By Bookmasters